ROS Robotics Projects
Second Edition

Build and control robots powered by the Robot Operating System, machine learning, and virtual reality

Ramkumar Gandhinathan
Lentin Joseph

BIRMINGHAM - MUMBAI

ROS Robotics Projects
Second Edition

Commissioning Editor: Vijin Boricha
Acquisition Editor: Meeta Rajani
Content Development Editor: Pratik Andrade
Senior Editor: Rahul Dsouza
Technical Editor: Dinesh Pawar
Copy Editor: Safis Editing
Project Coordinator: Anish Daniel
Proofreader: Safis Editing
Indexer: Rekha Nair
Production Designer: Alishon Mendonsa

First published: March 2017
Second edition: December 2019

Production reference: 1181219

Published by Packt Publishing Ltd.
Livery Place
35 Livery Street
Birmingham
B3 2PB, UK.

ISBN 978-1-83864-932-6

www.packt.com

To my dear mother and father, for their sacrifices and for exemplifying the power of determination.
To my influential sister, for her constant love and motivation.
To my loving wife, for being supportive, understanding my passion, and accepting me the way I am throughout our life and beyond.
To all my friends and colleagues who've been by my side, encouraging and inspiring me to do more.
To Meeta, Pratik, and other Packt employees for working closely with me, bringing the best out of me, and shaping the book.

– Ramkumar Gandhinathan

I dedicate this book to my parents, C. G. Joseph and Jancy Joseph, for giving me strong support in making this project happen.

– Lentin Joseph

Packt.com

Subscribe to our online digital library for full access to over 7,000 books and videos, as well as industry leading tools to help you plan your personal development and advance your career. For more information, please visit our website.

Why subscribe?

- Spend less time learning and more time coding with practical eBooks and Videos from over 4,000 industry professionals
- Improve your learning with Skill Plans built especially for you
- Get a free eBook or video every month
- Fully searchable for easy access to vital information
- Copy and paste, print, and bookmark content

Did you know that Packt offers eBook versions of every book published, with PDF and ePub files available? You can upgrade to the eBook version at www.packt.com and as a print book customer, you are entitled to a discount on the eBook copy. Get in touch with us at customercare@packtpub.com for more details.

At www.packt.com, you can also read a collection of free technical articles, sign up for a range of free newsletters, and receive exclusive discounts and offers on Packt books and eBooks.

Contributors

About the authors

Ramkumar Gandhinathan is a roboticist and researcher by profession. He started building robots in the sixth grade and has been in the robotics field for over 15 years through personal and professional connections. He has personally built over 80 robots of different types. With 7 years' overall professional experience (4 years full-time and 3 years part-time/internship) in the robotics industry, he has 5 years of ROS experience. As a part of his professional career, he has built over 15 industrial robot solutions using ROS. He is also fascinated by building drones and is a drone pilot. His research interests and passion are in the fields of SLAM, motion planning, sensor fusion, multi-robot communication, and systems integration.

Lentin Joseph is an author, roboticist, and robotics entrepreneur from India. He runs a robotics software company called Qbotics Labs in Kochi, Kerala. He has 8 years of experience in the robotics domain, primarily in ROS, OpenCV, and PCL.

He has authored several books on ROS, namely *Learning Robotics Using Python*, first and second edition; *Mastering ROS for Robotics Programming*, first and second edition; *ROS Robotics Projects*, first edition; and *Robot Operating System for Absolute Beginners*.

He completed his master's degree in robotics and automation in India and has worked at Robotics Institute, CMU, USA. He is also a TEDx speaker.

Packt is searching for authors like you

If you're interested in becoming an author for Packt, please visit `authors.packtpub.com` and apply today. We have worked with thousands of developers and tech professionals, just like you, to help them share their insight with the global tech community. You can make a general application, apply for a specific hot topic that we are recruiting an author for, or submit your own idea.

Table of Contents

Preface

Robot Operating System (**ROS**) is one of the most popular robotics middleware and is used by universities and industries for robot-specific applications. Ever since its introduction, many robots have been introduced to the market and users have been able to use them with ease within their applications. One of its main draws is its open source nature. ROS does not need a user to reinvent the wheel; instead, standardizing robot operations and applications is simple.

This book is an upgrade to the previous edition and introduces you to newer ROS packages, interesting projects, and some added features. This book targets projects in the latest (at the time of writing) ROS distribution—ROS Melodic Morenia with Ubuntu Bionic version 18.04.

Here, you will understand how robots are used in industries and will learn the step-by-step procedure of building heterogeneous robot solutions. Unlike the service call and action features in ROS, you will be introduced to cooler techniques that let robots handle intricate tasks in a smart way. This knowledge should pave the way to far more intelligent and self-performing autonomous robots. Additionally, we will also introduce ROS-2, so you can learn the differences between this version and the previous ROS version and find help in choosing a specific middleware for your application.

Industries and research institutes are focusing primarily on the fields of computer vision and natural language processing. While the previous edition of this book introduced you to some simple vision applications such as object detection and face tracking, this edition will introduce you to one of the most widely used smart speaker platforms on the market, Amazon's Alexa, and how to control robots using it. In parallel, we will introduce new hardware, such as Nvidia Jetson, Asus Tinker Board, and BeagleBone Black and explore their capabilities with ROS.

While people may know how to control robots individually, one of the most common problems faced by users in the ROS community is the use of multiple robots working in synchronization, whether they are of the same type or not. This becomes complicated, as robots may follow similar topic names and may possibly lead to confusion in a sequence of operations. This book helps in highlighting the possible conflicts and suggests solutions.

This book also touches on reinforcement learning, including how it can be used with robotics and ROS. Furthermore, you will find the most interesting projects for building a self-driving car, deep learning with ROS, and building teleoperation solutions using VR headsets and Leap Motion, as they're currently trending and are being researched continuously.

Who this book is for

This book is for students, hobbyists, professionals, and individuals with a passion for learning robotics technology. Additionally, it is aimed at those individuals who are most interested in learning about and writing algorithms, motion control, and perception capabilities from scratch. This might even help a start-up build a new product or help researchers utilize what's already available and create something new and innovative. This book is also intended for those people who would like to work in the software domain or who want to have a career as a robotics software engineer.

What this book covers

`Chapter 1`, *Getting Started with ROS*, is a basic introductory chapter on ROS for beginners. This chapter will help you get an idea of the ROS software framework and its concepts.

`Chapter 2`, *Introduction to ROS-2 and Its Capabilities*, introduces you to ROS-2, the newest upgraded framework that helps us use ROS in real-time applications. This chapter is organized in a similar manner to `Chapter 1`, *Getting Started with ROS*, such that users are able to differentiate between both ROS versions and understand their capabilities and limitations.

`Chapter 3`, *Building an Industrial Mobile Manipulator*, is where you will learn how to build a mobile robot and a robot arm and combine them both to be used in a virtual environment and control them through ROS.

`Chapter 4`, *Handling Complex Robot Tasks Using State Machines*, introduces you to techniques in ROS that could be adapted while using robots for continuous and complicated task management.

`Chapter 5`, *Building an Industrial Application*, is where you will combine the skills acquired in `Chapters 3`, *Building an Industrial Mobile Manipulator* and `Chapter 4`, *Handling Complex Robot Tasks Using State Machines*, effectively, to create a user application. Here, we will demonstrate how to use the mobile manipulator to deliver products to houses in a neighborhood.

`Chapter 6`, *Multi-Robot Collaboration*, teaches you how to communicate between multiple robots of the same or different category and control them separately and together in groups.

`Chapter 7`, *ROS on Embedded Platforms and Their Control*, helps you understand the latest embedded controller and processor boards, such as STM32-based controllers, Tinker Board, Jetson Nano, and many more. We will also look at how to control their GPIOs via ROS and control them via voice-based commands through Alexa.

`Chapter 8`, *Reinforcement Learning and Robotics*, introduces you to one of the most commonly used learning techniques in robotics called reinforcement learning. In this chapter, you will understand what reinforcement learning is and the math behind it using examples. Additionally, we will discover how to incorporate this learning technique with ROS by means of simple projects.

`Chapter 9`, *Deep Learning Using ROS and TensorFlow*, is a project made using a trending technology in robotics. Using the TensorFlow library and ROS, we can implement interesting deep learning applications. You can implement image recognition using deep learning, and an application using SVM can be found in this chapter.

`Chapter 10`, *Creating a Self-Driving Car Using ROS*, is one of the more interesting projects in this book. In this chapter, we will build a simulation of a self-driving car using ROS and Gazebo.

`Chapter 11`, *Teleoperating Robots Using a VR Headset and Leap Motion*, shows you how to control a robot's actions using a VR headset and Leap Motion sensor. You can play around with VR, which is a trending technology these days.

`Chapter 12`, *Face Detection and Tracking Using ROS, OpenCV, and Dynamixel Servos*, takes you through a cool project that you can make with ROS and the OpenCV library. This project basically creates a face tracker application in which your face will be tracked in such a way that the camera will always point to your face. We will use intelligent servos such as Dynamixel to rotate the robot on its axis.

To get the most out of this book

- You should have a powerful PC running a Linux distribution, preferably Ubuntu 18.04 LTS.
- You can use a laptop or desktop with a graphics card, and RAM of at least 4 to 8 GB is preferred. This is actually for running high-end simulations in Gazebo, as well as for processing point clouds and computer vision.

- You should have the sensors, actuators, and I/O boards mentioned in the book and should be able to connect them all to your PC. You also need Git installed to clone the package files.
- If you are a Windows user, it will be good to download VirtualBox and set up Ubuntu on it. However, do note that you may have issues while you try to interface real hardware to ROS when working with VirtualBox.

Download the example code files

You can download the example code files for this book from your account at `www.packt.com`. If you purchased this book elsewhere, you can visit `www.packtpub.com/support` and register to have the files emailed directly to you.

You can download the code files by following these steps:

1. Log in or register at `www.packt.com`.
2. Select the **Support** tab.
3. Click on **Code Downloads**.
4. Enter the name of the book in the **Search** box and follow the onscreen instructions.

Once the file is downloaded, please make sure that you unzip or extract the folder using the latest version of:

- WinRAR/7-Zip for Windows
- Zipeg/iZip/UnRarX for Mac
- 7-Zip/PeaZip for Linux

The code bundle for the book is also hosted on GitHub at `https://github.com/PacktPublishing/ROS-Robotics-Projects-SecondEdition`. In case there's an update to the code, it will be updated on the existing GitHub repository.

We also have other code bundles from our rich catalog of books and videos available at `https://github.com/PacktPublishing/`. Check them out!

Download the color images

We also provide a PDF file that has color images of the screenshots/diagrams used in this book. You can download it here: `http://www.packtpub.com/sites/default/files/downloads/9781838649326_ColorImages.pdf`.

Code in Action

Visit the following link to check out videos of the code being run:
`http://bit.ly/34p6hL0`

Conventions used

There are a number of text conventions used throughout this book.

`CodeInText`: Indicates code words in text, database table names, folder names, filenames, file extensions, pathnames, dummy URLs, user input, and Twitter handles. Here is an example: "Remove the `CMakelists.txt` file."

A block of code is set as follows:

```
def talker_main():
    rospy.init_node('ros1_talker_node')
     pub = rospy.Publisher('/chatter', String)
     msg = String()
     i = 0
```

Any command-line input or output is written as follows:

```
$ sudo apt-get update
$ sudo rosdep init
```

Bold: Indicates a new term, an important word, or words that you see on screen. For example, words in menus or dialog boxes appear in the text like this. Here is an example: "Click on **Software & Updates** and enable all of the Ubuntu repositories."

Warnings or important notes appear like this.

Tips and tricks appear like this.

Get in touch

Feedback from our readers is always welcome.

General feedback: If you have questions about any aspect of this book, mention the book title in the subject of your message and email us at `customercare@packtpub.com`.

Errata: Although we have taken every care to ensure the accuracy of our content, mistakes do happen. If you have found a mistake in this book, we would be grateful if you would report this to us. Please visit `www.packtpub.com/support/errata`, selecting your book, clicking on the Errata Submission Form link, and entering the details.

Piracy: If you come across any illegal copies of our works in any form on the internet, we would be grateful if you would provide us with the location address or website name. Please contact us at `copyright@packt.com` with a link to the material.

If you are interested in becoming an author: If there is a topic that you have expertise in, and you are interested in either writing or contributing to a book, please visit `authors.packtpub.com`.

Reviews

Please leave a review. Once you have read and used this book, why not leave a review on the site that you purchased it from? Potential readers can then see and use your unbiased opinion to make purchase decisions, we at Packt can understand what you think about our products, and our authors can see your feedback on their book. Thank you!

For more information about Packt, please visit `packt.com`.

1
Getting Started with ROS

Robotics is one of the upcoming technologies that can change the world. Robots can replace people in many ways, and we are all afraid of them stealing our jobs. One thing is for sure: robotics will be one of the influential technologies of the future. When a new technology gains momentum, the opportunities in that field also increase. This means that robotics and automation can generate a lot of job opportunities in the future.

One of the main areas in robotics that can provide mass job opportunities is robotics software development. As we all know, software gives life to a robot or any machine. We can expand a robot's capabilities through software. If a robot exists, its capabilities, such as control, sensing, and intelligence, are realized using software.

Robotics software involves a combination of related technologies, such as computer vision, artificial intelligence, and control theory. In short, developing software for a robot is not a simple task; it can require expertise in many fields.

If you're looking for mobile application development in iOS or Android, there is a **Software Development Kit** (**SDK**) available to build applications in it, but what about robots? Is there any generic software framework available? Yes. One of the more popular robotics software frameworks is called the **Robot Operating System** (**ROS**).

In this chapter, we will take a look at an abstract concept of ROS and how to install it, along with an overview of simulators and its use on virtual systems. We will then cover the basic concepts of ROS, along with the different robots, sensors, and actuators that support ROS. We will also look at ROS with respect to industries and research. This entire book is dedicated to ROS projects, so this chapter will be a kick-start guide for those projects and help you set up ROS.

The following topics are going to be covered in this chapter:

- Getting started with ROS
- Fundamentals of ROS
- ROS client libraries

- ROS tools
- ROS simulators
- Installing ROS
- Setting up ROS on VirtualBox
- Introduction to Docker
- Setting up the ROS workspace
- Opportunities for ROS in industries and research

So, let's get started with ROS.

Technical requirements

Let's look into the technical requirements for this chapter:

- ROS Melodic Morenia on Ubuntu 18.04 (Bionic)
- VMware and Docker
- Timelines and test platform:
    - **Estimated learning time**: On average, 65 minutes
    - **Project build time (inclusive of compile and run time)**: On average, 60 minutes
    - **Project test platform**: HP Pavilion laptop (Intel® Core™ i7-4510U CPU @ 2.00 GHz × 4 with 8 GB Memory and 64-bit OS, GNOME-3.28.2)

Getting started with ROS

ROS is an open source, flexible software framework for programming robots. ROS provides a hardware abstraction layer in which developers can build robotics applications without worrying about the underlying hardware. ROS also provides different software tools to visualize and debug robot data. The core of the ROS framework is a message-passing middleware in which processes can communicate and exchange data with each other, even when they're running from different machines. ROS message passing can be synchronous or asynchronous.

Software in ROS is organized as packages, and it offers good modularity and reusability. Using the ROS message-passing middleware and hardware abstraction layer, developers can create tons of robotic capabilities, such as mapping and navigation (in mobile robots). Almost all the capabilities in ROS will be robot agnostic so that all kinds of robots can use it. New robots can directly use this capability package without modifying any code inside the package.

ROS has widespread collaborations in universities, and lots of developers contribute to it. We can say that ROS is a community-driven project supported by developers worldwide. This active developer ecosystem distinguishes ROS from other robotic frameworks.

In short, ROS is the combination of **Plumbing** (or communication), **Tools**, **Capabilities**, and **Ecosystem**. These capabilities are demonstrated in the following diagram:

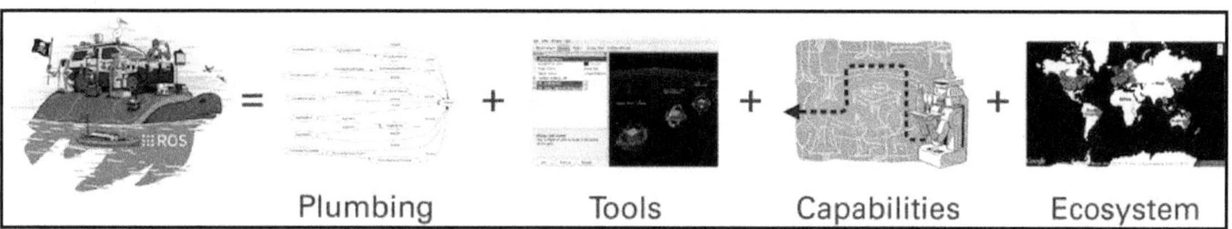

The ROS equation (source: ros.org. licensed under Creative Commons CC-BY-3.0: https://creativecommons.org/licenses/by/3.0/us/legalcode)

The ROS project was started in 2007 at Stanford University under the name Switchyard. Later on, in 2008, the development was undertaken by a robotic research startup called **Willow Garage**. The major development in ROS happened in Willow Garage. In 2013, the Willow Garage researchers formed the **Open Source Robotics Foundation** (**OSRF**). ROS is actively maintained by OSRF now. Now, let's look at a few ROS distributions.

Here are links to their websites: Willow Garage: `http://www.willowgarage.com/`
OSRF: `http://www.osrfoundation.org/`.

ROS distributions

The ROS distributions are very similar to Linux distributions, that is, a versioned set of ROS packages. Each distribution maintains a stable set of core packages, up to the **End Of Life** (**EOL**) of the distribution.

The ROS distributions are fully compatible with Ubuntu, and most of the ROS distributions are planned according to their respective Ubuntu versions.

The following are some of the latest ROS distributions (at the time of writing) that are recommended for use from the ROS website (`http://wiki.ros.org/Distributions`):

Distro	Release date	Poster	*Tuturtle*, turtle in tutorial	EOL date
ROS Melodic Morenia **(Recommended)**	May 23rd, 2018			May, 2023 (Bionic EOL)
ROS Lunar Loggerhead	May 23rd, 2017			May,2019
ROS Kinetic Kame	May 23rd, 2016			April, 2021 (Xenial EOL)

Latest ROS distributions (source: ros.org. licensed under Creative Commons CC-BY-3.0: https://creativecommons.org/licenses/by/3.0/us/legalcode)

The latest ROS distribution is Melodic Morenia. We will get support for this distribution up until May 2023. One of the problems with this latest ROS distribution is that most of the packages will not be available on it because it will take time to migrate them from the previous distribution. If you are looking for a stable distribution, you can go for ROS Kinetic Kame because the distribution started in 2016, and most of the packages are available on this distribution. The ROS Lunar Loggerhead distribution will stop being supported in May 2019, so I do not recommend that you use it.

Supported OSes

The main OS ROS is tuned for is Ubuntu. ROS distributions are planned according to Ubuntu releases. Other than Ubuntu, it is partially supported by Ubuntu ARM, Debian, Gentoo, macOS, Arch Linux, Android, Windows, and OpenEmbedded.

This table shows new ROS distributions and the specific versions of the supporting OSes:

ROS distribution	Supporting OSes
Melodic Morenia (LTS)	Ubuntu 18.04 (LTS) and 17.10, Debian 8, macOS (Homebrew), Gentoo, and Ubuntu ARM
Kinetic Kame (LTS)	Ubuntu 16.04 (LTS) and 15.10, Debian 8, macOS (Homebrew), Gentoo, and Ubuntu ARM

Jade Turtle	Ubuntu 15.04, 14.10, and 14.04, Ubuntu ARM, macOS (Homebrew), Gentoo, Arch Linux, Android NDK, and Debian 8
Indigo Igloo (LTS)	Ubuntu 14.04 (LTS) and 13.10, Ubuntu ARM, macOS (Homebrew), Gentoo, Arch Linux, Android NDK, and Debian 7

ROS Melodic and Kinetic are **Long-Term Support** (**LTS**) distributions that come with the LTS version of Ubuntu. The advantage of using LTS distribution is that we will get maximum lifespan and support.

We will look at a few robots and sensors that are supported by ROS in the next section.

Robots and sensors supported by ROS

The ROS framework is one of the most successful robotics frameworks, and universities around the globe contribute to it. Because of its active ecosystem and open source nature, ROS is being used in a majority of robots and is compatible with major robotic hardware and software. Here are some of the most famous robots completely running on ROS:

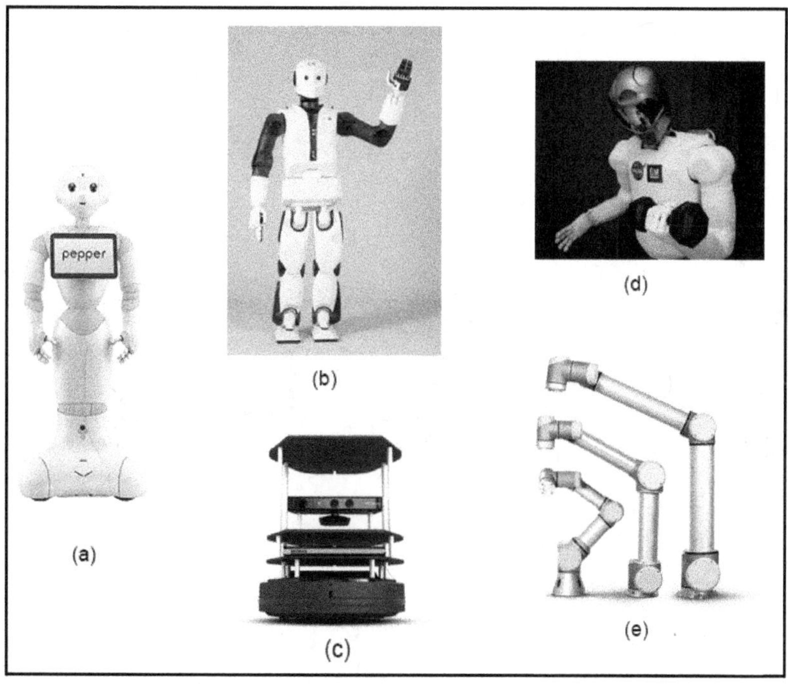

Popular robots supported by ROS (Source: ros.org. Licensed under Creative Commons CC-BY-3.0: https://creativecommons.org/licenses/by/3.0/us/legalcode)

The names of the robots listed in preceding images are Pepper (**a**), REEM-C (**b**), Turtlebot (**c**), Robonaut (**d**), and Universal Robots (**e**).

The robots supported by ROS are listed at the following link: `http://wiki.ros.org/Robots`.

The following are the links where you can get the ROS packages of these robots:

- **Pepper**: `http://wiki.ros.org/Robots/Pepper`
- **REEM-C**: `http://wiki.ros.org/Robots/REEM-C`
- **Turtlebot 2**: `http://wiki.ros.org/Robots/TurtleBot`
- **Robonaut**: `http://wiki.ros.org/Robots/Robonaut2`
- **Universal robotic arms**: `http://wiki.ros.org/universal_robot`

Some popular sensors that support ROS are as follows:

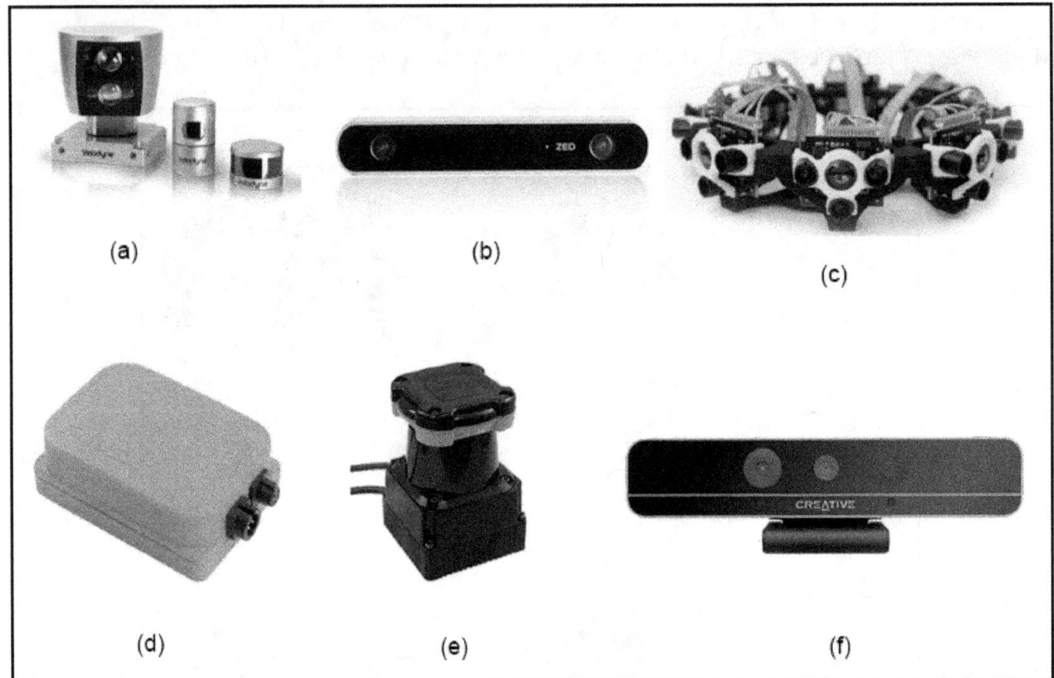

Popular robot sensors supported in ROS

The names of the sensors in the preceding image are Velodyne (**a**), ZED Camera (**b**), Teraranger (**c**), Xsens (**d**), Hokuyo Laser range finder (**e**), and Intel RealSense (**f**).

The list of sensors supported by ROS is available at the following link: `http://wiki.ros.org/Sensors`.

The following are the links to the ROS wiki pages of these sensors:

- **Velodyne (a)**: `http://wiki.ros.org/velodyne`
- **ZED Camera (b)**: `http://wiki.ros.org/zed-ros-wrapper`
- **Teraranger (c)**: `http://wiki.ros.org/teraranger`
- **Xsens (d)**: `http://wiki.ros.org/xsens_driver`
- **Hokuyo Laser range finder (e)**: `http://wiki.ros.org/hokuyo_node`
- **Intel real sense (f)**: `http://wiki.ros.org/realsense_camera`

Now, let's look at the advantages of using ROS.

Why use ROS?

The main intention behind building the ROS framework is to become a generic software framework for robots. Even though there was robotics research happening before ROS, most of the software was exclusive to their own robots. Their software may be open source, but it is very difficult to reuse.

Compared to existing robotic frameworks, ROS is outperforming in the following aspects:

- **Collaborative development**: As we've already discussed, ROS is open source and free to use for industries and research. Developers can expand the functionalities of ROS by adding packages. Almost all ROS packages work on a hardware abstraction layer, so it can be reused easily for other robots. So, if one university is good in mobile navigation and another is good in robotic manipulators, they can contribute that to the ROS community and other developers can reuse their packages and build new applications.
- **Language support**: The ROS communication framework can be easily implemented in any modern programming language. It already supports popular languages such as C++, Python, and Lisp, and it has experimental libraries for Java and Lua.
- **Library integration**: ROS has an interface to many third-party robotics libraries, such as **Open Source Computer Vision** (**OpenCV**), **Point Cloud Library** (**PCL**), Open-NI, Open-Rave, and Orocos. Developers can work with any of these libraries without much hassle.

- **Simulator integration**: ROS also has ties to open source simulators such as Gazebo and has a good interface with proprietary simulators such as Webots and V-REP.
- **Code testing**: ROS offers an inbuilt testing framework called **rostest** to check code quality and bugs.
- **Scalability**: The ROS framework is designed to be scalable. We can perform heavy computation tasks with robots using ROS, which can either be placed on the cloud or on heterogeneous clusters.
- **Customizability**: As we have already discussed, ROS is completely open source and free, so we can customize this framework as per the robot's requirements. If we only want to work with the ROS messaging platform, we can remove all of the other components and use only that. We can even customize ROS for a specific robot for better performance.
- **Community**: ROS is a community-driven project, and it is mainly led by OSRF. The large community support is a great plus for ROS and means we can easily start robotics application development.

The following are the URLs of libraries and simulators that can be integrated with ROS:

- **Open-CV:** `http://wiki.ros.org/vision_opencv`
- **PCL:** `http://wiki.ros.org/pcl_ros`
- **Open-NI:** `http://wiki.ros.org/openni_launch`
- **Open-Rave:** `http://openrave.org/`
- **Orocos:** `http://www.orocos.org/`
- **V-REP:** `http://www.coppeliarobotics.com/`

Let's go through some of the basic concepts of ROS; these can help you get started with ROS projects.

Fundamentals of ROS

Understanding the basic working of ROS and its terminology can help you understand existing ROS applications and build your own. This section will teach you important concepts that we are going to use in the upcoming chapters. If you find that a topic is missing in this chapter, then rest assured that it will be covered in a corresponding chapter later.

There are three different concepts in ROS. Let's take a look at them.

The filesystem level

The filesystem level explains how ROS files are organized on the hard disk:

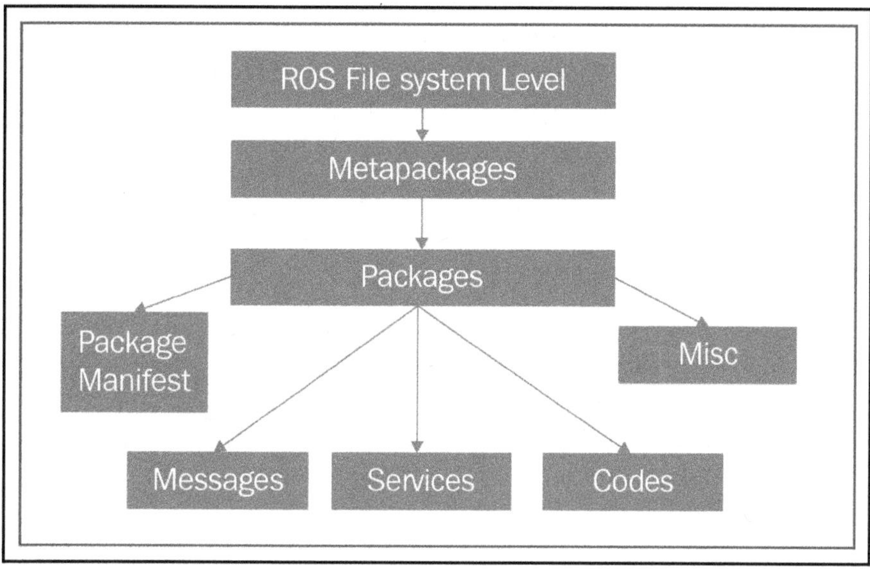

The ROS filesystem level

As you can see from the preceding diagram, the filesystem in ROS can be categorized mainly as metapackages, packages, package manifest, messages, services, codes, and miscellaneous files. The following is a short description of each component:

- **Metapackages**: Metapackages group a list of packages for a specific application. For example, in ROS, there is a metapackage called navigation for mobile robot navigation. It can hold the information of related packages and helps install those packages during its own installation.
- **Packages**: The software in ROS is mainly organized as ROS packages. We can say that ROS packages are the atomic build units of ROS. A package may consist of ROS nodes/processes, datasets, and configuration files, all organized in a single module.
- **Package manifest**: Inside every package will be a manifest file called `package.xml`. This file consists of information such as the name, version, author, license, and dependencies that are required of the package. The `package.xml` file of a metapackage consists of the names of related packages.

- **Messages (msg)**: ROS communicates by sending ROS messages. The type of message data can be defined inside a file with the `.msg` extension. These files are called message files. Here, we are going to follow a convention where we put the message files under `our_package/msg/message_files.msg`.
- **Service (srv)**: One of the computation graph level concepts is services. Similar to ROS messages, the convention is to put service definitions under `our_package/srv/service_files.srv`.

This sums up the ROS filesystem.

The computation graph level

The ROS computation graph is a peer-to-peer based network that processes all the information together. The ROS graph concept constitutes nodes, topics, messages, master, parameter server, services, and bags:

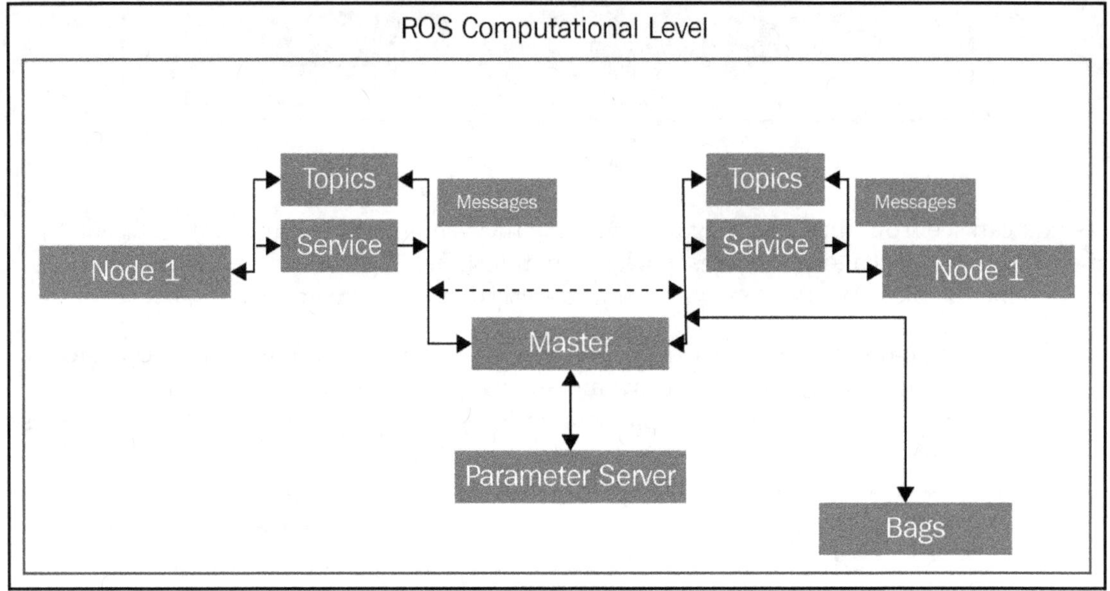

The ROS computational graph concept diagram

The preceding diagram shows the various concepts in the ROS computational graph. Here is a short description of each concept:

- **Nodes**: ROS nodes are simply processes that use ROS APIs to communicate with each other. A robot may have many nodes to perform its computations. For example, an autonomous mobile robot may have a node each for hardware interfacing, reading laser scan, and localization and mapping. We can create ROS nodes using ROS client libraries such as `roscpp` and `rospy`, which we will be discussing in the upcoming sections.
- **Master**: The ROS master works as an intermediate node that aids connections between different ROS nodes. The master has all of the details about all the nodes running in the ROS environment. It will exchange details of one node with another to establish a connection between them. After exchanging this information, communication will start between the two ROS nodes.
- **Parameter server**: The parameter server is a pretty useful thing in ROS. A node can store a variable in the parameter server and set its privacy, too. If the parameter has a global scope, it can be accessed by all other nodes. The ROS parameter runs along with the ROS master.
- **Messages**: ROS nodes can communicate with each other in many ways. In all the methods, nodes send and receive data in the form of ROS messages. The ROS message is a data structure that's used by ROS nodes to exchange data.
- **Topics**: One of the methods to communicate and exchange ROS messages between two ROS nodes is called ROS topics. Topics are named buses in which data is exchanged using ROS messages. Each topic will have a specific name, and one node will publish data to a topic and another node can read from the topic by subscribing to it.
- **Services**: Services are another kind of communication method, similar to topics. Topics use publish or subscribe interaction, but in services, a request or reply method is used. One node will act as the service provider, which has a service routine running, and a client node requests a service from the server. The server will execute the service routine and send the result to the client. The client node should wait until the server responds with the results.
- **Bags**: Bags are a useful utility in ROS for the recording and playback of ROS topics. While working on robots, there may be some situations where we need to work without actual hardware. Using `rosbag`, we can record sensor data and copy the bag file to other computers to inspect data by playing it back.

This sums up the computational graph concept.

The ROS community level

The ROS community has grown more now compared to the time it was introduced. You can find at least 2,000+ packages being supported, altered, and used by the community actively. The community level comprises the ROS resources for sharing software and knowledge:

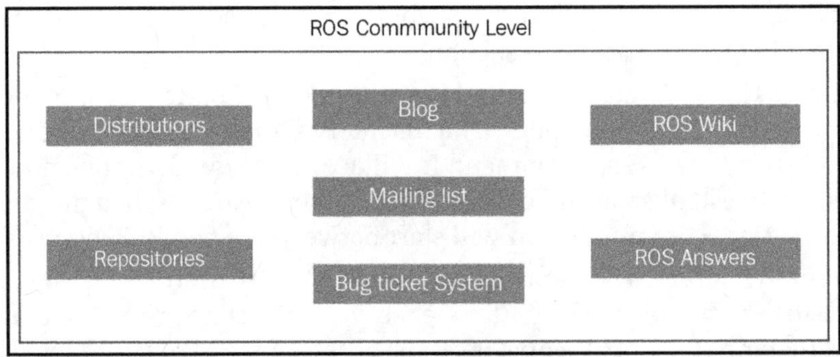

ROS community level diagram

Here is a brief description of each section:

- **Distributions**: ROS distributions are versioned collections of ROS packages, such as Linux distributions.
- **Repositories**: ROS-related packages and files depend on a **Version Control System** (**VCS**) such as Git, SVN, and Mercurial, using which developers around the world can contribute to the packages.
- **ROS Wiki**: The ROS community wiki is the knowledge center of ROS, in which anyone can create documentation of their packages. You can find standard documentation and tutorials about ROS on the ROS wiki.
- **Mailing lists:** Subscribing to ROS mailing lists enables users to get new updates regarding ROS packages and gives them a place to ask questions about ROS (`http://wiki.ros.org/Mailing%20Lists?action=show`).
- **ROS Answers**: The ROS Answers website is the stack overflow of ROS. Users can ask questions regarding ROS and related areas (`http://answers.ros.org/questions/`).
- **Blog**: The ROS blog provides regular updates about the ROS community with photos and videos (`http://www.ros.org/news`).

Now, let's learn how communication is carried out in ROS in the next section.

Communication in ROS

Let's learn how two nodes communicate with each other using ROS topics. The following diagram shows how this happens:

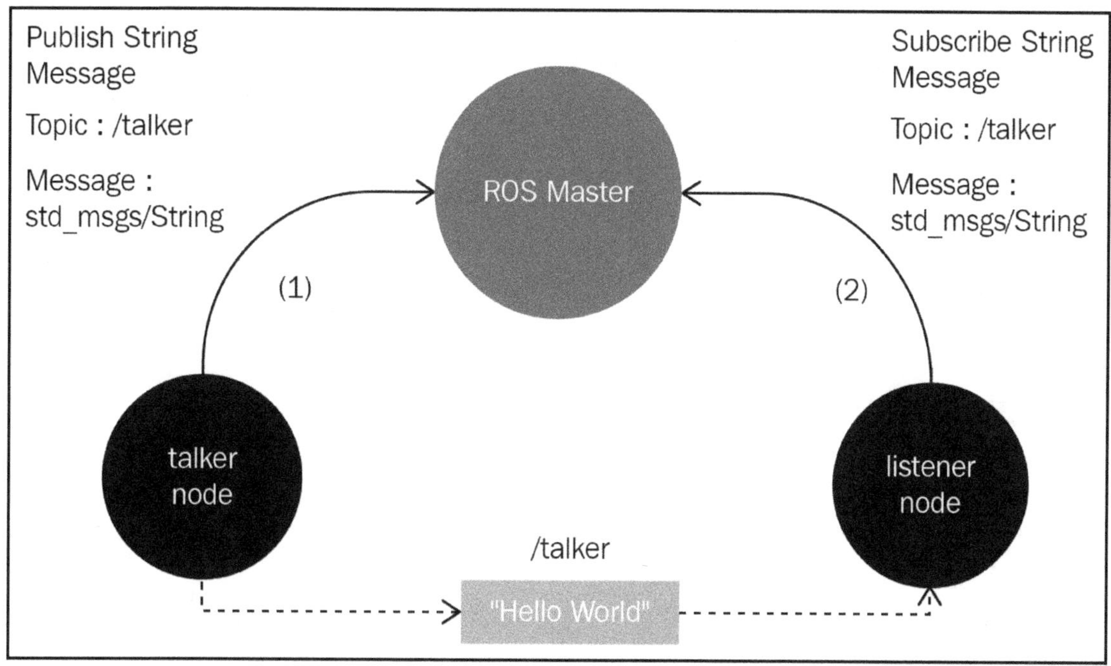

Communication between ROS nodes using topics

As you can see, there are two nodes named talker and listener. The **talker node** publishes a string message called **Hello World** into a topic called **/talker**, while the **listener node** is subscribed to this topic. Let's see what happens at each stage, marked **(1)**, **(2)**, and **(3)** in the preceding diagram:

1. Before running any nodes in ROS, we should start the **ROS Master**. After it has been started, it will wait for nodes. When the **talker node** (publisher) starts running, it will connect to the **ROS Master** and exchange the publishing topic details with the master. This includes the topic name, message type, and publishing node URI. The URI of the master is a global value, and all the nodes can connect to it. The master maintains the tables of the publisher connected to it. Whenever a publisher's details change, the table updates automatically.

2. When we start the **listener node** (subscriber), it will connect to the master and exchange the details of the node, such as the topic going to be subscribed to, its message type, and the node URI. The master also maintains a table of subscribers, similar to the publisher.
3. Whenever there is a subscriber and publisher for the same topic, the master node will exchange the publisher URI with the subscriber. This will help both nodes to connect and exchange data. After they've connected, there is no role for the master. The data is not flowing through the master; instead, the nodes are interconnected and exchange messages.

More information on nodes, their namespaces, and usage can be found here: `http://wiki.ros.org/Nodes`.

Now that we know the fundamentals of ROS, let's look at a few ROS client libraries.

ROS client libraries

ROS client libraries are used to write ROS nodes. All of the ROS concepts are implemented in client libraries. So, we can just use it without implementing everything from scratch. We can implement ROS nodes with a publisher and subscriber, write service callbacks, and so on using client libraries.

The main ROS client libraries are in C++ and Python. Here is a list of popular ROS client libraries:

- `roscpp`: This is one of the most recommended and widely used ROS client libraries for building ROS nodes. This client library has most of the ROS concepts implemented and can be used in high-performance applications.
- `rospy`: This is a pure implementation of the ROS client library in Python. The advantage of this library is its ease of prototyping, which means that development time isn't as long. It is not recommended for high-performance applications, but it is perfect for non-critical tasks.
- `roslisp`: This is the client library for LISP and is commonly used to build robot planning libraries.

Details of all the client ROS libraries can be found at the following link: `http://wiki.ros.org/Client%20Libraries`. The next section will give us an overview of different ROS tools.

ROS tools

ROS has a variety of GUI and command-line tools to inspect and debug messages. These tools come in handy when you're working in a complex project involving a lot of package integrations. It would help to identify whether the topics and messages are published in the right format and are available to the user as desired. Let's look at some commonly used ones.

ROS Visualizer (RViz)

RViz (`http://wiki.ros.org/rviz`) is one of the 3D visualizers available in ROS that can visualize 2D and 3D values from ROS topics and parameters. RViz helps to visualize data such as robot models, robot 3D transform data (TF), point cloud, laser and image data, and a variety of different sensor data:

Point cloud data visualized in RViz

The preceding screenshot shows a 3D point cloud scan from a Velodyne sensor placed on an autonomous car.

rqt_plot

The `rqt_plot` program (`http://wiki.ros.org/rqt_plot`) is a tool for plotting scalar values that are in the form of ROS topics. We can provide a topic name in the **Topic** box:

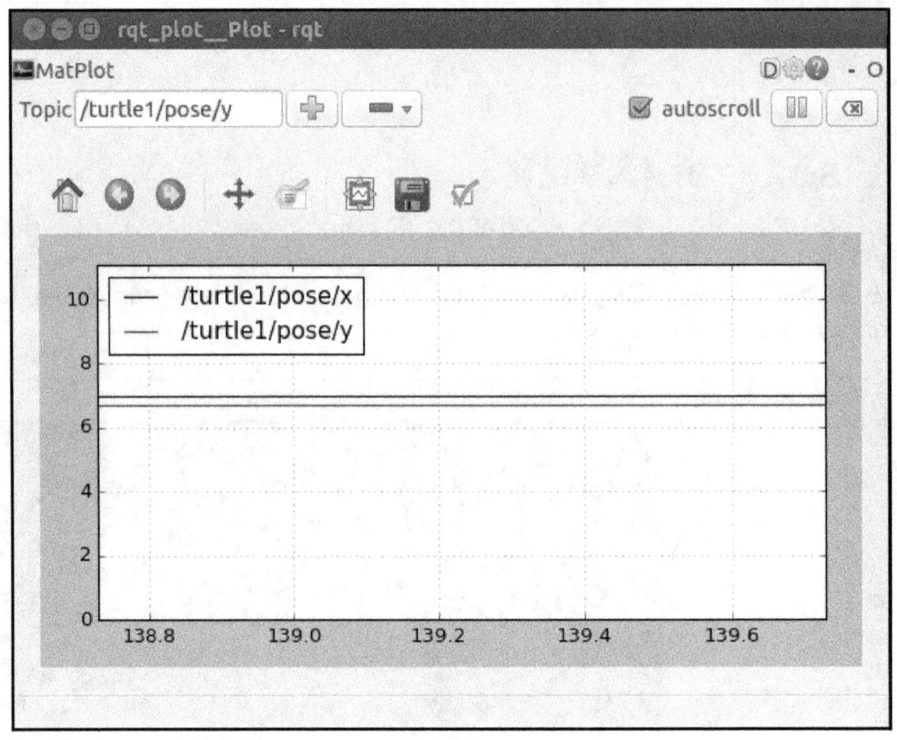

rqt_plot

The preceding screenshot is a plot of a pose from the `turtle_sim` node.

rqt_graph

The rqt_graph (http://wiki.ros.org/rqt_graph) ROS GUI tool can visualize the graph of interconnection between ROS nodes:

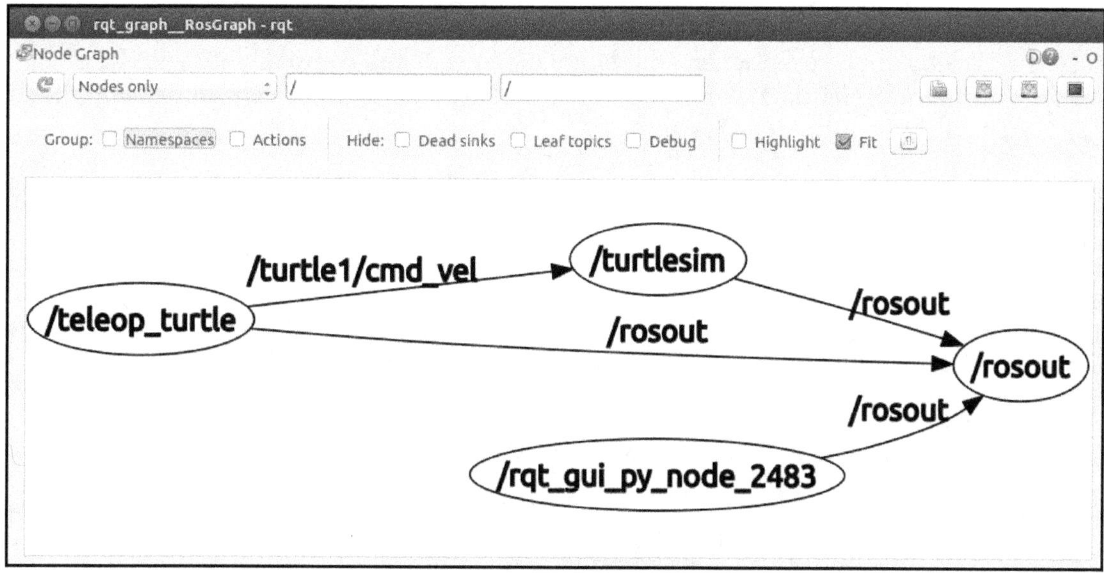

rqt_graph

The complete list of ROS tools is available at the following link: http://wiki.ros.org/Tools.

Since we now have a brief idea of the ROS tools, we can cover different ROS simulators.

ROS simulators

One of the open source robotic simulators tightly integrated with ROS is Gazebo (`http://gazebosim.org`). Gazebo is a dynamic robotic simulator that has a wide variety of robot models and extensive sensor support. The functionalities of Gazebo can be added via plugins. The sensor values can be accessed by ROS through topics, parameters, and services. Gazebo can be used when your simulation needs full compatibility with ROS. Most of the robotics simulators are proprietary and expensive; if you can't afford it, you can use Gazebo directly without any issues:

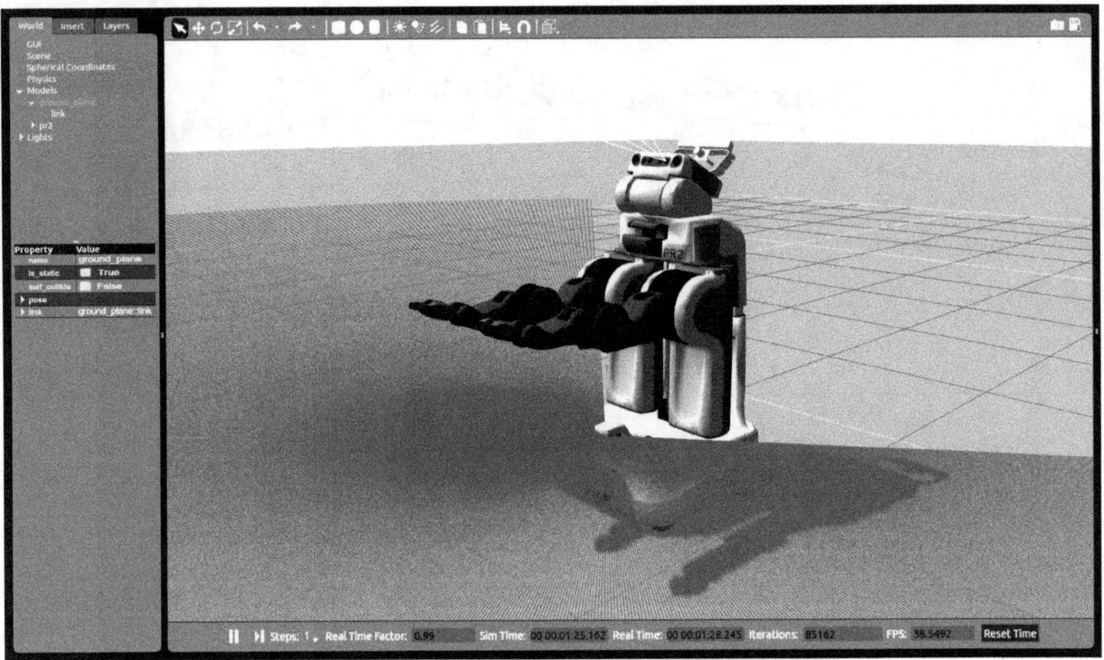

Gazebo simulator

The preceding is a PR2 robot model from OSRF. You can find the model at `https://github.com/pr2/pr2_common`, in the description folder.

The ROS interface of Gazebo is available at the following link: `http://wiki.ros.org/gazebo`.

Now that we know about the simulators of ROS, we can begin installing ROS Melodic on Ubuntu.

Installing ROS Melodic on Ubuntu 18.04 LTS

As we have already discussed, there are a variety of ROS distributions available to download and install, so choosing the exact distribution for our needs may be confusing. The following are the answers to some of the questions that are asked frequently while choosing a distribution:

- Which distribution should I choose to get maximum support?

  Answer: If you are interested in getting maximum support, choose an LTS release. It will be good if you choose the second-most recent LTS distribution.

- I need the latest features of ROS; which should I choose?

  Answer: Go for the latest version; you may not get the latest complete packages immediately after the release. You may have to wait a few months after the release. This is because of the migration period from one distribution to another.

In this book, we are dealing with two LTS distributions: ROS Kinetic, which is a stable release, and ROS Melodic, the latest one. Our chapters will use ROS Melodic Morenia.

Getting started with the installation

Go to the ROS installation website (`http://wiki.ros.org/ROS/Installation`). You will see a screen listing the latest ROS distributions:

Latest ROS distributions on the website

You can get the complete installation instructions for each distribution if you click on ROS Kinetic Kame or ROS Melodic Morenia.

We'll now step through the instructions to install the latest ROS distribution.

Configuring Ubuntu repositories

We are going to install ROS Melodic on Ubuntu 18.04 from the ROS package repository. The repository will have prebuilt binaries of ROS in `.deb` format. To be able to use packages from the ROS repository, we have to configure the repository options of Ubuntu first.

The details of the different kinds of Ubuntu repositories can be found at `https://help.ubuntu.com/community/Repositories/Ubuntu`.

To configure the repository, perform the following steps:

1. First, search for **Software & Updates** in the Ubuntu search bar:

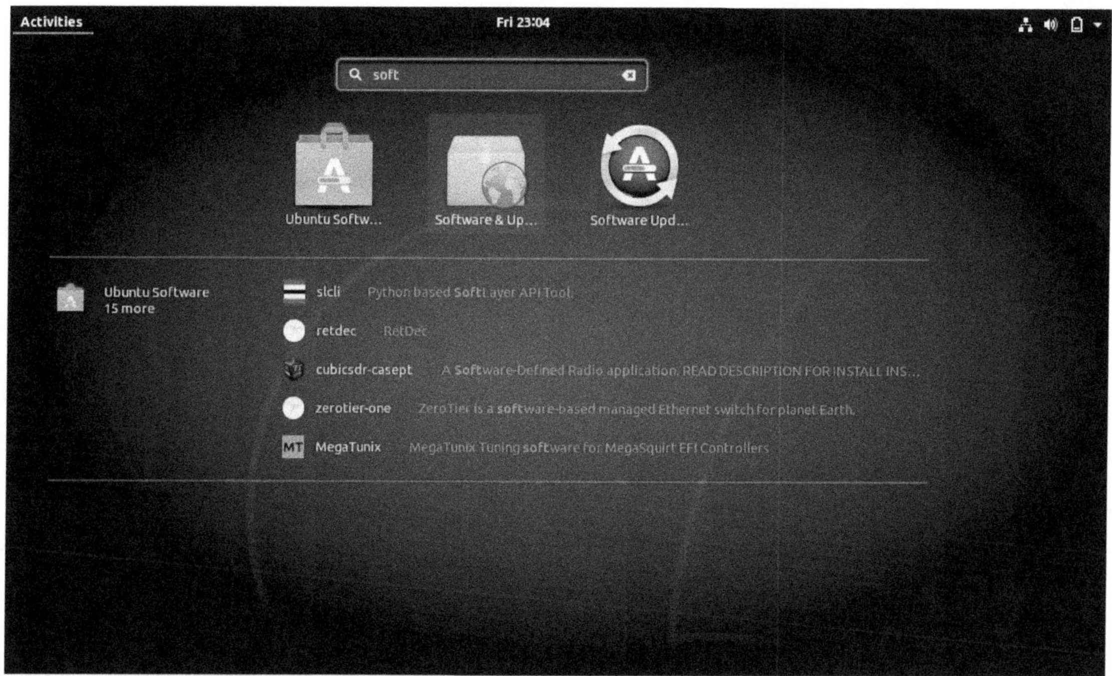

Ubuntu Software & Updates

2. Click on **Software & Updates** and enable all of the Ubuntu repositories, as shown in the following screenshot:

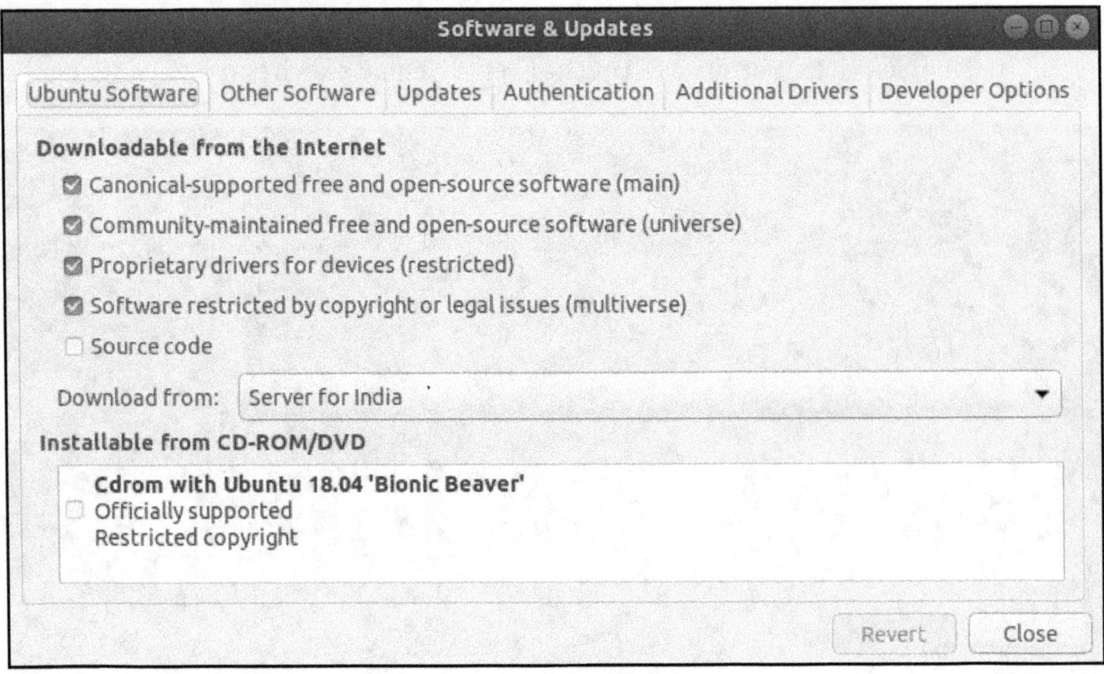

The Ubuntu Software & Updates center

Now that we've enabled the preceding set conditions, we can move on to the next step.

Setting up source.list

The next step is to allow ROS packages from the ROS repository server, called `packages.ros.org`. The ROS repository server details have to be fed into `source.list`, which is in `/etc/apt/`.

The following command will do this for ROS Melodic:

```
$ sudo sh -c 'echo "deb http://packages.ros.org/ros/ubuntu $(lsb_release -sc) main" > /etc/apt/sources.list.d/ros-latest.list'
```

Let's set up the keys now.

Setting up keys

When a new repository is added to Ubuntu, we should add the keys to make it trusted and to be able to validate the origin of the packages. The following key should be added to Ubuntu before starting installation:

```
$ sudo apt-key adv --keyserver hkp://ha.pool.sks-keyservers.net:80 --recv-key 421C365BD9FF1F717815A3895523BAEEB01FA116
```

Now, we are sure that we are downloading from an authorized server.

Installing ROS Melodic

Now, we are ready to install ROS packages on Ubuntu. Follow these steps to do so:

1. The first step is to update the list of packages on Ubuntu. You can use the following command to update the list:

   ```
   $ sudo apt-get update
   ```

   This will fetch all the packages from the servers that are in `source.list`.

2. After getting the package list, we have to install the entire ROS package suite using the following command:

   ```
   $ sudo apt-get install ros-melodic-desktop-full
   ```

   This will install most of the important packages in ROS. You will need at least 15 GB of space in your root Ubuntu partition to install and work with ROS.

Initializing rosdep

The `rosdep` tool in ROS helps us easily install the dependencies of the packages that we are going to compile. This tool is also necessary for some core components of ROS.

This command launches `rosdep`:

```
$ sudo rosdep init
$ rosdep update
```

Here, while the first command was called, a file called `20-default.list` was created in `/etc/ros/rosdep/sources.list.d/`, with a list of links that connect to the respective `ros-distros`.

Setting up the ROS environment

Congratulations! We are done with the ROS installation, but what next?

The ROS installation mainly consists of scripts and executables, which are mostly installed to `/opt/ros/<ros_version>`.

To get access to these commands and scripts, we should add ROS environment variables to the Ubuntu Terminal. It's easy to do this. To access ROS commands from inside the Terminal, we have to source the `/opt/ros/<ros_version>/setup.bash` file.

Here's the command to do so:

```
$ source /opt/ros/melodic/setup.bash
```

But in order to get the ROS environment in multiple Terminals, we should add the command to the `.bashrc` script, which is in the home folder. The `.bashrc` script will be sourced whenever a new Terminal opens:

```
$ echo "source /opt/ros/melodic/setup.bash" >> ~/.bashrc
$ source ~/.bashrc
```

We can install multiple ROS distributions on Ubuntu. If there are multiple distributions, we can switch to each ROS distribution by changing the distribution name in the preceding command.

Getting rosinstall

Last but not least, there is the ROS command-line tool, called `rosinstall`, for installing source trees for particular ROS packages. The tool is based on Python, and you can install it using the following command:

```
$ sudo apt-get install python-rosinstall
```

We are done with the ROS installation. Just check whether the installation is successful by running the following commands:

- Open a Terminal window and run the `roscore` command:

  ```
  $ roscore
  ```

- Run a `turtlesim` node in another Terminal:

  ```
  $ rosrun turtlesim turtlesim_node
  ```

If everything is correct, you will see the following output:

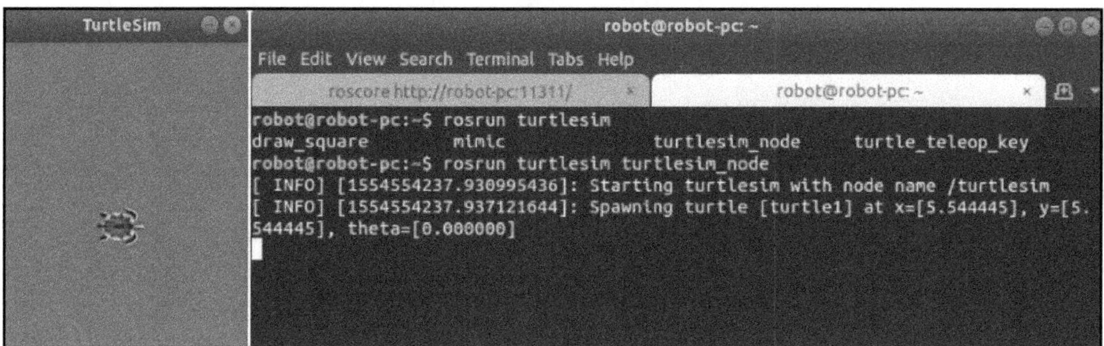

The turtlesim node GUI and Terminal with pose information

If you respawn the `turtlesim` node a couple of times, you should see the turtle changing. We have now successfully installed ROS on Ubutu. Now, let's learn how to set up ROS on VirtualBox.

Setting up ROS on VirtualBox

As you know, complete ROS support is only present on Ubuntu. So, what about Windows and macOS users? Can't they use ROS? Yes, they can, using a tool called VirtualBox (`https://www.virtualbox.org/`). VirtualBox allows us to install a guest OS without affecting the host OS. The virtual OS can work along with the host OS in a given specification of a virtual computer, such as the number of processors and RAM and hard disk size.

You can download VirtualBox for popular OSes from the following link: `https://www.virtualbox.org/wiki/Downloads`.

The complete installation procedure for Ubuntu on VirtualBox is shown in the following tutorial video on YouTube: `https://www.youtube.com/watch?v=QbmRXJJKsvs`.

The following is a screenshot of the VirtualBox GUI. You can see the virtual OS list on the left-hand side and the virtual PC configuration on the right-hand side. The buttons for creating a new virtual OS and starting the existing VirtualBox are on the top panel. The optimal virtual PC configuration is shown in the following screenshot:

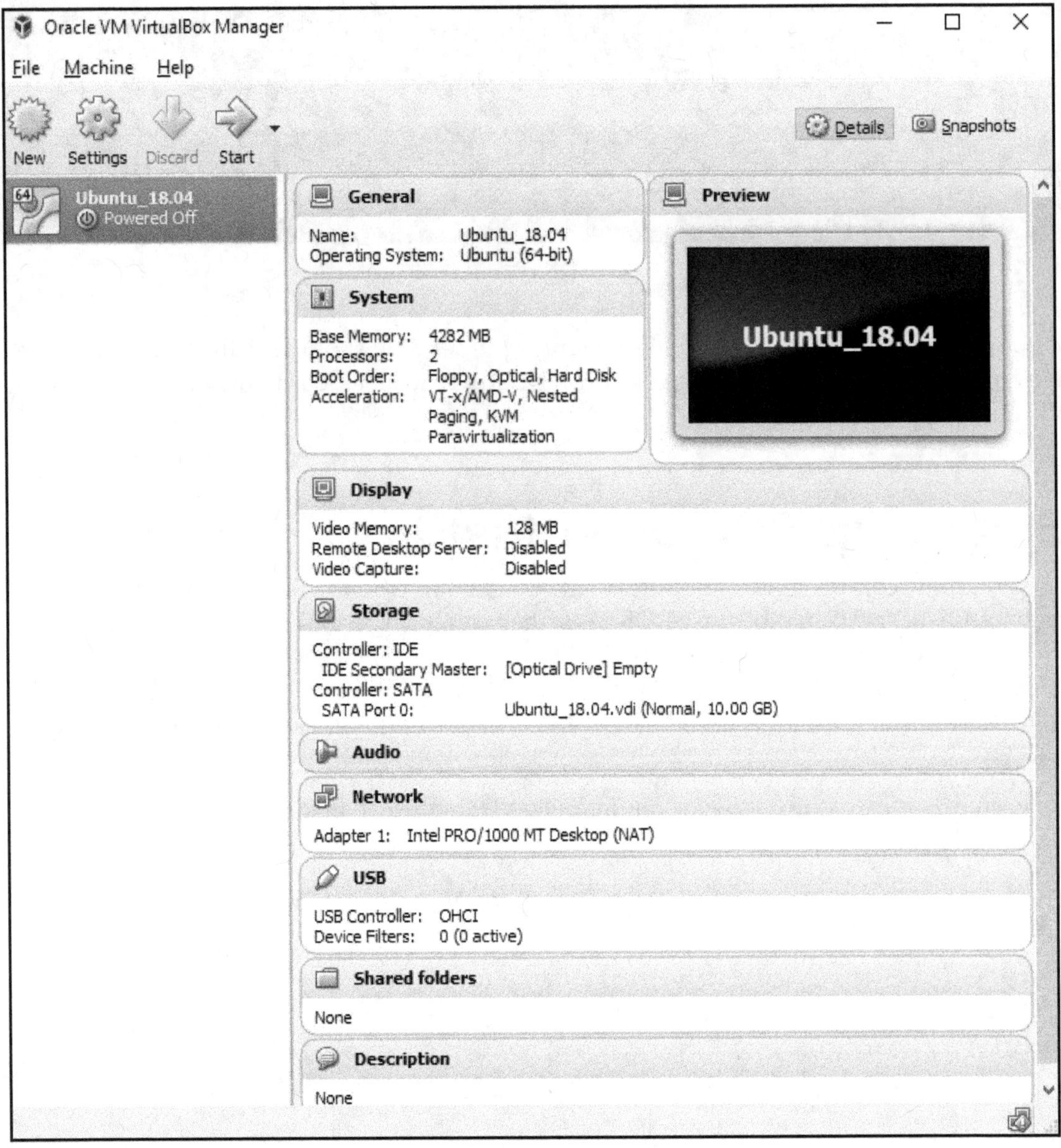

The VirtualBox configuration

Here are the main specifications of the virtual PC:

- **Number of CPUs**: 1
- **RAM**: 4 GB
- **Display | Video Memory**: 128 MB
- **Acceleration**: 3D
- **Storage**: 20 GB to 30 GB
- **Network adapter**: NAT

In order to have hardware acceleration, you should install drivers from the VirtualBox Guest addons disc. After booting into the Ubuntu desktop, navigate to **Devices | Insert Guest Addition CD Image**. This will mount the CD image in Ubuntu and ask the user to run the script to install drivers. If we allow it, it will automatically install all of the drivers. After a reboot, you will get full acceleration on the Ubuntu guest.

There is no difference in ROS installation on VirtualBox. If the virtual network adapter is in NAT mode, the internet connection of the host OS will be shared with the guest OS, so the guest can work the same as the real OS. We now have ROS set up on VirtualBox.

The next section is an introduction to Docker.

Introduction to Docker

Docker is a piece of free software and the name of the company that introduced it to open source community. You might have heard of virtual environments in Python, where you can create isolated environments for your projects and install dedicated dependencies that do not cause any trouble with other projects in other environments. Docker is similar, where we can create isolated environments for our projects called **containers**. Containers work like virtual machines but aren't actually similar to virtual machines. While virtual machines need a separate OS on top of the hardware layer, containers do not and work independently on top of the hardware layer, sharing the resources of the host machine.

This helps us consume less memory and it is often speedier than virtual machines. The best example to show the difference between both is shown here:

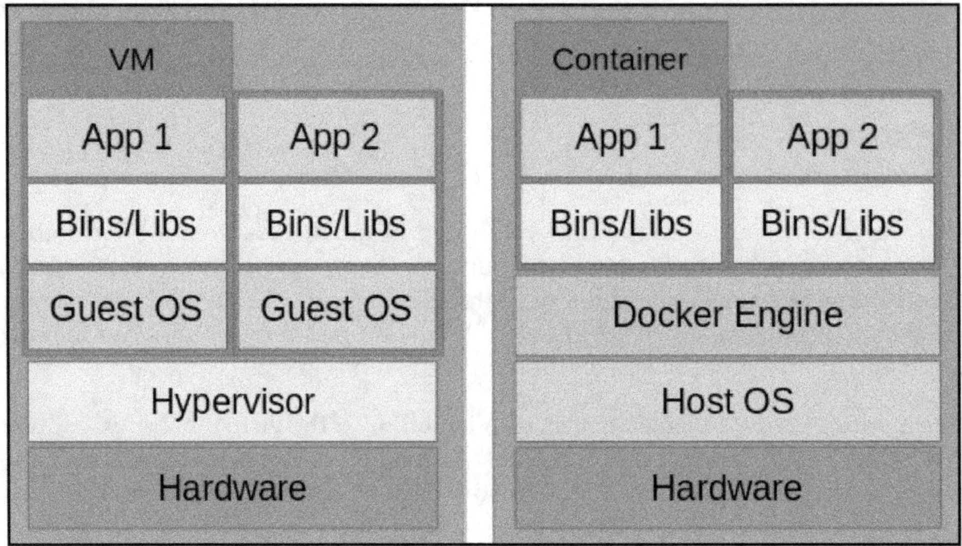

Differences between a virtual machine and Docker

Now that we know the difference between a virtual machine and Docker, let's understand why we use Docker.

Why Docker?

In ROS, a project may consist of several metapackages that contain subpackages, and those would need dependencies to work on. It could be quite annoying for a developer to set up packages in ROS as it is quite common that different packages would use different or the same dependencies of different versions and those could lead to compilation issues. The best example would be when we want to use OpenCV3 with ROS Indigo while working with vision algorithms or `gazebo_ros_controller` packages with different plugin versions, causing the famous gravity error (`https://github.com/ros-simulation/gazebo_ros_pkgs/issues/612`). By the time the developer tries to rectify them, he/she might end up losing other working projects due to replaced packages or dependency version changes. While there might be different ways to handle this problem, a practical way to go about this problem in ROS would be to use Docker containers. Containers are fast and can start or stop, unlike any process in an OS in a matter of seconds. Any upgrades or updates on the OS, or packages would not affect the containers inside or other containers in place.

Installing Docker

Docker can be installed in two ways: using the Ubuntu repositories or using the official Docker repository:

- If you would just like to explore and save a couple of minutes with just a single line installation, go ahead and install from the Ubuntu repository.
- If you would like to explore more options with Docker other than what is intended in this book, I would suggest that you go ahead and install from official Docker repositories as they will have the most stable and bug-fixed packages with added features.

Before going ahead with either of the following installations, ensure you update the `apt` package index using `$ sudo apt-get update`.

Installing from the Ubuntu repository

To install Docker from the Ubuntu repository, use the following command:

```
$ sudo apt-get install docker.io
```

If you've changed your mind and would like to try out installing from the Docker repository or if you wish to remove the Docker version you installed via the preceding step, move on to the next step.

Removing Docker

If you're not interested in the old Docker version and want to install the latest stable version, remove Docker using the following command and install it from the Docker repository:

```
$ sudo apt-get remove docker docker-engine docker.io containerd runc
```

The preceding is a general command to remove Docker, `docker-engine`, `docker.io` (as these are the older version names), and runtime containers, if any were pulled or created.

Installing from the Docker repository

To install Docker from the official repository, follow these steps:

1. First, we use the following command:

```
$ sudo apt-get install apt-transport-https ca-certificates curl
gnupg-agent software-properties-common
```

2. Then, we add the official GPG key from Docker:

```
$ curl -fsSL https://download.docker.com/linux/ubuntu/gpg | sudo
apt-key add -
```

3. Set up the Docker repository using the following command:

```
$ sudo add-apt-repository "deb [arch=amd64]
https://download.docker.com/linux/ubuntu bionic stable"
```

There are three types of update channels called the stable, nightly, and test channels. The test channel provides the prereleases that are ready for testing before availability, the nightly channel is the work in progress or beta version, and stable is the finalized bug-fixed channel. The best suggestion from the Docker team is the stable channel; however, you're free to test either channels by replacing the term `stable` with either `nightly` or `test`.

4. Update the `apt` package index once again:

```
$ sudo apt-get update
```

5. Now, install the Docker package using the following command:

```
$ sudo apt install docker-ce
```

6. After installing via either method, you could check the versions of Docker for both types of installation using the following command:

```
$ docker --version
```

The current version that is available in the Ubuntu repository is 17.12, while the latest release version at the time of writing this book is 18.09 (stable version).

Docker can only be run as a root user by default. Hence, add your username to the Docker group using the following command:

```
$ sudo usermod -aG docker ${USER}
```

Ensure you reboot the system for the preceding to take effect; otherwise, you will face a **permission denied** error, as shown here:

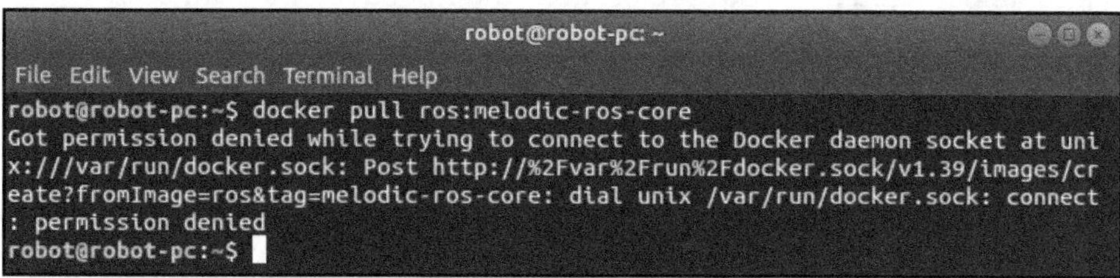

Permission denied error

A quick fix to the preceding error would be to use `sudo` before any Docker commands.

Working with Docker

Containers are built from Docker images and these images can be pulled from Docker Hub (`https://hub.docker.com/`). We can pull ROS containers from the `ros` repository using the following command:

```
$ sudo docker pull ros:melodic-ros-core
```

If everything is successful, you see the output shown here:

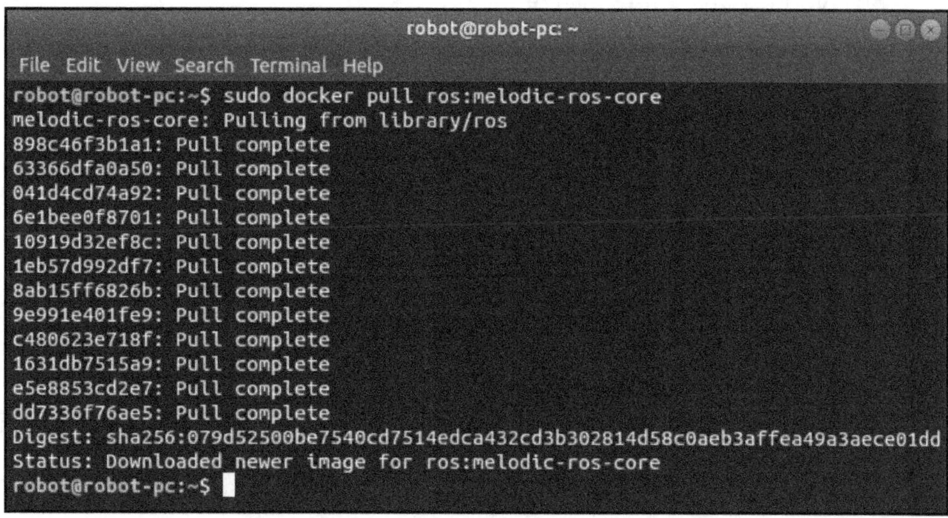

Successful Docker pull

You can choose the specific version of ROS you want to work with. The best suggestion for any application is to start with `melodic-core`, where you would continue to work and update the container related to your project goal and not have other unnecessary components installed. You can view Docker images using this command:

```
$ sudo docker images
```

By default, all of the containers are saved in `/var/lib/docker`. Using the preceding command, you can identify the repository name and tag. In my case, for the `ros` repository name, my tag was `melodic-ros-core`; hence, you could run the `ros` container using the following command:

```
$ sudo docker run -it ros:melodic-ros-core
```

Other information the `$ docker images` command gives is the container ID, which is `7c5d1e1e5096` in my case. You will need it when you want to remove the container. Once you're inside Docker, you can check the ROS packages that are available using the following command:

```
$ rospack list
```

When you run and exit Docker, you would've created another container, so for beginners, it's quite common to create a list of containers unknowingly. You could use `$ docker ps -a` or `$ docker ps -l` to view all active/inactive containers or the latest container and remove containers using `$ docker rm <docker_name>`. To continue working in the same container, you could use the following command:

```
$ sudo docker start -a -i silly_volhard
```

Here, `silly_volhard` is the default name created by Docker.

Now that you've opened the same container, let's install an ROS package and commit changes to the Docker. Let's install the `actionlib_tutorials` package using the following command:

```
$ apt-get update
$ apt-get install ros-melodic-actionlib-tutorials
```

Now, when you check the ROS packages list once again, you should be able to view a few extra packages. Since you have modified the container, you would need to commit it to experience the modifications while reopening the Docker image. Exit the container and commit using the following command:

```
$ sudo docker commit 7c5d1e1e5096 ros:melodic-ros-core
```

Now that we have installed ROS on Ubuntu and VirtualBox, let's learn how to set up the ROS workspace.

Setting up the ROS workspace

After setting up ROS on a real PC, VirtualBox, or Docker, the next step is to create a workspace in ROS. The ROS workspace is a place where we keep ROS packages. In the latest ROS distribution, we use a catkin-based workspace to build and install ROS packages. The catkin system (`http://wiki.ros.org/catkin`) is the official build system of ROS, which helps us build the source code into a target executable or libraries inside the ROS workspace.

Building an ROS workspace is an easy task; just open a Terminal and follow these instructions:

1. The first step is to create an empty workspace folder and another folder called `src` to store the ROS package in. The following command will do this for us. The workspace folder name here is `catkin_ws`:

   ```
   $ mkdir -p catkin_ws/src
   ```

2. Switch to the `src` folder and execute the `catkin_init_workspace` command. This command will initialize a catkin workspace in the current `src` folder. We can now start creating packages inside the `src` folder:

   ```
   $ cd ~/catkin_ws/src
   $ catkin_init_workspace
   ```

3. After initializing the catkin workspace, we can build the packages inside the workspace using the `catkin_make` command. We can also build the workspace without any packages:

   ```
   $ cd ~/catkin_ws/
   $ catkin_make
   ```

4. This will create additional folders called `build` and `devel` inside the ROS workspace:

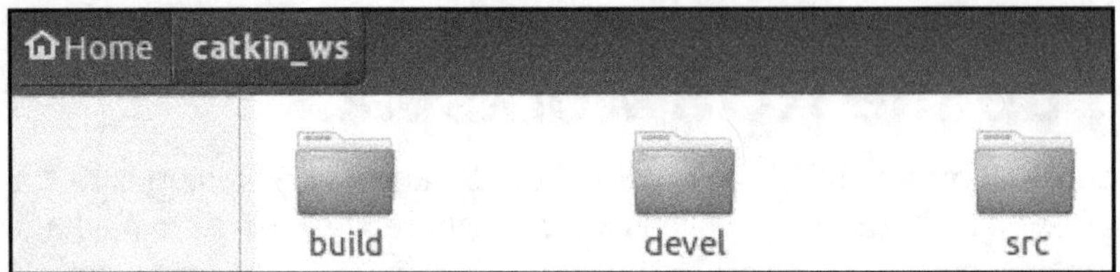

The catkin workspace folders

5. Once you've built the workspace, to access packages inside the workspace, we should add the workspace environment to our `.bashrc` file using the following command:

```
$ echo "source ~/catkin_ws/devel/setup.bash" >> ~/.bashrc
$ source ~/.bashrc
```

6. When everything is done, you can verify that everything is correct by executing the following command:

```
$ echo $ROS_PACKAGE_PATH
```

This command will print the entire ROS package path. If your workspace path is in the output, you are done:

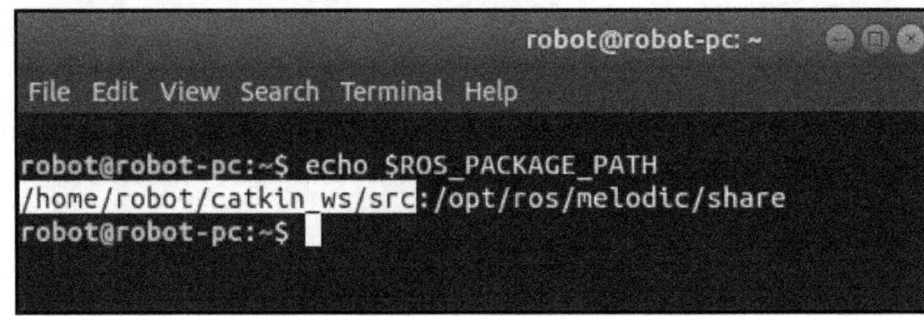

The ROS package path

You will see that two locations are sourced as `ROS_PACKAGE_PATH`. The former is the recent edition we made in *step 5* and the latter is the actual ROS installed packages folder. With this, we have set up the ROS workspace. We will now look at the different opportunities for ROS in industries and research.

Opportunities for ROS in industries and research

Now that we've installed ROS and set up our ROS workspace, we can discuss the advantages of using it. Why is learning ROS so important for robotics researchers? The reason is that ROS is becoming a generic framework for programming all kinds of robots. So, robots in universities and industries mainly use ROS.

Here are some famous robotics companies using ROS for their robots:

- **Fetch Robotics:** `http://fetchrobotics.com/`
- **Clearpath Robotics:** `https://www.clearpathrobotics.com/`
- **PAL Robotics:** `http://www.pal-robotics.com/en/home/`
- **Yujin Robot:** `http://yujinrobot.com/`
- **DJI:** `http://www.dji.com/`
- **ROBOTIS:** `http://www.robotis.com/html/en.php`

Knowledge of ROS will help you land a robotics application engineering job easily. If you go through the skillset of any job related to robotics, you're bound to find ROS on it.

There are independent courses and workshops in universities and industries to teach ROS development in robots. Knowing ROS will help you to get an internship and MS, Ph.D., and postdoc opportunities from prestigious robotic institutions such as CMU's Robotics Institute (`http://www.ri.cmu.edu/`) and UPenn's GRAP lab (`https://www.grasp.upenn.edu/`).

The following chapters will help you build a practical foundation and core skills in ROS.

Summary

This chapter was an introductory chapter to get you started with robotics application development using ROS. The main aim of this chapter was to get you started with ROS by installing and understanding it. This chapter can be used as a kick-start guide for ROS application development and can help you understand the following chapters, which mainly demonstrate ROS-based applications. At the end of this chapter, we saw job and research opportunities related to ROS and saw that a lot of companies and universities are looking for ROS developers for different robotics applications.

From the next chapter onward, we will discuss ROS-2 and its capabilities.

2
Introduction to ROS-2 and Its Capabilities

ROS or, to be more specific, ROS-1, has helped robotics reach a different milestone in the open source community. While there were difficulties connecting the hardware and software and synchronizing them, ROS-1 paved the way for a simple communication strategy that has helped the community to connect any sophisticated sensor to a microcomputer or microcontroller with ease. Over the past decade, ROS-1 has grown and has a huge package list, where each and every package solves a problem either in bits or in full, and has eliminated the concept of reinventing the wheel. These packages have led to a whole new way of looking at robotics and providing intelligence to the currently available systems. Connecting several such smaller packages could create a new complex autonomous system on its own.

Though ROS-1 has given us the liberty of communicating with complex hardware and software components with ease, there are some intricacies involved while producing them as a product. For instance, let's say that, in the manufacturing industry, there is a swarm of heterogeneous robots (robots of different types—say, mobile robots, robot arms, and so on). It is quite difficult to establish communication between them due to the way ROS-1 is architectured. This is because ROS-1 doesn't support the multi master concept.

Although there are other ways (which we will explore in `Chapter 6`, *Multi-Robot Collaboration*) we can use to communicate between nodes across the network, there is no secure way of communicating between them. Anyone connected to the master node would easily be able to gain access to a list of available topics and use or modify them. People often use ROS-1 for working on proof-of-concepts or as a quick workaround for research interests.

A marginal line has been drawn between using ROS-1 for prototyping and creating final products, and this gap isn't likely to be reduced. This is mainly because ROS-1 isn't real-time. Network connectivity varies between system components while using a wired connection (Ethernet) over a wireless connection (Wi-Fi). This could lead to a delay in data reception or even loss of data.

Keeping this in mind, OSRF began its journey toward improving and building the next-generation ROS, called ROS-2. It is being developed to fix the risks faced by ROS-1. In this chapter, you will be introduced to ROS-2, its differences with ROS-1, and its features. This chapter is organized in the same manner as the previous chapter for better understanding and comparison:

- Getting started with ROS-2
- Fundamentals of ROS-2
- ROS-2 client libraries
- ROS-2 tools
- Installing ROS-2
- Setting up the ROS-2 workspace
- Writing ROS-2 nodes
- Bridging ROS-1 and ROS-2

Technical requirements

Let's look into the technical requirements for this chapter:

- ROS 2 (source installation) on Ubuntu 18.04 (Bionic)
- Timelines and test platform:
    - **Estimated learning time**: On average, 90 minutes
    - **Project build time (inclusive of compile and run time)**: On average, 60 minutes
    - **Project test platform**: HP Pavilion laptop (Intel® Core™ i7-4510U CPU @ 2.00 GHz × 4 with 8 GB Memory and 64-bit OS, /GNOME-3.28.2)

The code for this chapter is available at `https://github.com/PacktPublishing/ROS-Robotics-Projects-SecondEdition/tree/master/chapter_2_ws/src/ros2_talker`.

Now, let's get started with ROS-2.

Getting started with ROS-2

ROS-2 is a deterministic effort toward improving the communication network framework that will be used with real-time systems and production-ready solutions. ROS-2 aims to do the following:

- Provide secure communication between components
- Communicate in real time
- Connect multiple robots with ease
- Improve the quality of communication irrespective of the communication medium
- Provide an ROS layer directly onto hardware, such as sensors and embedded boards
- Use the latest software improvements

Remember the ROS equation diagram from the previous chapter? ROS-2 follows the exact same equation but in a different manner:

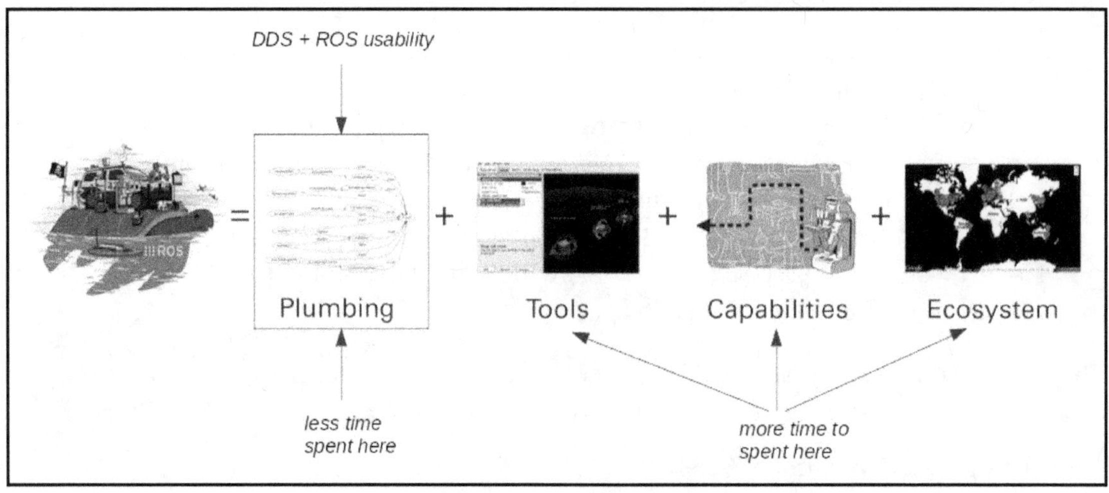

ROS equation with DDS implementation (source: ros.org. Licensed under Creative Commons CC-BY-3.0: https://creativecommons.org/licenses/by/3.0/us/legalcode)

ROS-2 follows an industrial standard and implements real-time communication through a concept called DDS implementation. **DDS** (short for **Data Distributed Services**) is a proven industry standard by **Object Management Groups** (**OMGs**) and it is implemented by many vendors such as RTI's implementation, called Connext (`https://www.rti.com/products/`), ADLink's implementation, called OpenSplice RTPS (`https://github.com/ADLINK-IST/opensplice`), and eProsima's implementation, called Fast RTPS (`http://www.eprosima.com/index.php/products-all/eprosima-fast-rtps`).

These standards are used in time-critical applications such as airline systems, hospitals, financial services, and space exploration systems. This implementation aims to simplify the plumbing strategy (the publish-subscribe infrastructure) and make it deterministic across different hardware and software components. This ensures that the user can concentrate on the capabilities and ecosystem more.

ROS-2 distributions

After a couple of years of alpha and beta version releases, the first official stable version of ROS-2 was released in December 2017. The distro was named Ardent Apalone, a typical alphabetical order naming style OSRF usually follows to name their ROS distributions. The second release was in June 2018, called Bouncy Bolson. This version came with added features and bug fixes and support for Ubuntu 18.04 and Windows 10 with Visual Studio 2017. The third version was called Crystal Clemmys and was released in December 2018. All of the ROS distributions are code-named. For instance, this release is code-named crystal, while the former two are named bouncy and ardent, respectively.

At the time of writing this book, the latest version is Dashing Diademata, which was released on May 31, 2019 and code-named dashing.

Ardent and bouncy have already reached their **End Of License** (**EOL**) deadlines and hence aren't supported. The third stable release, crystal, has an EOL deadline of December 2019 and the only long-term stable release is dashing, which has an EOL deadline of May 2021.

Supported operating systems

ROS-2 supports Linux, Windows, macOS, and **Real-Time Operating Systems** (**RTOS**) OS layers, while ROS-1 only supported Linux and macOS layers. Though there is support available from the ROS community on Windows, it wasn't officially supported by OSRF. The following table shows the ROS distributions and specific versions of supported operating systems:

ROS distributions	**Supported operating systems**
Dashing Diademata	Ubuntu 18.04 (Bionic—arm64 and amd64), Ubuntu 18.04 (Bionic-arm32), macOS 10.12 (Sierra), Windows 10 with Visual Studio 2019, Debian Stretch(9)—arm64, amd64 and arm32, and OpenEmbedded Thud (2.6)
Crystal Clemmys	Ubuntu 18.04 (Bionic); Ubuntu 16.04 (Xenial)—source build available, not Debian package; macOS 10.12 (Sierra); and Windows 10
Bouncy Bolson	Ubuntu 18.04 (Bionic); Ubuntu 16.04 (Xenial)—source build available, not Debian package; macOS 10.12 (Sierra); and Windows 10 with Visual Studio 2017
Ardent Apalone	Ubuntu 16.04 (Xenial), macOS 10.12 (Sierra),and Windows 10

The ROS distro that's targeted in this book is Dashing Diademata, the third ROS-2 release.

Robots and sensors supported in ROS-2

ROS-2 is being widely used and supported by research institutes and industries, especially the robot manufacturing industry. The following are links to the robots and sensors that are supported in ROS-2:

- Turtlebot 2 (a): `https://github.com/ros2/turtlebot2_demo`
- Turtlebot 3 (b): `https://github.com/ROBOTIS-GIT/turtlebot3/tree/ros2`
- Mara robot arm (c): `https://github.com/AcutronicRobotics/MARA`
- Dr. Robot's Jaguar 4x4 (d): `https://github.com/TRI-jaguar4x4/jaguar4x4`
- Intel Realsense camera (e): `https://github.com/intel/ros2_intel_realsense`
- Ydlidar (f): `https://github.com/Adlink-ROS/ydlidar_ros2`

For now, the ROS-2 page doesn't have a dedicated page of robots and sensors that are supported on the ROS-2 platform. The preceding robots and sensor packages are efforts from manufacturers and researchers to use ROS-2 with their hardware and provide it to the community.

Why ROS-2?

ROS-2 is designed to work similarly to ROS-1, keeping in mind the robotics community. It is standalone and is not a part of ROS-1. It does, however, interpolate and work alongside ROS-1 packages. ROS-2 is being developed to overcome the challenges faced by ROS-1 by using modern dependencies and tools. Unlike ROS-1, whose feature stack was written in C++, and the client libraries were written in C++ and Python (where Python was built using a ground-up approach and was written based on C++ libraries), the components in ROS-2 are written in the C language. There is a separate layer written in C that connects to ROS-2 client libraries such as `rclcpp`, `rclpy`, and `rcljava`.

ROS-2 has better support for various network configurations and provides reliable communication. ROS-2 also eliminates the concept of nodelets (`http://wiki.ros.org/nodelet`) and supports multi-node initialization. Unlike ROS-1, a couple of pretty interesting features in ROS-2 are the heartbeat detection of a node through a state machine cycle and notifications when nodes and topics are added or removed. These could help design fault-tolerant systems. Also, ROS-2 is soon expected to support different platforms and architectures.

Now that we've understood how ROS-2 sets itself apart from ROS-1, let's look at its fundamentals for a detailed explanation.

Fundamentals of ROS-2

In ROS-1, the user code would connect to the ROS client libraries (such as `rospy` or `roscpp`) and they would communicate directly with other nodes from within the network, whereas in ROS-2, the ROS client libraries act like an abstraction layer and connect to another layer that communicates with the network using other nodes through DDS implementation. A simple comparison is shown here:

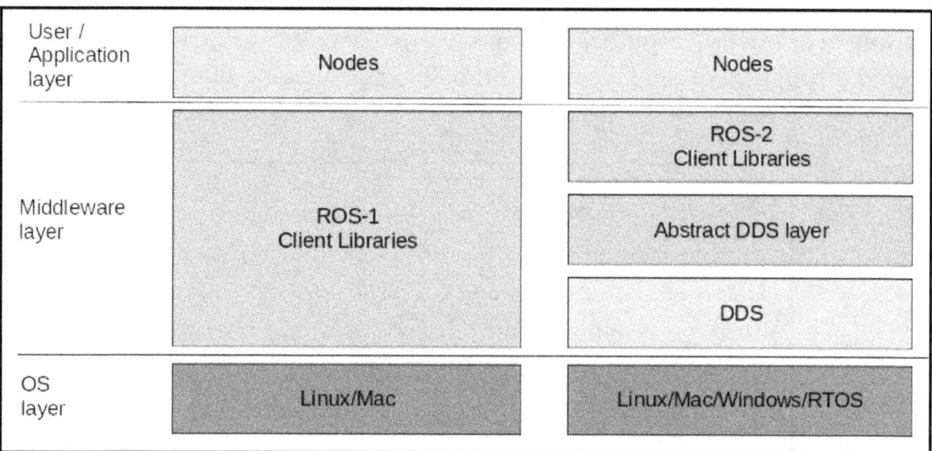

Comparison between ROS-1 and ROS-2

As you can see, in ROS-2, the communication with the OS layer and further down to the hardware layer is done through DDS implementation. The DDS component in the preceding diagram is vendor-specific and implemented by the vendors.

The abstract DDS layer component is connected via ROS-2 and helps the user connect their code through DDS implementation. This way, the user need not be explicitly aware of the DDS APIs as they come through ROS-2. Also, ROS-1 uses a custom transport protocol, as well as a custom central discovery mechanism, hence the usage of the master node. ROS-2, on the other hand, has the abstract DDS layer, through which serialization, transport, and discovery are being provided.

What is DDS?

As we mentioned earlier, DDS is a standard defined by OMG. It is a publish-subscribe transport technique similar to the one used in ROS-1. The DDS implements a distributed discovery system technique that helps two or more DDS programs communicate with each other without the use of the ROS master, unlike ROS-1. The discovery system need not be dynamic and vendors who implement DDS provide options for static discovery.

How is DDS implemented?

There are multiple DDS implementations supported in ROS-2 since different vendors offer different features. For instance, RTI's Connext may be specifically implemented either for microcontrollers or toward applications requiring safety certifications. In all cases, DDS is implemented through a special layer called the ROS middleware interface layer (or `rmw`), as shown here:

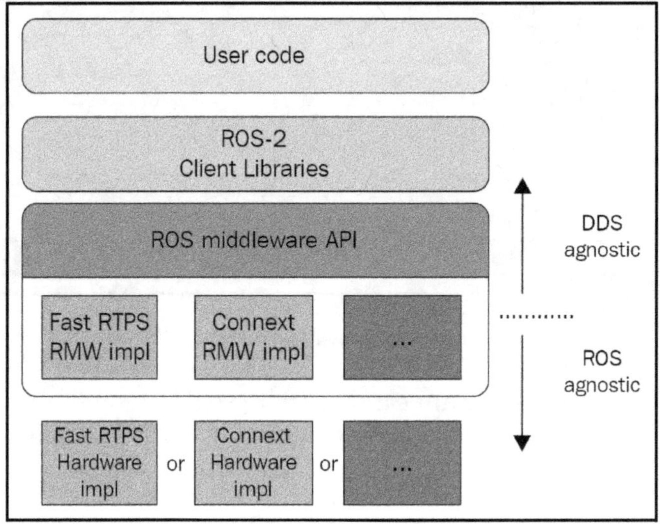

ROS-2 middleware layer

The user code is the topmost block and contains the user logic or the algorithm. In ROS-1, the user code usually sits on top of the ROS client library (such as `roscpp` or `rospy`) and these libraries help the user connect their code with other components in ROS (such as nodes, topics, or services). Unlike the ROS-client library in ROS-1, the ROS-client library in ROS-2 is split into two layers:

- One specific to the programming language (such as `rclcpp`, `rclpy`, or `rcljava`) so that it can handle threading processes such as rosspin and intra-process communication like memory management
- One common layer called `rcl`, implemented in C, that handles the names, services, parameters, time, and console logging

The ROS middleware layer, `rmw`, is also a C implementation and connects with the DDS implementations on top of the hardware layer. This layer is responsible for the service calls, node discovery, graph events, and publish/subscribe calls with **Quality of Services** (**QoS**). QoS in ROS-2 is a measure of performance across nodes in the network. By default, ROS-2 follows eProsima's Fast RTPS implementation since it is open source.

Computational graph

ROS-2 follows the same computational graph concept as ROS-1 but with some changes:

- **Nodes**: Nodes are called participants in ROS-2. Apart from how nodes are defined in the ROS-1 computational graph, there is a possibility to initialize more than one node in a process. They may be located in the same process, different processes, or different machines.
- **Discovery**: While there is a master concept in ROS-1 that aids in communicating between nodes, there is no concept of a master in ROS-2. Don't panic! By default, the DDS standard implementation provides a distributed discovery method where nodes automatically discover themselves in the network. This aids in communication between multiple robots of different types.

Apart from this, the rest of the concepts that we looked at in the previous chapter, such as messages, topics, the parameter server, services, and bags, are the same for ROS-2.

ROS-2 community level

Unlike the ROS-1 community, the ROS-2 community has begun to have some insights. There are good contributions from several research institutes and industries. Since ROS-2 only started its development in 2014, there is lots more information regarding ROS-2 research and tools to watch out for since implementing real-time systems—especially in an open source community—is one huge mountain to climb.

OSRF has been communicating really well with vendors that offer DDS implementation and contribute to the community. While ROS-1 has around 2,000+ packages in its repository, ROS-2 has around 100+ packages. The latest ROS-2 page is `https://index.ros.org/doc/ros2/`.

Communication in ROS-2

If you have completed the previous chapter successfully, you must be familiar with how ROS-1 uses a simple publish-subscribe model where a master node is used to establish a connection between nodes and communicate data. As we mentioned previously, ROS-2 works in a slightly different way. Since DDS implementation is used in ROS-2, the **DDS Interoperability Real-Time Publish-Subscribe Protocol** (**DDSI-RTPS**) is used. The idea is to establish secure and efficient communication between nodes, even in heterogeneous platforms:

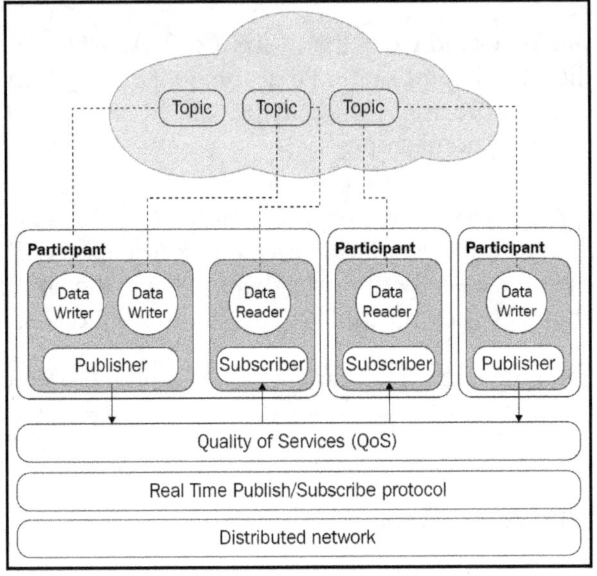

The DCPS model

As you can see, there are additional components that are involved in this communication method. As we mentioned previously, a node is called a participant in ROS-2. Each participant can have single or multiple DDS topics and these topics are neither publishers nor subscribers, like in ROS-1, but are referred to as code objects in ROS-2.

These DDS topics are available in the global data space. From these topics, DDS publishers and subscribers are created but they do not directly publish to or subscribe to a topic. These are the responsibilities of components called data writers and data readers. Data writers and data readers write or read data in a specialized message type, and this is how communication is achieved in ROS-2. These levels of abstraction are necessary to ensure the secure and efficient transfer of data. A user can set QoS settings at each level to provide the highest granularity of configuration.

Changes between ROS-1 and ROS-2

This section specifies the differences between ROS-1 and ROS-2 so that you can understand what upgrades are targeted in ROS-2. For ease of understanding, they're represented in the following table:

	ROS-1	**ROS-2**
Platforms	Continuous Integration for Ubuntu 16.04 Community support: macOS.	Continuous Integration for Ubuntu 16.04 and 18.04, OS X EL Capitan, and Windows 10.
OS layers	Linux and macOS.	Linux, macOS, Windows, and RTOS.
Languages	C++ 03 Python 2.	C++ 11, 14, and 17 Python 3.5.
Build system	catkin.	ament and colcon.
Environment setup	Here, the build tool generates scripts that would need to be sourced into the environment to use the packages built in the workspace.	Here, the build tool generates package-specific and workspace-specific scripts so that only those specific packages are sourced into the environment for usage.
Build multiple packages	Multiple packages are built in a single CMake context, hence the possibility of colliding target names.	Isolated builds are supported where each package is built separately.
Node initiation	Only one node per process.	Multiple nodes allowed per process.

Apart from the preceding changes, there are other changes, all of which can be found at `http://design.ros2.org/articles/changes.html`.

Now that we have covered the fundamentals of ROS-2, let's look at the ROS-2 client libraries.

ROS-2 client libraries (RCL)

As we saw in the previous chapter, ROS client libraries are nothing but APIs for implementing ROS concepts. Hence, we can directly use them in our user code to access ROS concepts such as nodes, services, and topics. ROS client libraries come with the ease of connecting with multiple programming languages.

Since each programming language has its own advantages and disadvantages, it's left to the users to decide which to choose from. For instance, if a system is concerned with efficiency and faster response rates, you could choose `rclcpp`, and if the system demands prototyping as a priority with respect to the development time that's utilized, you could choose `rclpy`.

The ROS client libraries in ROS-2 are split into two sections: one is language-specific (such as `rclcpp`, `rclpy`, and `rcljava`) and the other contains common functionalities that are implemented in the C language. This keeps the client library thin, lightweight, and easy to develop.

While a developer writes code in ROS, there is a high chance that the code could go for various iterations and changes with respect to the way it might be needed to connect with other nodes or participants in the network. There may be a need to change the logic in which ROS concepts are implemented in the code. Hence, the developer may need to worry only about the intra-process communication and threading processes in their code as they are language-specific implementations; `ros::spin()`, for instance, may have different methods of implementation in C++ and Python.

The changes that are made in the language-specific layer would not directly affect the ROS concepts layer and hence maintaining several client libraries is easy when it comes to bug fixing:

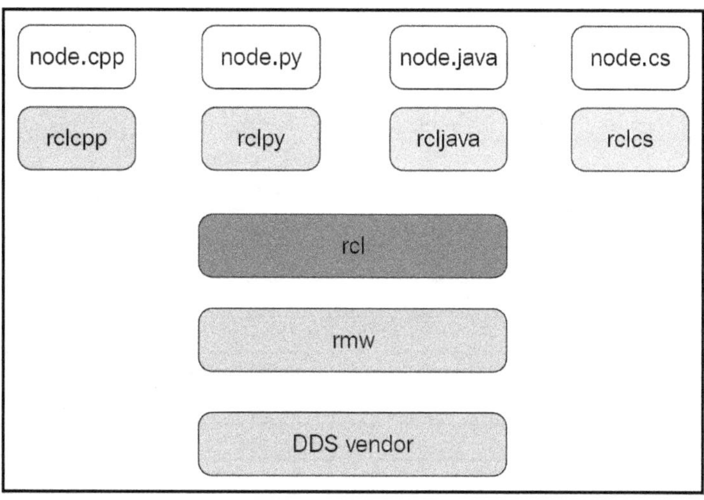

RCL

The preceding diagram shows the RCL structure, where all the language-specific ROS components sit on the layers below the user code (`rclcpp`, `rclpy`, `rcljava`, and `rclcs`). The next common layer, `rcl`, is a C implementation and consists of ROS-specific functions such as names, namespaces, services, topics, time, parameters, and logging information. This allows any language-specific layer to connect with the `rcl` layer and establish communication between different nodes or participants with ease.

This section gave us an overview of the fundamentals of ROS-2. Now, let's look at ROS-2 tools.

ROS-2 tools

Unlike ROS-1, ROS-2 provides tools for debugging message logs and topic information as well. ROS-2 supports visual and command-line tools. However, since there is heavy development in ROS-2, not all the tools have been ported yet. You can still use ROS-1 tools using a ROS-1 to ROS-2 bridge (shown in the upcoming sections) for development.

Rviz2

Rviz is the exact same as how it was defined in ROS-1. The following table shows the features that have currently been ported from `ros-visualization/rviz` to `ros2/rviz`:

Displays	Tools	View controller	Panels
Camera	Move Camera	Orbit	Displays
Fluid Pressure	Focus Camera	XY Orbit	Help
Grid	Measure	First Person	Selections
Grid Cells	Select	Third Person Follower	Tool Properties
Illuminance	2D Nav Goal	Top-Down Orthographic	Views
Image	Publish Point		
Laser Scan	Initial Pose		
Map			
Marker			
Marker Array			
Odometry			
Point Cloud (1 and 2)			
Point			
Polygon			
Pose			
Pose Array			
Range			
Relative Humidity			
Robot Model			
Temperature			
TF			

Other features that are yet to be ported include image transport filters, filtering of topic lists by topic type, message filters, and the features shown in the following list:

Displays	Tools	Panels
Axes	Interact	Time
DepthCloud		
Effort		
Interactive Marker		
Oculus		
Pose With Covariance		
Wrench		

You can find more information about these features here: `http://wiki.ros.org/rviz/DisplayTypes`. A simple default Rviz2 window looks follows:

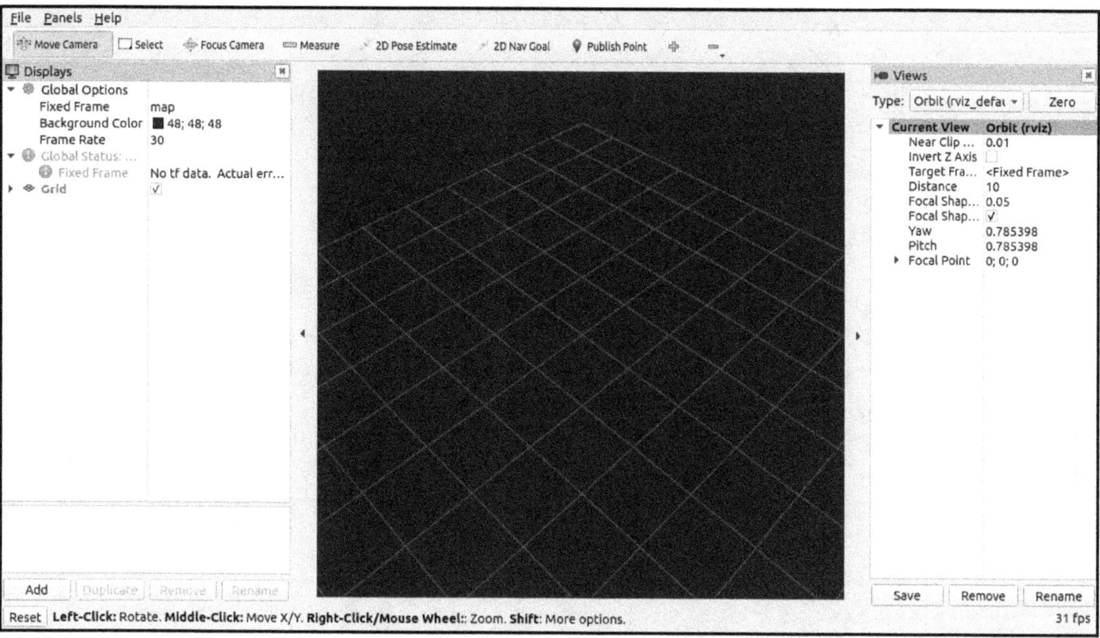

Rviz2 visualizer

Now, let's move on to the next tool, which is Rqt.

Rqt

Like in ROS-1, the `rqt` console is available in ROS-2 as well. Most of the plugins from ROS-1 `rqt` are ported and reused in ROS-2 `rqt`. You can find an available list of plugins here: `http://wiki.ros.org/rqt/Plugins`. Some of the notable advantages of rqt are as follows:

- There's a GUI console, which is interactive and controls process startup and shutdown states.
- More than one widget can be docked in one window.

- You can use already available Qt plugins and convert them into custom `rqt` plugins.
- It supports multiple languages (such as Python and C++) and multiple platforms (on which ROS runs).

Now, let's learn how to install ROS-2 since we have basic knowledge of the ROS-2 client libraries and tools and have covered ROS-2 fundamentals.

Installing ROS-2

In the previous chapter, we used the package manager to set up the ROS-1 environment, desktop, pull keys, and startup installation. ROS-2 is no different and the same method can be followed (but obviously using different setup keys and commands). However, unlike ROS-1, where we followed the Debian package installation technique, we will try the source installation technique in this chapter for installing ROS-2, although you can use the Debian package installation technique as well. It is best to build ROS-2 manually on Linux (that is, via source installation) so that we can add any new changes or releases directly into the package and compile them. Let's get started with the source installation.

Getting started with the installation

The ROS distro we would like to install is Dashing Diademata, code-named dashing, and it will be installed in Ubuntu 18.04 (our setup). You can find out more about how to install ROS-2 at `https://index.ros.org/doc/ros2/Installation/`. To get started with the installation from the source, let's set up the environment.

Setting up the system locale

You need to ensure your environment supports the UTF-8 format. If you're in Docker, the environment will be set to POSIX. Set the locale to the specified format using the following commands:

```
$ sudo locale-gen en_US en_US.UTF-8
$ sudo update-locale LC_ALL=en_US.UTF-8 LANG=en_US.UTF-8
$ export LANG=en_US.UTF-8
```

If all went well you should see the following output:

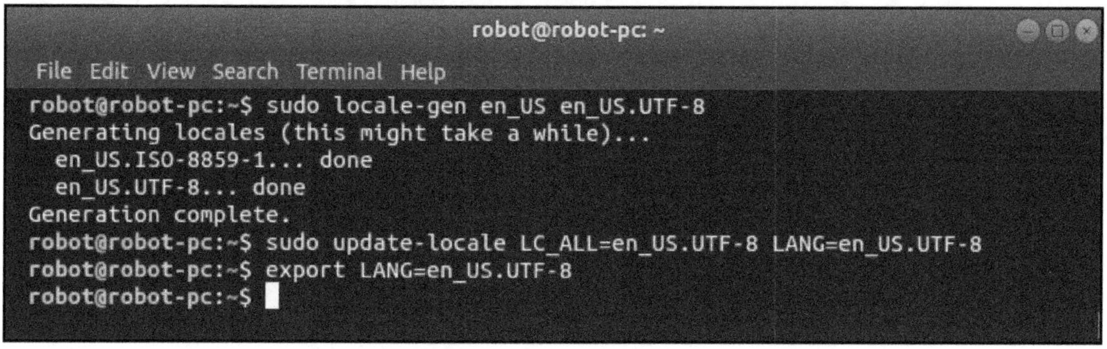
robot@robot-pc: ~
File Edit View Search Terminal Help
robot@robot-pc:~$ sudo locale-gen en_US en_US.UTF-8
Generating locales (this might take a while)...
en_US.ISO-8859-1... done
en_US.UTF-8... done
Generation complete.
robot@robot-pc:~$ sudo update-locale LC_ALL=en_US.UTF-8 LANG=en_US.UTF-8
robot@robot-pc:~$ export LANG=en_US.UTF-8
robot@robot-pc:~$

Setting up the system locale

It should be fine if you're using a different UTF-8 supported locale.

Adding ROS-2 repositories

Let's follow these steps to add ROS-2 repositories:

1. Ensure that you have added ROS-2 apt repositories to your system. You will need to authorize their keys with the package manager using the following command:

```
$ sudo apt update && sudo apt install curl gnupg2 lsb-release

$ curl -s
https://raw.githubusercontent.com/ros/rosdistro/master/ros.asc |
sudo apt-key add -
```

2. Then, add the repository to your sources list using the following command:

```
$ sudo sh -c 'echo "deb [arch=amd64,arm64]
http://packages.ros.org/ros2/ubuntu `lsb_release -cs` main" >
/etc/apt/sources.list.d/ros2-latest.list'
```

Now that you have added the repositories, let's install the other necessities.

Installing development and ROS tools

Follow these steps to install the ROS tools:

1. You need to install the following dependencies and tools using the package manager using the following command:

```
$ sudo apt update && sudo apt install -y build-essential cmake git
python3-colcon-common-extensions python3-lark-parser python3-pip
python-rosdep python3-vcstool wget
```

2. Using `pip3`, install the packages for testing using the following command:

```
$ python3 -m pip install -U argcomplete flake8 flake8-blind-except
flake8-builtins flake8-class-newline flake8-comprehensions flake8-
deprecated flake8-docstrings flake8-import-order flake8-quotes
pytest-repeat pytest-rerunfailures pytest pytest-cov pytest-runner
setuptools
```

3. Install the FAST-RTPS dependencies using the following command:

```
$ sudo apt install --no-install-recommends -y libasio-dev
libtinyxml2-dev
```

Now that all of the dependencies and tools are installed, let's move on to creating and building our workspace.

Getting the ROS-2 source code

We need to create a workspace and clone the ROS-2 repository into it. You can do this using the following commands:

```
$ mkdir -p ~/ros2_ws/src
$ cd ~/ros2_ws
$ wget
https://raw.githubusercontent.com/ros2/ros2/release-latest/ros2.repos
```

The output would be `ros2.repos` being saved, as shown here:

```
robot@robot-pc:~$ cd ~/ros2_ws
robot@robot-pc:~/ros2_ws$ wget https://raw.githubusercontent.com/ros2/ros2/relea
se-latest/ros2.repos
--2019-04-28 10:23:29--  https://raw.githubusercontent.com/ros2/ros2/release-lat
est/ros2.repos
Resolving raw.githubusercontent.com (raw.githubusercontent.com)... 151.101.8.133
Connecting to raw.githubusercontent.com (raw.githubusercontent.com)|151.101.8.13
3|:443... connected.
HTTP request sent, awaiting response... 200 OK
Length: 8469 (8.3K) [text/plain]
Saving to: 'ros2.repos'

ros2.repos          100%[===================>]   8.27K  --.-KB/s    in 0.007s

2019-04-28 10:23:30 (1.19 MB/s) - 'ros2.repos' saved [8469/8469]

robot@robot-pc:~/ros2_ws$
```

Getting the latest repositories

If you're an advanced user and would like to use the developmental release, simply replace `release-latest` with `master` in the preceding command. If you would like to continue experimenting, the best suggestion is to use the preceding command as is because the preceding tree would have gone through intense testing before release. The last step would be to import the repository information using the following command:

```
$ vcs import src < ros2.repos
```

Now that the workspace is set, let's install the dependencies.

Installing dependencies using rosdep

Like we did in ROS-1, we install dependencies using `rosdep`. You will have initialized `rosdep list` in the previous chapter. If not, use the following command; otherwise, skip to the next command:

```
$ sudo rosdep init
```

Now, update `rosdep`:

```
$ rosdep update
```

Then, install the following dependencies:

```
$ rosdep install --from-paths src --ignore-src --rosdistro dashing -y --
skip-keys "console_bridge fastcdr fastrtps libopensplice67 libopensplice69
rti-connext-dds-5.3.1 urdfdom_headers"
```

If everything is successful, you should see the following window:

```
robot@robot-pc: ~/ros2_ws
File Edit View Search Terminal Help
Unpacking python3-snowballstemmer (1.2.1-1) ...
Selecting previously unselected package python3-pydocstyle.
Preparing to unpack .../python3-pydocstyle_2.0.0-1_all.deb ...
Unpacking python3-pydocstyle (2.0.0-1) ...
Selecting previously unselected package pydocstyle.
Preparing to unpack .../pydocstyle_2.0.0-1_all.deb ...
Unpacking pydocstyle (2.0.0-1) ...
Setting up python3-snowballstemmer (1.2.1-1) ...
Processing triggers for man-db (2.8.3-2ubuntu0.1) ...
Setting up python3-pydocstyle (2.0.0-1) ...
Setting up pydocstyle (2.0.0-1) ...
executing command [sudo -H apt-get install -y python3-numpy]
Reading package lists... Done
Building dependency tree
Reading state information... Done
python3-numpy is already the newest version (1:1.13.3-2ubuntu1).
python3-numpy set to manually installed.
0 upgraded, 0 newly installed, 0 to remove and 184 not upgraded.
#All required rosdeps installed successfully
robot@robot-pc:~/ros2_ws$
```

Successfully installed ros dependencies

If you would like to install additional DDS implementations, go to the next section; otherwise, skip them.

Installing DDS implementations (optional)

This is an optional section, in case you would like to install DDS implementations. Follow these steps to install them:

1. As you should know by now, ROS-2 runs on top of DDS and there are multiple DDS vendors. The default DDS implementation middleware that comes with ROS-2 is FAST RTPS. You can install DDS implementations from other vendors such as OpenSplice or Connext using the following commands:

    - For Openslice, use this:

    ```
    $ sudo apt install libopensplice69
    ```

    - For Connext, use this:

    ```
    $ sudo apt install -q -y rti-connext-dds-5.3.1
    ```

Note that, for Connext, you will need a license from RTI. If you need to test RTI's Connext implementation, you can try their 30-day evaluation license, which can be found at `https://www.rti.com/free-trial`.

2. Now that you have acquired either the trial or official license, you need to indicate the license path. You can use the `$RTI_LICENSE_FILE` env variable to point to the license path. You can use the `export` command to point to the license file, as shown here:

```
$ export RTI_LICENSE_FILE=path/to/rti_license.dat
```

3. Ensure that, after downloading the license, you grant permissions to the file using `$chmod +x` on the `$.run` file and run it. Also, you need to source the setup file to set the `$ NDDSHOME` env variable. Go to the following directory, as shown here:

```
$ cd /opt/rti.com/rti_connext_dds-5.3.1/resource/scripts
```

Then, source the following file:

```
$ source ./rtisetenv_x64Linux3gcc5.4.0.bash
```

Now, you can go ahead with building the code. RTI support will be built along with it.

Building code

In ROS-1, you would build packages using the `catkin` tool. In ROS-2, we use an upgraded tool that's an iteration of `catkin_make`, `catkin_make_isolated`, `catkin_tools`, and `ament_tools` (a build system used in the first ROS-2 distro ardent) called `colcon`. This means that `colcon` can also build ROS-1 packages and those without manifests alongside ROS-2 packages.

We shall look into building individual packages using `colcon` in the upcoming sections. For now, let's build ROS-2. Install `colcon` using the following command:

```
$ sudo apt install python3-colcon-common-extensions
```

Now that `colcon` is installed, let's build ROS-2 by going back to the workspace:

```
$ cd ~/ros2_ws/
```

Now, build the package using the following command:

```
$ colcon build --symlink-install
```

It usually takes around 40 minutes to an hour or more to build. In my case, it took around 1 hour 14 minutes. If everything went successfully, you should see something similar to the following:

```
robot@robot-pc: ~/ros2_ws
File Edit View Search Terminal Help
[Processing: rviz_default_plugins]
[Processing: rviz_default_plugins]
[Processing: rviz_default_plugins]
[Processing: rviz_default_plugins]
[Processing: rviz_default_plugins]
[Processing: rviz_default_plugins]
[Processing: rviz_default_plugins]
[Processing: rviz_default_plugins]
[Processing: rviz_default_plugins]
[Processing: rviz_default_plugins]
[Processing: rviz_default_plugins]
Finished <<< rviz_default_plugins [15min 6s]
Starting >>> rviz2
Finished <<< rviz2 [10.9s]

Summary: 232 packages finished [1h 13min 50s]
  9 packages had stderr output: qt_gui_cpp rmw_connext_cpp rmw_connext_shared_cp
p rmw_opensplice_cpp rosidl_typesupport_connext_c rosidl_typesupport_connext_cpp
 rosidl_typesupport_opensplice_c rosidl_typesupport_opensplice_cpp rqt_gui_cpp
robot@robot-pc:~/ros2_ws$
```

Successful compilation

As you can see, nine packages weren't built and they were skipped since I hadn't installed the optional DDS implementations. Now, let's try out some ROS-2 examples from the package we built.

Setting up ROS-1, ROS-2, or both environments

Now that we have both ROS-1 and ROS-2, let's learn how to use both together and individually with ease.

In case you installed ROS-2 via the Debian package installation, try running `roscore` and see whether things are fine with the ROS-1 setup. You will receive an error. This is simply because your bash got confused with, ROS-1 and ROS-2 environments. One of the best practices to use here is to source either the ROS-1 environment when we're working with ROS-1 packages and source ROS-2 if we're working with ROS-2 packages.

Instead of typing the source command every time, we could make use of the `alias` command and add them to the bash script using the following steps:

1. Invoke your bash script using the following command:

```
$ sudo gedit ~/.bashrc
```

2. Add the following two lines to it:

```
$ alias initros1='source /opt/ros/melodic/setup.bash'
$ alias initros2='source ~/ros2_ws/install/local_setup.bash'
```

3. Please delete or comment the following lines (that we added in the previous chapter) in the bash script:

```
$ source /opt/ros/melodic/setup.bash
$ source ~/catkin_ws/devel/setup.bash
```

Your bash file should look like this:

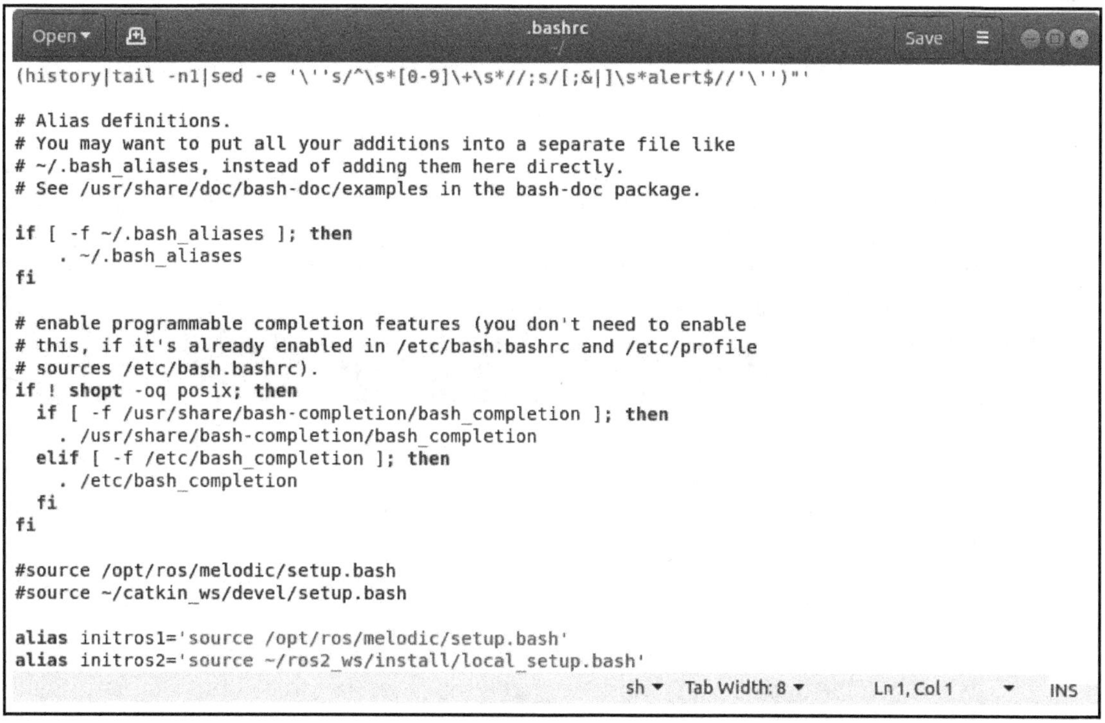

```
Open ▾    .bashrc    Save ☰
(history|tail -n1|sed -e '\''s/^\s*[0-9]\+\s*//;s/[;&|]\s*alert$//'\'')"'

# Alias definitions.
# You may want to put all your additions into a separate file like
# ~/.bash_aliases, instead of adding them here directly.
# See /usr/share/doc/bash-doc/examples in the bash-doc package.

if [ -f ~/.bash_aliases ]; then
    . ~/.bash_aliases
fi

# enable programmable completion features (you don't need to enable
# this, if it's already enabled in /etc/bash.bashrc and /etc/profile
# sources /etc/bash.bashrc).
if ! shopt -oq posix; then
  if [ -f /usr/share/bash-completion/bash_completion ]; then
    . /usr/share/bash-completion/bash_completion
  elif [ -f /etc/bash_completion ]; then
    . /etc/bash_completion
  fi
fi

#source /opt/ros/melodic/setup.bash
#source ~/catkin_ws/devel/setup.bash

alias initros1='source /opt/ros/melodic/setup.bash'
alias initros2='source ~/ros2_ws/install/local_setup.bash'
sh ▾  Tab Width: 8 ▾  Ln 1, Col 1  ▾  INS
```

The bash file

Now, save and close the bash script. This is to ensure that, when you open your Terminal, neither the ROS-1 or ROS-2 workspace is invoked. Now that these changes have been made to the bash file, source the bash script once again to ensure we're able to use the alias commands:

```
$ source ~/.bashrc
```

Now, you can peacefully invoke the ROS-1 or ROS-2 environments using the `initros1` or `initros2` command in the Terminal to work with the ROS-1 or ROS-2 packages, respectively.

Running test nodes

Now that your ROS environments have been set up properly, let's try testing the ROS-2 nodes. Follow these steps to test the nodes:

1. Open a Terminal and source the ROS-2 environment workspace using the following command:

   ```
   $ initros2
   ```

   You should see a message, as shown here:

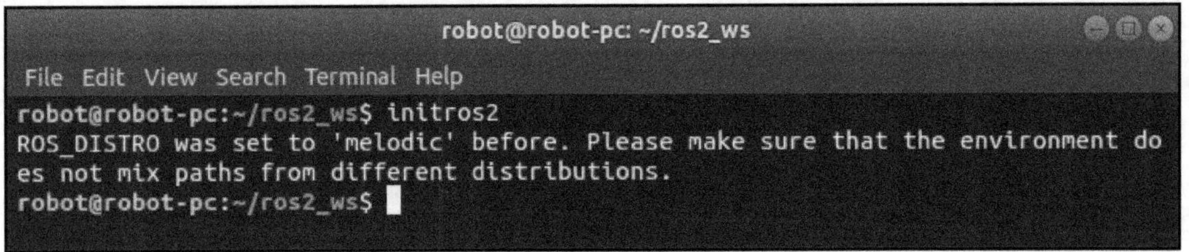

Distro warning

2. Let's run the traditional talker node that we saw, similar to ROS-1, using the following command:

   ```
   $ ros2 run demo_nodes_cpp talker
   ```

   As you may have noticed, there isn't much difference between the way ROS-1 and ROS-2 run the nodes. You could very well say from the preceding command that the package name is `demo_nodes_cpp` and that the node name is talker. And instead of `rosrun` in ROS-1, it is `ros2 run` in ROS-2.

Notice the space between `ros2` and `run`.

3. In another Terminal, let's initialize ROS-2 once again using the following command:

```
$ initros2
```

4. Let's run the traditional listener node that we saw, similar to ROS-1, using the following command:

```
$ ros2 run demo_nodes_py listener
```

You should see `talker` saying that it's `Publishing: 'Hello World: 1,2...'` and `listener` saying `I heard: [Hello World: <respective count>]`. The output should look like this:

```
[INFO] [talker]: Publishing: 'Hello World: 29'
[INFO] [talker]: Publishing: 'Hello World: 30'
[INFO] [talker]: Publishing: 'Hello World: 31'
[INFO] [talker]: Publishing: 'Hello World: 32'
[INFO] [talker]: Publishing: 'Hello World: 33'
[INFO] [talker]: Publishing: 'Hello World: 34'
[INFO] [talker]: Publishing: 'Hello World: 35'
[INFO] [talker]: Publishing: 'Hello World: 36'
[INFO] [talker]: Publishing: 'Hello World: 37'
[INFO] [talker]: Publishing: 'Hello World: 38'
[INFO] [talker]: Publishing: 'Hello World: 39'
[INFO] [talker]: Publishing: 'Hello World: 40'
[INFO] [talker]: Publishing: 'Hello World: 41'
```

```
[INFO] [listener]: I heard: [Hello World: 29]
[INFO] [listener]: I heard: [Hello World: 30]
[INFO] [listener]: I heard: [Hello World: 31]
[INFO] [listener]: I heard: [Hello World: 32]
[INFO] [listener]: I heard: [Hello World: 33]
[INFO] [listener]: I heard: [Hello World: 34]
[INFO] [listener]: I heard: [Hello World: 35]
[INFO] [listener]: I heard: [Hello World: 36]
[INFO] [listener]: I heard: [Hello World: 37]
[INFO] [listener]: I heard: [Hello World: 38]
[INFO] [listener]: I heard: [Hello World: 39]
[INFO] [listener]: I heard: [Hello World: 40]
[INFO] [listener]: I heard: [Hello World: 41]
```

ROS-2 publisher and subscriber output

You could use the `ros2 topic list` command to see the list of topics that are available as well, as shown here:

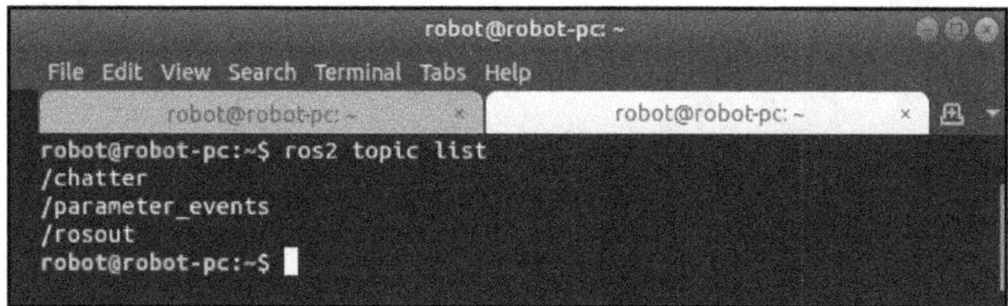

ROS-2 topic list

This brings us to the end of ROS-2's installation. Now, let's learn to set up our ROS-2 workspace.

Setting up the ROS-2 workspace

A ROS workspace is a directory where we keep the ROS packages. As you saw in the preceding setup, you know that the build technique that's used in ROS-2 is `colcon` instead of `catkin`, which was used in ROS-1. The workspace layout is a bit different in ROS-2. Colcon does out of source builds and creates the following folders:

- The `build` folder is where intermediate files are stored.
- The `install` folder is where each package will be installed.
- The `log` folder is where all the logging information is available.
- The `src` folder is where the source code is placed.

Note that there is no `devel` folder like there is in ROS-1.

As the build steps were already explained and tried out while building the ROS-2 packages, let's quickly look at the commands that are needed to build any ROS-2 workspace.

Let's consider the `ros2_examples_ws` package for demonstration purposes:

```
$ initros2
$ mkdir ~/ros2_workspace_ws && cd ~/ros2_workspace_ws
$ git clone https://github.com/ros2/examples src/examples
$ cd ~/ros2_workspace_ws/src/examples
```

Now, check out the branch that is compatible with our ROS distro version, that is, dashing:

```
$ git checkout dashing
$ cd ..
$ colcon build --symlink-install
```

To run tests for the packages we built, use the following command:

```
$ colcon test
```

Now, you can run the package nodes after sourcing the package environment:

```
$ . install/setup.bash
$ ros2 run examples_rclcpp_minimal_subscriber subscriber_member_function
```

In the other Terminal, run the following commands:

```
$ initros2
$ cd ~/ros2_workspace_ws
$ . install/setup.bash
$ ros2 run examples_rclcpp_minimal_publisher publisher_member_function
```

You should see a similar output of publishing and subscribing that we saw earlier.

Now, let's learn how to write ROS-2 nodes using the concepts we covered in the previous sections.

Writing ROS-2 nodes

Writing nodes in ROS-2 is comparatively different from ROS-1 due to the introduction of the additional software layers we saw in the *Fundamentals of ROS-2* section. However, OSRF has ensured there isn't a huge difference while writing the ROS-2 code in order to save time and effort. In this section, we'll compare ROS-1 code and ROS-2 code and find out the differences in their usage.

Note that this is just an example section and if you want to write a ROS-2-specific package for your project, I would recommend that you go through the ROS-2 tutorials page (`https://index.ros.org/doc/ros2/Tutorials/`), which contains the necessary information.

ROS-1 example code

Let's consider the traditional publish-subscribe code, that is, the talker-listener code written in Python. I'm assuming that you're familiar with package creation in ROS-1 by now using the `catkin_create_pkg` command.

Let's create a simple package and run our ROS-1 node from that:

1. Open a fresh Terminal and use the following commands:

```
$ initros1
$ mkdir -p ros1_example_ws/src
$ cd ~/ros1_example_ws/src
$ catkin_init_workspace
$ catkin_create_pkg ros1_talker rospy
$ cd ros1_talker/src/
$ gedit talker.py
```

2. Now, copy the following code into the text editor and save the file:

```
#!/usr/bin/env python

import rospy
from std_msgs.msg import String
from time import sleep

def talker_main():
   rospy.init_node('ros1_talker_node')
    pub = rospy.Publisher('/chatter', String)
    msg = String()
    i = 0
    while not rospy.is_shutdown():
        msg.data = "Hello World: %d" % i
        i+=1
        rospy.loginfo(msg.data)
        pub.publish(msg)
       sleep(0.5)

if __name__ == '__main__':
        talker_main()
```

3. After you've saved the file, close the file and give permissions to the file:

```
$ chmod +x talker.py
```

4. Now, go back to the workspace and compile the package:

```
$ cd ~/ros1_example_ws
$ catkin_make
```

5. To run the node, source the workspace and run the node using the following commands:

```
$ source devel/setup.bash
$ rosrun ros1_talker talker.py
```

Ensure you have `roscore` running in another Terminal to check the preceding node. Do that by calling `$ initros1` and then `$ roscore`.

You should now see the node publishing information. Let's see how this is done for ROS-2.

ROS-2 example code

As you know, ROS-2 uses the `colcon` build technique, and package creation is slightly different.

Let's create a simple package and run our ROS-2 node from that:

1. Open a fresh Terminal and use the following commands:

```
$ initros2
$ mkdir -p ros2_example_ws/src
$ cd ~/ros2_example_ws/src
$ ros2 pkg create ros2_talker
$ cd ros2_talker/
```

2. Now, remove the `CMakelists.txt` file:

```
$ rm CMakelists.txt
```

3. Modify the `package.xml` file using the `$ gedit package.xml` command:

```
<?xml version="1.0"?>
<?xml-model
href="http://download.ros.org/schema/package_format2.xsd"
schematypens="http://www.w3.org/2001/XMLSchema"?>
<package format="2">
  <name>ros2_talker</name>
  <version>0.0.0</version>
  <description>Examples of minimal publishers using
rclpy.</description>

  <maintainer email="ram651991@gmail.com">Ramkumar
Gandhinathan</maintainer>
  <license>Apache License 2.0</license>

  <exec_depend>rclpy</exec_depend>
  <exec_depend>std_msgs</exec_depend>

  <!-- These test dependencies are optional
  Their purpose is to make sure that the code passes the linters -
->
  <test_depend>ament_copyright</test_depend>
  <test_depend>ament_flake8</test_depend>
  <test_depend>ament_pep257</test_depend>
  <test_depend>python3-pytest</test_depend>

  <export>
    <build_type>ament_python</build_type>
  </export>
</package>
```

4. Now, create another file, `setup.py`, using `$ gedit setup.py` and copy the following code into it:

```
from setuptools import setup
setup(
    name='ros2_talker',
    version='0.0.0',
    packages=[],
    py_modules=['talker'],
    install_requires=['setuptools'],
    zip_safe=True,
    author='Ramkumar Gandhinathan',
    author_email='ram651991@gmail.com',
    maintainer='Ramkumar Gandhinathan',
    maintainer_email='ram651991@gmail.com',
    keywords=['ROS'],
```

```
        classifiers=[
            'Intended Audience :: Developers',
            'License :: OSI Approved :: Apache Software License',
            'Programming Language :: Python',
            'Topic :: Software Development',
        ],
        description='Example to explain ROS-2',
        license='Apache License, Version 2.0',
        entry_points={
            'console_scripts': [
                'talker = talker:talker_main'
            ],
        },
    )
```

5. Now, create another file and name it `talker.py`, using `$ gedit talker.py`, and paste the following code into it:

```
#!/usr/bin/env python3

import rclpy
from std_msgs.msg import String
from time import sleep

def talker_main():
    rclpy.init(args=None)
    node = rclpy.create_node('ros2_talker_node')
    pub = node.create_publisher(String, '/chatter')
    msg = String()
    i = 0
    while rclpy.ok():
        msg.data = 'Hello World: %d' % i
        i += 1
        node.get_logger().info('Publishing: "%s"' % msg.data)
        pub.publish(msg)
        sleep(0.5)

if __name__ == '__main__':
    talker_main()
```

If you followed the preceding steps carefully, this is what your folder tree should look like:

```
robot@robot-pc: ~/ros2_example_ws
File Edit View Search Terminal Help
.
├── build
│   └── ros2_talker
│       ├── __pycache__
│       ├── ros2_talker.egg-info
│       └── share
│           └── ros2_talker
│               └── hook
├── install
│   └── ros2_talker
│       ├── lib
│       │   ├── python3.6
│       │   │   └── site-packages
│       │   │       └── __pycache__
│       │   └── ros2_talker
│       └── share
│           ├── ament_index
│           │   └── resource_index
│           │       └── packages
│           ├── colcon-core
│           │   └── packages
│           └── ros2_talker
│               └── hook
├── log
│   ├── build_2019-04-29_07-49-32
│   │   └── ros2_talker
│   ├── latest -> latest_build
│   └── latest_build -> build_2019-04-29_07-49-32
└── src
    └── ros2_talker
        ├── include
        │   └── ros2_talker
        └── src
```

ROS-2 folder tree

6. Change back to the workspace folder and build the package using the `colcon` command:

```
$ cd ~/ros2_example_ws
$ colcon build --symlink-install
```

7. Now, run the node using the following command:

```
$ ros2 run ros2_talker talker
```

You should see the same information being published that was published in ROS-1. Let's look at the major differences between them.

Differences between ROS-1 and ROS-2 talker nodes

For ease of understanding, let's consider the following table, which differentiates between both nodes, side by side:

ROS-1 talker	ROS-2 talker
`!/usr/bin/env python` ROS-1 follows Python 2, hence the preceding shebang line.	`!/usr/bin/env python3` ROS-2 follows Python 3, hence the preceding shebang line.

`import rospy`
The ROS-1 client library is `rospy`. There are no changes in the following import statements:
`from std_msgs.msg import String`
`from time import sleep`
	`import rclpy`
	The ROS-2 client library is `rclpy`. There are no changes in the following import statements:
	`from std_msgs.msg import String`
	`from time import sleep`

`def talker():` There's no change in the function call.	`def talker():` There's no change in the function call.

`rospy.init_node('ros1_talker_node')`
This is ROS-1's way of initializing a node. Only the node can be initialized in a file. If you need to initialize more than one node in ROS-1, you should use the `nodelet` concept.
	`rclpy.init(args=None)`
	`node =`
	`rclpy.create_node('ros2_talker_node')`
	This is ROS-2's way of initializing a node. In this method, more than one node can be initialized in a single file.

`pub = rospy.Publisher('/chatter',`
`String)`
This is how a topic is published in ROS-1.
	`pub = node.create_publisher(String,`
	`'/chatter')`
	This is how a topic is published in ROS-2.

`msg = String()`
`    i = 0`
`    while not rospy.is_shutdown():`
`        msg.data = "Hello World:`
`%d" % i`
`        i+=1`
Message declaration and counter logic.
	`msg = String()`
	`    i = 0`
	`    while rclpy.ok():`
	`        msg.data = 'Hello World: %d'`
	`% i`
	`        i += 1`
	Message declaration and counter logic. Unlike `is_shutdown()`, there is `shutdown()` in `rclpy`, but it cannot be used in the preceding method of publishing data using Python's sleep, hence why `ok()` is used.

`rospy.loginfo(msg.data)` This is how ROS-1 log information is displayed.	`node.get_logger().info('Publishing: "%s"' % msg.data)` This is how ROS-2 log information is displayed.

`pub.publish(msg)`
`        sleep(0.5)`
 	
`if __name__ == '__main__':`
`        talker_main()`
No change.
	`pub.publish(msg)`
	`        sleep(0.5)`
	
	`if __name__ == '__main__':`
	`    talker_main()`
	No change.

The preceding table sums up the differences between the ROS-1 and ROS-2 nodes.

Please note that the preceding ROS-2 node representation is only indicative in reference to our ROS-1 node. There are different ways of representing them in ROS-2. In fact, we have spoken about the QoS measure-based publish-subscribe that is available to ensure nodes communicate efficiently in a network with disturbances. These won't be explained in this book. If you would like to learn more about ROS-2, please have a look at the ROS-2 tutorials page (`https://index.ros.org/doc/ros2/Tutorials/`).

Now, let's learn how to bridge ROS-1 and ROS-2.

Bridging ROS-1 and ROS-2

As we already know, ROS-2 is still under heavy development and there isn't a proper long-term release. Hence, writing nodes in ROS-2 directly or porting them to ROS-2 immediately would be an arduous task. There is an option where you could still use ROS-1 packages with ROS-2 packages to develop your projects. This is done through a package called `ros1_bridge`.

It is actually an ROS-2 package, which establishes the automatic or manual mapping of messages, topics, and services and communicates between ROS-1 and ROS-2 nodes. It could subscribe to messages in a ROS version and then publish them in the other ROS version. This package comes in handy when you would like to use a simulator such as Gazebo with ROS-2 to test your project. There's only a limited number of messages supported in the `ros1_bridge` package at the time of writing. Hopefully, other packages will soon be ported into ROS-2 or their message support will be available in the preceding package so that we can run them in parallel.

Testing the ros1_bridge package

The `ros1_bridge` package comes installed by default in our `distro-crystal`. Let's run a node in ROS-1, open the bridge, and test the published topics in ROS-2. You need to open four separate Terminals for this experiment:

1. In the first Terminal, let's source the ROS-1 and ROS-2 workspaces simultaneously and start the `ros1_bridge` package:

```
$ initros1
$ initros2
$ ros2 run ros1_bridge simple_bridge_1_to_2
```

You should see the `Trying to connect to the ROS master` error.

2. In the second Terminal, initialize ROS-1 workspace and run `roscore`:

```
$ initros1
$ roscore
```

You should now see the connected to ROS master information.

3. In the third Terminal, start the ROS-1 talker node:

```
$ initros1
$ rosrun rospy_tutorials talker.py
```

You should now see that Terminals 1 and 3 show the published information.

4. In the fourth Terminal, source the ROS-2 workspace and start the ROS-2 listener node:

```
$ initros2
$ ros2 run demo_nodes_py listener
```

You should now see that Terminals 1, 3, and 4 show the published information.

The output should look like this:

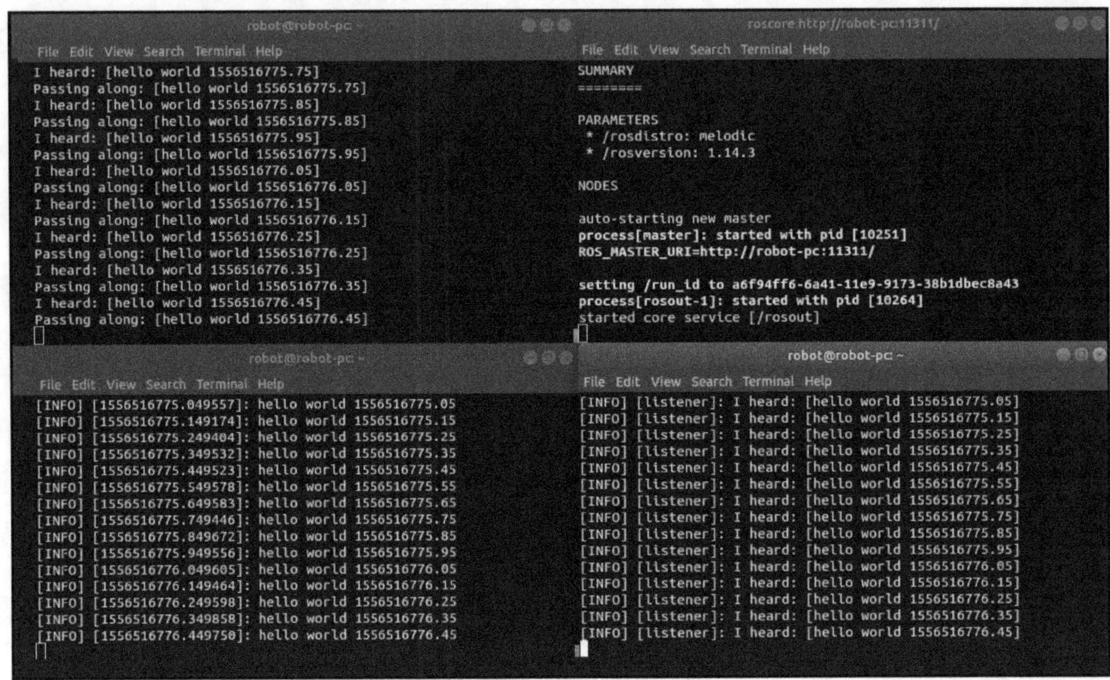

ROS-1 to ROS-2 bridge—from top left to bottom right—Terminals 1 to 4

By doing this, you can easily communicate with the ROS-1 and ROS-2 packages with ease. Communication between ROS-2 to ROS-1 can be done in the same way using the `simple_bridge_2_to_1` node.

Try the `dynamic_bridge` node. It should automatically establish communication between ROS-1 and/or ROS-2 nodes.

Summary

This chapter was an introductory chapter to ROS-2 and introduced us to its features and capabilities compared to ROS-1. We understood how different the build systems are and how the nodes are written. We also wrote custom nodes and compared them to how they're written in ROS-1. Since ROS-2 is under development, we also learned how to make use of `ros1_bridge`, which helps us use ROS-1 packages and tools alongside ROS-2. From the next chapter onward, we shall begin working with interesting robot projects using ROS-1.

In the next chapter, we will learn how to build an industrial mobile manipulator.

3
Building an Industrial Mobile Manipulator

Robotics is a broad interdisciplinary engineering field with immense application today. Unlike the standard classification of mobile and immobile robots, they're also classified based on applications for mobile robotics, aerial robotics, marine, and space robotics. The most common robots among them are the mobile robots that maneuver around the environment by trying to understand and learn the changes in it. They equally enjoy the environment, performing necessary actions that the application demands. Another common set of static robots that are used and are of huge demand in industries are robot arms. They were initially used to carry out lifting heavy payloads where the tasks were usually automatic and repetitive. But these manipulators have grown intelligent enough to understand the environment and objects of interest in it and perform tasks alongside humans. This category of robot arms is now called a **cobot**. Imagine putting together such a robot arm and a mobile robot with enough payload capabilities; this class of robots are known as **mobile manipulators**.

Mobile manipulators are useful in industries as they can go around carrying objects and deliver them to the respective work cells or stations as needed. This gives the robot liberty to reach almost all points in space, unlike a constrained robot arm's work volume. They are a very good addition to the flexible management systems that most industries follow for their plant automation. Apart from industries, they are useful in other fields, such as space exploration, mining, retail, and military applications. They could be either autonomously controlled, semi-autonomously controlled, or manually controlled based on the application's needs.

This is the first chapter where we will get into building robots using ROS. A list of available mobile manipulators will be introduced to provide better insight into how they can be used. Then, we will look into the prerequisites and approaches toward building the Mobile Manipulator project. Furthermore, you may model and simulate a robot base and robot arm individually and then put them together and see them in action.

In this chapter, the following topics will be covered:

- Understanding available mobile manipulators
- Applications of mobile manipulators
- Getting started building mobile manipulators
- Units and coordinate system
- Building the robot base
- Building the robot arm
- Putting things together

Technical requirements

Let's look into the technical requirements for this chapter:

- ROS Melodic Morenia on Ubuntu 18.04, with Gazebo 9 (preinstalled)
- ROS packages: `ros_control`, `ros_controllers`, and `gazebo_ros_control`
- Timelines and test platform:
    - **Estimated learning time**: On average, 120 mins
    - **Project build time (inclusive of compile and run time)**: On average, 90 mins
    - **Project Test platform**: HP Pavilion laptop (Intel® Core™ i7-4510U CPU @ 2.00 GHz × 4 with 8 GB Memory and 64-bit OS, GNOME-3.28.2)

The code for this chapter is available at `https://github.com/PacktPublishing/ROS-Robotics-Projects-SecondEdition/tree/master/chapter_3_ws/src/robot_description`.

Let's begin this chapter by understanding mobile manipulators.

Understanding available mobile manipulators

Mobile manipulators have been in the market for quite a while. Universities and research institutes initially began reusing their mobile robots and robot arms to improve dexterity. When ROS was gaining popularity in early 2007, PR2, a mobile manipulator (`http://www.willowgarage.com/pages/pr2/overview`) from Willow Garage (shown in the following photograph), was the testbed for testing a variety of ROS packages:

PR2 robot from OSRF (source: https://www.flickr.com/photos/jiuguangw/5136649984. Image by Jiuguang Wang. Licensed under Creative Commons CC-BY-SA 2.0: https://creativecommons.org/licenses/by-sa/2.0/legalcode)

PR2 has the mobility to navigate like a human being rationally and the dexterity to manipulate objects in an environment. However, industries didn't prefer PR2 initially as it cost $400k to own one. Soon, mobile robot manufacturers began building robot arms onto their available mobile robot bases and this began gaining popularity due to its lower cost compared to PR2. Some of these well-known manufacturers are Fetch Robotics, Pal Robotics, Kuka, and many more.

Fetch robotics has a mobile manipulator called Fetch (`https://fetchrobotics.com/robotics-platforms/fetch-mobile-manipulator/`). Fetch is a combination of a 7 degrees of freedom robot arm with an additional degree of freedom—the torso lift mounted on a mobile robot base that was targeted to carry a payload of 100 kg. Fetch is 5 foot tall and can understand the environment through its sensors, such as depth cameras and laser scanners. It is robust and low in cost. Fetch uses a parallel jaw gripper as its end effector and has an additional pan and tilt-based head (with the depth sensor). The arm can carry a payload of 6 kg.

Pal robotics introduced a mobile manipulator called Tiago (`http://tiago.pal-robotics.com/`), which is similar in design to Fetch. It has good features too, unlike Fetch, and is used by various research institutes across Europe and the world. It also has a 7 degrees of freedom robot arm, but it also has a force-torque sensor on its wrist to carefully monitor manipulation. However, the payload capabilities are slightly lower than Fetch as the arm can only lift around 3 kg of payload and the base can only carry up to 50 kg. Tiago has lots of built-in software features such as NLP systems and face-detection packages that might be handy for ready deployment:

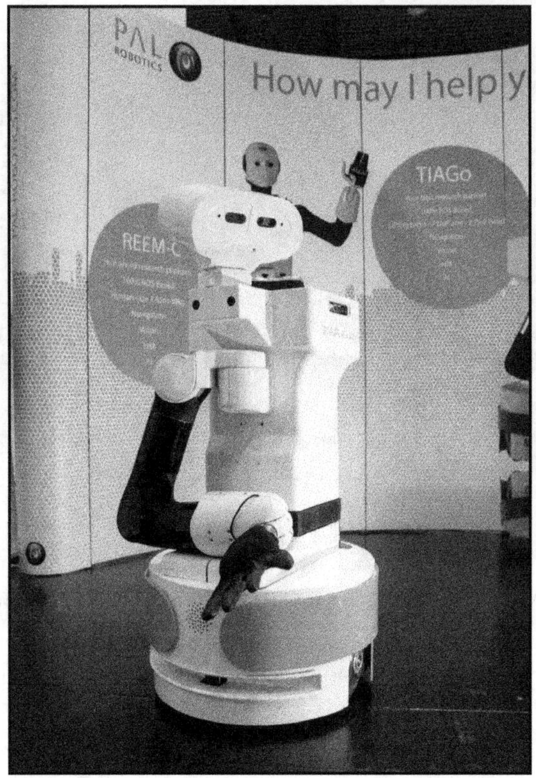

Tiago from Pal Robotics (source: https://commons.wikimedia.org/wiki/File:RoboCup_2016_Leipzig_-_TIAGo.jpg. Image by ubahnverleih. Licensed under public domain: https://creativecommons.org/publicdomain/zero/1.0/legalcode)

Kinova Robotics also has a mobile manipulator called MOVO (`https://www.kinovarobotics.com/en/knowledge-hub/movo-mobile-manipulator`), which has two robot arms, like a PR2. Unlike the preceding two mobile manipulators, this robot operation is sophisticated, slow, and smooth.

Other mobile manipulators make use of already available mobile robots and robot arms. For instance, Clearpath robotics has a wide variety of mobile manipulators that make use of their mobile bases—husky and ridgebacks, in combination with Universal robot's UR5 robot arms and the Baxter platform, respectively. Robotnik, on the other hand, has a mobile manipulator that is a combination of its mobile base RB-Kairos and Universal robot's UR5.

Since we now know the basics of mobile manipulators, let's look at its applications.

Applications of mobile manipulators

Industries used to make use of articulated manipulators that were doing dull and dangerous repetitive tasks. Over time, those robots have grown modern and are capable of working alongside a human operator rather than alone in a work cell. Hence, a combination of such robots with industrial-grade ground vehicles help in certain industrial applications. One of the most common applications is **machine tending**. It is one of the most trending applications today and a lot of robots are getting deployed in this field. A machine tending robot is one in which certain tasks that help to "tend" to a machine are carried out by the robot. Some tending tasks are the loading and unloading of parts from a machine, assembly operations, meteorological inspections, and more.

In warehouses, mobile manipulators are common as well. These are used in material handling tasks such as the loading and unloading of materials. The robot also helps in stock monitoring and analysis where it could help to identify a sudden reduction in the quantity of stock either based on consumption or market demand. A lot of collaboration is carried out in this area with robots of different kinds. Aerial robots such as drones are popular in warehouses and are used alongside mobile manipulators to keep track of stock monitoring and other inspection activities.

In space exploration and hazardous environments, since the mobile manipulators could be either an autonomous or semi-autonomous operations, these robots could be used in places that are not suitable for human life sustainability. These robots are used both in space and nuclear reactor plants to perform human-like operations without humans. They use specialized end effectors that replicate human hands and are suited up according to the environment so that the robot's internal systems aren't fried or damaged. An expert-human takes control of such robots, visualizes the robot and environment around the robot through certain sensors, and performs manipulation operations with the help of sophisticated human interface devices.

In military operations, one area where mobile manipulator research is strongly focused and being researched and funded highly for development is in the military. Mobile manipulators are accompanied by smaller reconnaissance robots that help in surveillance operations. They are used to either carry heavy payloads or even pull injured people to a safe location. They're also used in manipulation activities through manual operations, where a soldier could either defuse a bomb or open doors of an unmanned locality. Now, let's learn to build mobile manipulators.

Getting started building mobile manipulators

By now, you know what mobile manipulators are, what they constitute, and where they are used. Let's get into building one in simulation. As you are very well aware by now, a mobile manipulator would need a good payload mobile robot base and a robot arm, so let's begin building our mobile manipulator in terms of its parts and then combine them. Let's also consider certain parameters and constraints for building and simulating one. To avoid complexities in robot types and to account for a simple and effective simulation, let's consider the following assumptions:

For a good payload mobile robot base, we want the following:

- The robot may move on a flat or inclined flat surface but not an irregular surface.
- The robot may be a differential drive robot with fixed steering wheels and all wheels driven.
- The target payload of the mobile robot is 50 kg.

For a robot arm, we want the following:

- 5 degrees of freedom
- The target payload of the robot arm is up to 5 kg

Now, let's cover the units and coordinate system conventions of ROS.

Units and coordinate system

Before you begin building the mobile manipulator in Gazebo and ROS, you need to keep the units of measurement and coordinate conventions ROS follows in mind. Information like this is defined in design documents called **ROS Enhancement Proposals** (**REPs**). They act as standard references to the community members who use ROS while building their projects. Any new feature that's introduced or is planned to be introduced in ROS would be available as a proposal document for the community. The standard units of measurement and coordinate conventions are defined in REP-0103 (`http://www.ros.org/reps/rep-0103.html`). You can find all the available lists of REPS in the REP index here: `http://www.ros.org/reps/rep-0000.html`.

As far as we're concerned, the following information is sufficient enough that we can go ahead with building our mobile manipulator.

For units of measurement, we have the following:

- **The base units**: Length is in meters; mass is in kilograms; time is in seconds
- **The derived units**: Angle is in radians; frequency is in hertz; force is in newtons
- **The kinematic-derived units**: Linear velocity is in meters per second; angular velocity is in radians per second

For coordinate system conventions, we have the following:

- The right-hand thumb rule is followed, where the thumb is the *z* axis, the middle finger is the *y* axis, and the index finger is the *x* axis. Also, positive rotation of the *z* axis is anti-clockwise and the negative rotation of *z* is clockwise.

We'll look at Gazebo and ROS assumptions in the next section.

Gazebo and ROS assumptions

As we know, Gazebo is a physics simulation engine with ROS support. It works as a standalone without ROS as well. Most models that are created in Gazebo are in an XML format called **Simulation Description Format** (**SDF**). ROS has a different approach to representing robot models. They are defined in an XML format called **Universal Robotic Description Format** (**URDF**). So, there is nothing to worry about here because, if the models are created in URDF, with some extra XML tags, they could be easily understood by Gazebo as they're converted automatically into SDF (because of those extra XML tags) under its hood. But if the models are defined in SDF, porting some of the robot's ROS-based features might be a bit tricky.

There is support for a variety of SDF-based plugins that work or provide message information to ROS, but they're limited to few sensors and controllers. We have Gazebo-9 installed alongside our ROS-1 Melodic Morenia. Although we have the latest Gazebo and ROS versions (at the time of writing this book), most `ros_controllers` are still not supported in SDF and we need to create custom controllers to make them work. Hence, we shall create robot models in URDF format and spawn it into Gazebo and allow the Gazebo's built-in APIs (URDF2SDF: `http://osrf-distributions.s3.amazonaws.com/sdformat/api/6.0.0/classsdf_1_1URDF2SDF.html`) do the job of conversion for us.

To achieve ROS integration with Gazebo, we need certain dependencies that would establish a connection between both and convert the ROS messages into Gazebo-understandable information. We also need a framework that implements real-time-like robot controllers that help the robot move kinematically. The former constitutes the `gazebo_ros_pkgs` package, which is a bunch of ROS wrappers written to help Gazebo understand the ROS messages and services, and while latter constitutes the `ros_control` and `ros_controller` packages, which provide robot joint and actuator space conversions and ready-made controllers that control position, velocity, or effort (force). You can install them through these commands:

```
$ sudo apt-get install ros-melodic-ros-control
$ sudo apt-get install ros-melodic-ros-controllers
$ sudo apt-get install ros-melodic-Gazebo-ros-control
```

We will be using the `hardware_interface::RobotHW` class from `ros_control` as it already has defined abstraction layers and `joint_trajectory_controller` and `diff_drive_controller` from `ros_controllers` for our robot arm and mobile base, respectively.

More information about `ros_control` and `ros_controllers` can be found in this article: `http://www.theoj.org/joss-papers/joss.00456/10.21105.joss.00456.pdf`.

Now that we know the basics of building mobile manipulators, let's start building the robot base.

Building the robot base

Let's begin by modeling our robot base. As we mentioned previously, ROS understands a robot in terms of URDF. URDF is a list of XML tags that contains all of the necessary information of the robot. Once the URDF for the robot base is created, we shall bring in the necessary connectors and wrappers around the code so that we can interact and communicate with a standalone physics simulator such as Gazebo. Let's see how the robot base is built step by step.

Robot base prerequisites

To build a robot base, we need the following:

- A good solid chassis with a good set of wheels with friction properties
- Powerful drives that can help carry the required payload
- Drive controls

In case you plan to build a real robot base, there are additional considerations you might need to look into, for instance, power management systems—to run the robot efficiently for as long as you wish—the necessary electrical and embedded characteristics, and mechanical power transmission systems. What can help you get there is building a robot in ROS. Why, exactly? You would be able to emulate (actually, simulate, but if you tweak some parameters and apply real-time constraints, you could definitely emulate) a real working robot, as in the following examples:

- Your chassis and wheels would be defined with physical properties in URDF.
- Your drives could be defined using `Gazebo-ros` plugins.
- Your drive controls could be defined using `ros-controllers`.

Hence, to build a custom robot, let's consider the specifications.

Robot base specifications

Our robot base might need to carry a robot arm and some additional payload along with it. Also, our robot base should ensure it is electromechanically stable so that it has enough torque to pull its own load, along with the rated payload, and move smoothly with fewer jerks and with marginal pose error.

A pose constitutes both the robot's translation and rotation coordinates in space with respect to the world/earth/environment.

Let's consider the following specifications for our robot base:

- **Size**: Somewhere within 600 x 450 x 200 (L x B x H, all in mm)
- **Type**: Four-wheel differential drive robot
- **Speed**: Up to 1 m/s
- **Payload**: 50 kg (excluding the robot arm)

Robot base kinematics

Our robot base has only 2 degrees of freedom: a translation along the *x* axis and rotation along the *z* axis. Our robot cannot move instantaneously in the *y* axis due to the fixed steering wheel assumption. Since our robot moves only on the ground, it cannot translate in the *z* axis as well. I guess it is understood that a rotation along the *x* or *y* axes would mean that the robot either summersaults or topples; hence, it is not possible.

In the case of Swedish or Mecanum wheels, the robot base would have 3 degrees of freedom where it could translate along the *x* and *y* axes instantaneously and rotate along the *z* axis.

Hence, our robot equation is as follows:

$$\begin{matrix} x' & & \cos(\omega\delta t) & -\sin(\omega\delta t) & 0 & & x - ICCx & & ICCx \\ y' & = & \sin(\omega\delta t) & \cos(\omega\delta t) & 0 & * & y - ICCy & + & ICCy \\ \theta' & & 0 & 0 & 1 & & \theta & & \omega\delta t \end{matrix}$$

Here, x', y', and θ' constitute the robot's final pose, ω is the angular velocity of the robot, and δt is the time interval. This is known as the forward kinematics equation since you're determining the robot's pose by knowing the robot's dimensions and velocities.

The unknown variables are as follows:

$$R = l/2 * (nl + nr)/(nr - nl)$$

Here, *nl* and *nr* are the encoder counts for the left and right wheels, and *l* is the length of the wheel axis:

$$ICC = [x - Rsin\theta, y + Rcos\theta]$$

We also have the following:

$$\omega\delta t = (nr - nl)step/l$$

Here, *step* is the distance covered by the wheel in each encoder tick.

Software parameters

Now that we have the robot specifications, let's learn about the ROS-related information we need to know of while building a robot arm. Let's consider the mobile robot base as a black box: if you give it specific velocity, the robot base should move and, in turn, give the position it has moved to. In ROS terms, the mobile robot takes in information through a topic called `/cmd_vel` (command velocity) and gives out `/odom` (odometery). A simple representation is shown as follows:

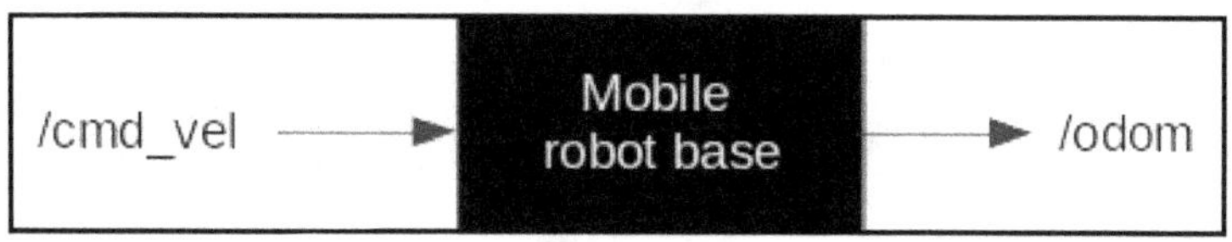

The mobile robot base as a black box

Now, let's look into the message format next.

ROS message format

`/cmd-vel` is of the `geometry_msgs/Twist` message format. The message structure can be found at `http://docs.ros.org/melodic/api/geometry_msgs/html/msg/Twist.html`.

`/odom` is of the `nav_msgs/Odometry` message format. The message structure can be found at `http://docs.ros.org/melodic/api/nav_msgs/html/msg/Odometry.html`.

Not all the fields are necessary in the case of our robot base since our robot is a 2 degrees of freedom robot.

ROS controllers

We would define the robot base's differential kinematics model using the `diff_drive_controller` plugin. This plugin defines the robot equation we saw earlier. It helps our robot to move in space. More information about this controller is available at the `http://wiki.ros.org/diff_drive_controller` website.

Modeling the robot base

Now that we have all the necessary information about the robot, let's get straight into modeling the robot. The robot model we are going to build is as follows:

Our mobile manipulator model

There is something you need to know before you come out with thoughts about modeling robots using URDF. You could make use of the geometric tags that define standard shapes such as cylinder, sphere, and boxes, but you cannot model complicated geometries or style them. These can be done using third-party software, for example, sophisticated **Computer Aided Design** (**CAD**) software such as Creo or Solidworks or using open source modelers such as Blender, FreeCAD, or Meshlab. Once they are modeled, they're are imported as meshes. The models in this book are modeled by such open source modelers and imported into URDFs as meshes. Also, writing a lot of XML tags sometimes becomes cumbersome and we might get lost while building intricate robots. Hence, we shall make use of macros in URDF called `xacro` (`http://wiki.ros.org/xacro`), which will help to reduce our lines of code for simplification and to avoid the repetition of tags.

Our robot base model will need the following tags:

- `<xacro>`: To help define macros for reuse
- `<links>`: To contain the geometric representations of the robot and visual information
- `<inertial>`: To contain the mass and moment of inertia of the links
- `<joints>`: To contain connections between the links with constraint definitions
- `<Gazebo>`: To contain plugins to establish a connection between Gazebo and ROS, along with simulation properties

For our robot base, there is a chassis and four wheels that are modeled. Have a look at the following diagram for representation information of the links:

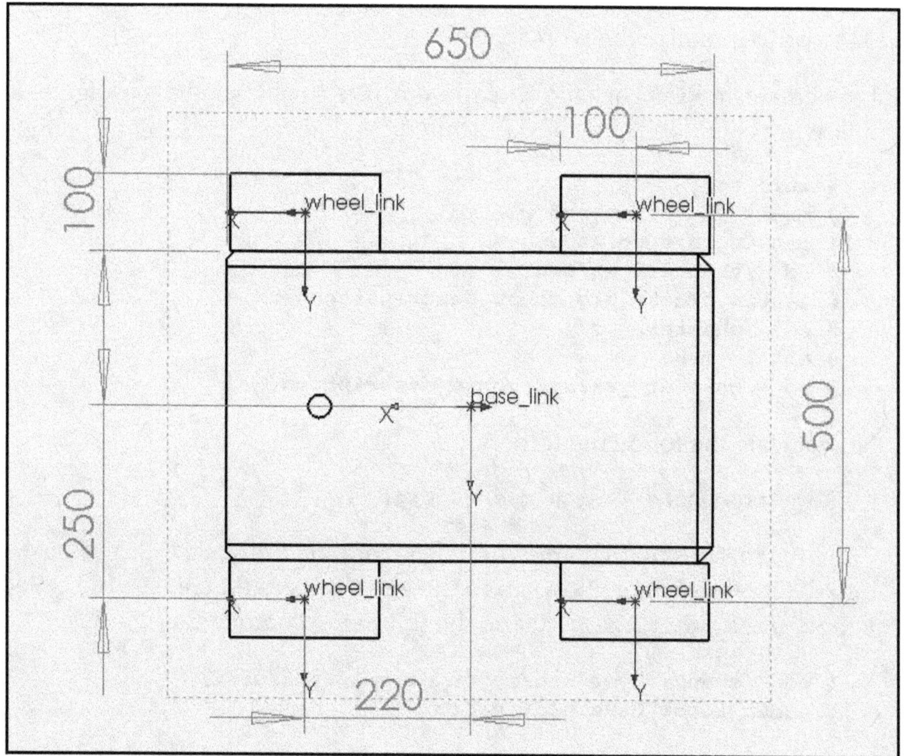

Representing the robot with link/coordinate system information

The chassis is named `base_link` and you can see the coordinate system in its center. Wheels (or `wheel_frames`) are placed with respect to the `base_link` frame. You can see that, as per our REP, the model follows the right-hand rule in the coordinate system. You can now make out that the forward direction of the robot will always be toward the *x* axis and that the rotation of the robot is around the *z* axis. Also, note that the wheels rotate around the *y* axis with respect to its frame of reference (you shall see this reference in the code in the next section).

Initializing the workspace

All we need to do now is define those meshes as `<link>` and `<joint>` tags for our robot. The mesh files are available at `https://github.com/PacktPublishing/ROS-Robotics-Projects-SecondEdition/tree/master/chapter_3_ws/src/robot_description/meshes`. Follow these steps to initialize the workspace:

1. Let's create an ROS package and add our files to it. Create a package using the following commands in a new Terminal:

```
$ initros1
$ mkdir -p ~/chapter3_ws/src
$ catkin_init_workspace
$ cd ~/chapter3_ws/src
$ catkin_create_pkg robot_description catkin
$ cd ~/chapter3_ws/
$ catkin_make
$ cd ~/chapter3_ws/src/robot_description/
```

2. Now, create the following folders:

```
$ mkdir config launch meshes urdf
```

3. Copy the meshes that you downloaded from the previous link and paste them into the meshes folder. Now, go to the `urdf` folder and create a file called `robot_base.urdf.xacro` using the following commands:

```
$ cd ~/chapter3_ws/src/robot_description/urdf/
$ gedit robot_base.urdf.xacro
```

4. Initialize the XML version tag and <robot> tag, as shown here, and begin copying the given XML code into it step by step:

```
<?xml version="1.0"?>
  <robot xmlns:xacro="http://ros.org/wiki/xacro" name="robot_base"
>
</robot>
```

Now that we have initialized the workspace, let's move on to the next step, which is defining the links.

Defining the links

Since you're defining the robot model by parts, copy all of the following code here into the <robot> tags (that is, the space between the <robot> tags). The chassis link is as follows:

```
<link name="base_link">
  <visual>
    <origin
      xyz="0 0 0"
      rpy="1.5707963267949 0 3.14" />
    <geometry>
      <mesh filename="package://robot_description/meshes/robot_base.stl"
/>
    </geometry>
    <material
      name="">
      <color
        rgba="0.79216 0.81961 0.93333 1" />
    </material>
  </visual>
</link>
```

The link defines the robot geometrically and helps in visualization. The preceding is the robot chassis, which we will call base_link.

The tags are pretty straightforward. If you wish to learn about them, have a look at http://wiki.ros.org/urdf/XML/link.

Four wheels need to connect to `base_link`. We could reuse the same model with different names and necessary coordinate information with the help of `xacro`. So, let's create another file called `robot_essentials.xacro` and define standard macros so that we can reuse them:

```
<?xml version="1.0"?>
<robot xmlns:xacro="http://ros.org/wiki/xacro" name="robot_essentials" >
<xacro:macro name="robot_wheel" params="prefix">
<link name="${prefix}_wheel">
<visual>
<origin
xyz="0 0 0"
rpy="1.5707963267949 0 0" />
<geometry>
<mesh filename="package://robot_description/meshes/wheel.stl" />
</geometry>
<material
name="">
<color
rgba="0.79216 0.81961 0.93333 1" />
</material>
</visual>
</link>
</xacro:macro>
</robot>
```

We have created a common macro for a wheel in this file. So, all you need to do now is call this macro in your actual robot file, `robot_base.urdf.xacro`, as shown here:

```
<xacro:robot_wheel prefix="front_left"/>
<xacro:robot_wheel prefix="front_right"/>
<xacro:robot_wheel prefix="rear_left"/>
<xacro:robot_wheel prefix="rear_right"/>
```

That's it. Can you see how quickly you have converted that many lines of code (for a link) into just one line of code for each link? Now, let's learn how to define joints.

If you wish to find out more about xacros, have a look at `http://wiki.ros.org/xacro`.

Defining the joints

As shown in the preceding representation diagram, the connections are only between the wheels and the chassis. The wheels are connected to `base_link` and they rotate around their *y* axis on their own frame of reference. Due to this, we can make use of continuous joint types. Since they're the same for all wheels, let's define them as `xacro` in the `robot_essentials.xacro` file:

```
<xacro:macro name="wheel_joint" params="prefix origin">
  <joint name="${prefix}_wheel_joint" type="continuous">
    <axis xyz="0 1 0"/>
    <parent link ="base_link"/>
    <child link ="${prefix}_wheel"/>
    <origin rpy ="0 0 0" xyz= "${origin}"/>
  </joint>
</xacro:macro>
```

As you can see, only the origin and the name needs to change in the preceding block of code. Hence, in our `robot_base.urdf.xacro` file, we'll define the wheel joints as follows:

```
<xacro:wheel_joint prefix="front_left" origin="0.220 0.250 0"/>
<xacro:wheel_joint prefix="front_right" origin="0.220 -0.250 0"/>
<xacro:wheel_joint prefix="rear_left" origin="-0.220 0.250 0"/>
<xacro:wheel_joint prefix="rear_right" origin="-0.220 -0.250 0"/>
```

Now that you have everything in your file, let's visualize this in `rviz` and see whether it matches our representation. You can do that by using the following commands in a new Terminal:

```
$ initros1
$ cd ~/chapter3_ws/
$ source devel/setup.bash
$ roscd robot_description/urdf/
$ roslaunch urdf_tutorial display.launch model:=robot_base.urdf.xacro
```

Add the robot model and, in the **Global** options, set **Fixed Frame** to `base_link`. Now, you should see our robot model if everything was successful. You can add the `tf` display and see whether the representation matches the *Representing the robot 2D with link/coordinate system information* diagram. You can also move the wheels using the sliders that were launched as a result of setting the `gui` argument to `true`.

Now, let's see what is needed to simulate the robot base.

Simulating the robot base

The first four steps were initially used to define the robot URDF model so that it could be understood by ROS. Now that we have a proper robot model that is understood by ROS, we need to add a few more tags to view the model in Gazebo.

Defining collisions

To visualize the robot in Gazebo, we need to add the `<collision>` tags, along with the `<visual>` tags defined in the `<link>` tag, as follows:

```
<collision>
<origin
xyz="0 0 0"
rpy="1.5707963267949 0 3.14" />
<geometry>
<mesh filename="package://robot_description/meshes/robot_base.stl" />
</geometry>
</collision>
```

For the base, add them to the `robot_base.urdf.xacro` file since we defined `base_link` there.

For all the wheel links, add them to the `robot_essentials.xacro` file since we defined the wheel link there:

```
<collision>
  <origin
   xyz="0 0 0"
   rpy="1.5707963267949 0 0" />
  <geometry>
    <mesh filename="package://robot_description/meshes/wheel.stl" />
  </geometry>
</collision>
```

Since Gazebo is a physics simulator, we define the physical properties using the `<inertial>` tags. We can acquire the mass and inertial properties from this third-party software. The inertial properties that are acquired from this software are added inside the `<link>` tag, along with suitable tags, as indicated here:

- For the base, the following is added:

```
<inertial>
 <origin
   xyz="0.0030946 4.78250032638821E-11 0.053305"
```

```
      rpy="0 0 0" />
    <mass value="47.873" />
    <inertia
      ixx="0.774276574699151"
      ixy="-1.03781944357671E-10"
      ixz="0.00763014265820928"
      iyy="1.64933255189991"
      iyz="1.09578155845563E-12"
      izz="2.1239326987473" />
  </inertial>
```

- For all of the wheels, the following is added:

```
  <inertial>
    <origin
      xyz="-4.1867E-18 0.0068085 -1.65658661799998E-18"
      rpy="0 0 0" />
    <mass value="2.6578" />
    <inertial>
      ixx="0.00856502765719703"
      ixy="1.5074118157338E-19"
      ixz="-4.78150098725052E-19"
      iyy="0.013670640432096"
      iyz="-2.68136447099727E-19"
      izz="0.00856502765719703" />
  </inertial>
```

Now that the Gazebo properties have been created, let's create the mechanisms.

Defining actuators

Now, we need to define the actuator information for our robot wheels in the `robot_base_essentials.xacro` file:

```
<xacro:macro name="base_transmission" params="prefix ">
  <transmission name="${prefix}_wheel_trans" type="SimpleTransmission">
    <type>transmission_interface/SimpleTransmission</type>
    <actuator name="${prefix}_wheel_motor">
<hardwareInterface>hardware_interface/VelocityJointInterface</hardwareInter
face>
    <mechanicalReduction>1</mechanicalReduction>
    </actuator>

  <joint name="${prefix}_wheel_joint">
<hardwareInterface>hardware_interface/VelocityJointInterface</hardwareInter
face>
  </joint>
```

```
  </transmission>
</xacro:macro>
```

Let's call them in our robot file as macros:

```
<xacro:base_transmission prefix="front_left"/>
<xacro:base_transmission prefix="front_right"/>
<xacro:base_transmission prefix="rear_left"/>
<xacro:base_transmission prefix="rear_right"/>
```

You can find the `robot_base_essentials.xacro` file in this book's GitHub repository: https://github.com/PacktPublishing/ROS-Robotics-Projects-SecondEdition/blob/master/chapter_3_ws/src/robot_description/urdf/robot_base_essentials.xacro.

Now that the mechanisms have been called, let's call the controllers that use them and make our robot dynamic.

Defining ROS_CONTROLLERS

Finally, we need to port the plugins that are needed to establish communication between Gazebo and ROD. For these, we need to create another file called `gazebo_essentials_base.xacro` that will contain the `<Gazebo>` tags.

In the created file, add the following `gazebo_ros_control` plugin:

```
<Gazebo>
  <plugin name="gazebo_ros_control" filename="libgazebo_ros_control.so">
   <robotNamespace>/</robotNamespace>
   <controlPeriod>0.001</controlPeriod>
   <legacyModeNS>false</legacyModeNS>
  </plugin>
</Gazebo>
```

The robot's differential drive plugin is as follows:

```
<Gazebo>

<plugin name="diff_drive_controller"
filename="libgazebo_ros_diff_drive.so">
 <legacyMode>false</legacyMode>
 <alwaysOn>true</alwaysOn>
 <updateRate>1000.0</updateRate>
 <leftJoint>front_left_wheel_joint, rear_left_wheel_joint</leftJoint>
 <rightJoint>front_right_wheel_joint, rear_right_wheel_joint</rightJoint>
 <wheelSeparation>0.5</wheelSeparation>
```

```
  <wheelDiameter>0.2</wheelDiameter>
  <wheelTorque>10</wheelTorque>
  <publishTf>1</publishTf>
  <odometryFrame>map</odometryFrame>
  <commandTopic>cmd_vel</commandTopic>
  <odometryTopic>odom</odometryTopic>
  <robotBaseFrame>base_link</robotBaseFrame>
  <wheelAcceleration>2.8</wheelAcceleration>
  <publishWheelJointState>true</publishWheelJointState>
  <publishWheelTF>false</publishWheelTF>
  <odometrySource>world</odometrySource>
  <rosDebugLevel>Debug</rosDebugLevel>
</plugin>

</Gazebo>
```

The friction property for the wheels as macros is as follows:

```
<xacro:macro name="wheel_friction" params="prefix ">
  <Gazebo reference="${prefix}_wheel">
    <mu1 value="1.0"/>
    <mu2 value="1.0"/>
    <kp value="10000000.0" />
    <kd value="1.0" />
    <fdir1 value="1 0 0"/>
  </Gazebo>
</xacro:macro>
```

The call macro in the robot file is as follows:

```
<xacro:wheel_friction prefix="front_left"/>
<xacro:wheel_friction prefix="front_right"/>
<xacro:wheel_friction prefix="rear_left"/>
<xacro:wheel_friction prefix="rear_right"/>
```

You can find the `gazebo_essentials_base.xacro` file in this book's GitHub repository: `https://github.com/PacktPublishing/ROS-Robotics-Projects-SecondEdition/blob/master/chapter_3_ws/src/robot_description/urdf/gazebo_essentials_base.xacro`.

Now that we have defined the macros for our robot, along with the Gazebo plugins, let's add them to our robot file. This can be easily done by just adding the following two lines in the robot file inside the `<robot>` macro tag:

```
<xacro:include filename="$(find
robot_description)/urdf/robot_base_essentials.xacro" />
<xacro:include filename="$(find
robot_description)/urdf/gazebo_essentials_base.xacro" />
```

Now that the URDFs are complete, let's configure the controllers. Let's create the following config file to define the controllers we are using. For this, let's go to our workspace, go inside the config folder we created, and create a controller config file, as shown here:

```
$ cd ~/chapter3_ws/src/robot_description/config/
$ gedit control.yaml
```

Now, copy the code that's available in this book's GitHub repository, `https://github.com/PacktPublishing/ROS-Robotics-Projects-SecondEdition/blob/master/chapter_3_ws/src/robot_description/config/control.yaml`, into the file. Now that we have completed our `robot_base` model, let's test it in Gazebo.

Testing the robot base

Now that we have the complete model for our robot base, let's put it into action and see how our base moves. Follow these steps:

1. Let's create a launch file that will spawn the robot and its controllers. Now, go into the `launch` folder and create the following launch file:

```
$ cd ~/chapter3_ws/src/robot_description/launch
$ gedit base_gazebo_control_xacro.launch
```

2. Now, copy the following code into it and save the file:

```
<?xml version="1.0"?>

<launch>
  <param name="robot_description" command="$(find xacro)/xacro --
inorder $(find robot_description)/urdf/robot_base.urdf.xacro" />
  <include file="$(find gazebo_ros)/launch/empty_world.launch"/>
  <node name="spawn_urdf" pkg="gazebo_ros" type="spawn_model"
args="-param robot_description -urdf -model robot_base" />
  <rosparam command="load" file="$(find
robot_description)/config/control.yaml" />
  <node name="base_controller_spawner" pkg="controller_manager"
type="spawner" args="robot_base_joint_publisher
robot_base_velocity_controller"/>

</launch>
```

3. Now, you can visualize your robot by running the following command:

```
$ cd ~/chapter3_ws
$ source devel/setup.bash
$ roslaunch robot_description base_gazebo_control_xacro.launch
```

Once the Gazebo environment launches, you should see something like the following without any error prompt:

```
/home/robot/chapter_3_ws/src/robot_description/launch/base_gazebo_control_xacro.launch http://localhost:11311
File Edit View Search Terminal Help
process[spawn_urdf-4]: started with pid [23271]
process[base_controller_spawner-5]: started with pid [23272]
[ INFO] [1557664801.100103700]: Finished loading Gazebo ROS API Plugin.
[ INFO] [1557664801.103041871]: waitForService: Service [/gazebo/set_physics_properties] has not been advertised, waiting...
[ INFO] [1557664801.239502410]: Finished loading Gazebo ROS API Plugin.
[ INFO] [1557664801.242657079]: waitForService: Service [/gazebo_gui/set_physics_properties] has not been advertised, waiting...
[ INFO] [1557664801.615825535, 0.023000000]: waitForService: Service [/gazebo/set_physics_properties] is now available.
[ INFO] [1557664801.666036358, 0.069000000]: Physics dynamic reconfigure ready.
[ INFO] [1557664802.009308761, 0.314000000]: Loading gazebo_ros_control plugin
[ INFO] [1557664802.009494434, 0.314000000]: Starting gazebo_ros_control plugin in namespace: /
[ INFO] [1557664802.010807312, 0.314000000]: gazebo_ros_control plugin is waiting for model URDF in parameter [robot_description] on the ROS p
aram server.
[ INFO] [1557664802.215271155, 0.314000000]: Loaded gazebo_ros_control.
[spawn_urdf-4] process has finished cleanly
log file: /home/robot/.ros/log/0dcc156c-74b3-11e9-8061-38b1dbec8a43/spawn_urdf-4*.log
[ INFO] [1557664802.251202646, 0.314000000]: Starting plugin DiffDrive(ns = //)
[ INFO] [1557664802.251419943, 0.314000000]: DiffDrive(ns = //): <rosDebugLevel> = Debug
[ INFO] [1557664802.252370583, 0.314000000]: DiffDrive(ns = //): <tf_prefix> =
[DEBUG] [1557664802.252506285, 0.314000000]: DiffDrive(ns = //): <commandTopic> = cmd_vel
[DEBUG] [1557664802.252539237, 0.314000000]: DiffDrive(ns = //): <odometryTopic> = odom
[DEBUG] [1557664802.252564858, 0.314000000]: DiffDrive(ns = //): <odometryFrame> = map
[DEBUG] [1557664802.252613353, 0.314000000]: DiffDrive(ns = //): <robotBaseFrame> = base_link
[DEBUG] [1557664802.252722087, 0.314000000]: DiffDrive(ns = //): <publishWheelTF> = false
[ WARN] [1557664802.252766247, 0.314000000]: DiffDrive(ns = //): missing <publishOdomTF> default is true
[DEBUG] [1557664802.252795882, 0.314000000]: DiffDrive(ns = //): <publishWheelJointState> = true
[DEBUG] [1557664802.252900507, 0.314000000]: DiffDrive(ns = //): <wheelSeparation> = 0.5
[DEBUG] [1557664802.252940285, 0.314000000]: DiffDrive(ns = //): <wheelDiameter> = 0.20000000000000001
[DEBUG] [1557664802.252969589, 0.314000000]: DiffDrive(ns = //): <wheelAcceleration> = 2.7999999999999998
[DEBUG] [1557664802.252998123, 0.314000000]: DiffDrive(ns = //): <wheelTorque> = 10
[DEBUG] [1557664802.253028607, 0.314000000]: DiffDrive(ns = //): <updateRate> = 1000
[DEBUG] [1557664802.253199561, 0.314000000]: DiffDrive(ns = //): <odometrySource> = world := 1
[DEBUG] [1557664802.253282733, 0.314000000]: DiffDrive(ns = //): <leftJoint> = front_left_wheel_joint, rear_left_wheel_joint
[ INFO] [1557664802.553809450, 0.600000000]: Controller state will be published at 50Hz.
[ INFO] [1557664802.556058217, 0.600000000]: Wheel separation will be multiplied by 1.
[ INFO] [1557664802.557208962, 0.601000000]: Left wheel radius will be multiplied by 1.
[ INFO] [1557664802.557322897, 0.601000000]: Right wheel radius will be multiplied by 1.
[ INFO] [1557664802.558400608, 0.602000000]: Velocity rolling window size of 10.
[ INFO] [1557664802.561558254, 0.605000000]: Velocity commands will be considered old if they are older than 0.5s.
```

Successful Gazebo launch

4. You can view the necessary ROS topics by opening another Terminal and running `rostopic list`:

```
$ initros1
$ rostopic list
```

The following screenshot shows the list of ROS topics:

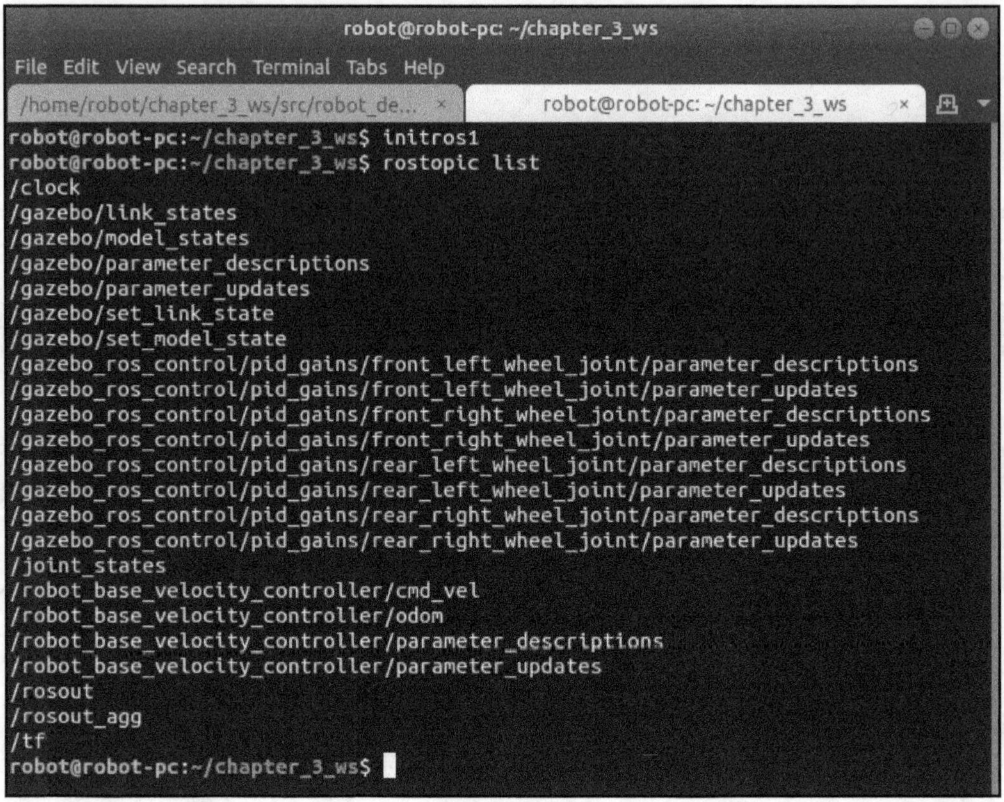

```
robot@robot-pc: ~/chapter_3_ws
File Edit View Search Terminal Tabs Help
/home/robot/chapter_3_ws/src/robot_de...    robot@robot-pc: ~/chapter_3_ws
robot@robot-pc:~/chapter_3_ws$ initros1
robot@robot-pc:~/chapter_3_ws$ rostopic list
/clock
/gazebo/link_states
/gazebo/model_states
/gazebo/parameter_descriptions
/gazebo/parameter_updates
/gazebo/set_link_state
/gazebo/set_model_state
/gazebo_ros_control/pid_gains/front_left_wheel_joint/parameter_descriptions
/gazebo_ros_control/pid_gains/front_left_wheel_joint/parameter_updates
/gazebo_ros_control/pid_gains/front_right_wheel_joint/parameter_descriptions
/gazebo_ros_control/pid_gains/front_right_wheel_joint/parameter_updates
/gazebo_ros_control/pid_gains/rear_left_wheel_joint/parameter_descriptions
/gazebo_ros_control/pid_gains/rear_left_wheel_joint/parameter_updates
/gazebo_ros_control/pid_gains/rear_right_wheel_joint/parameter_descriptions
/gazebo_ros_control/pid_gains/rear_right_wheel_joint/parameter_updates
/joint_states
/robot_base_velocity_controller/cmd_vel
/robot_base_velocity_controller/odom
/robot_base_velocity_controller/parameter_descriptions
/robot_base_velocity_controller/parameter_updates
/rosout
/rosout_agg
/tf
robot@robot-pc:~/chapter_3_ws$
```

List of ROS topics

The Gazebo view of the robot should be as follows:

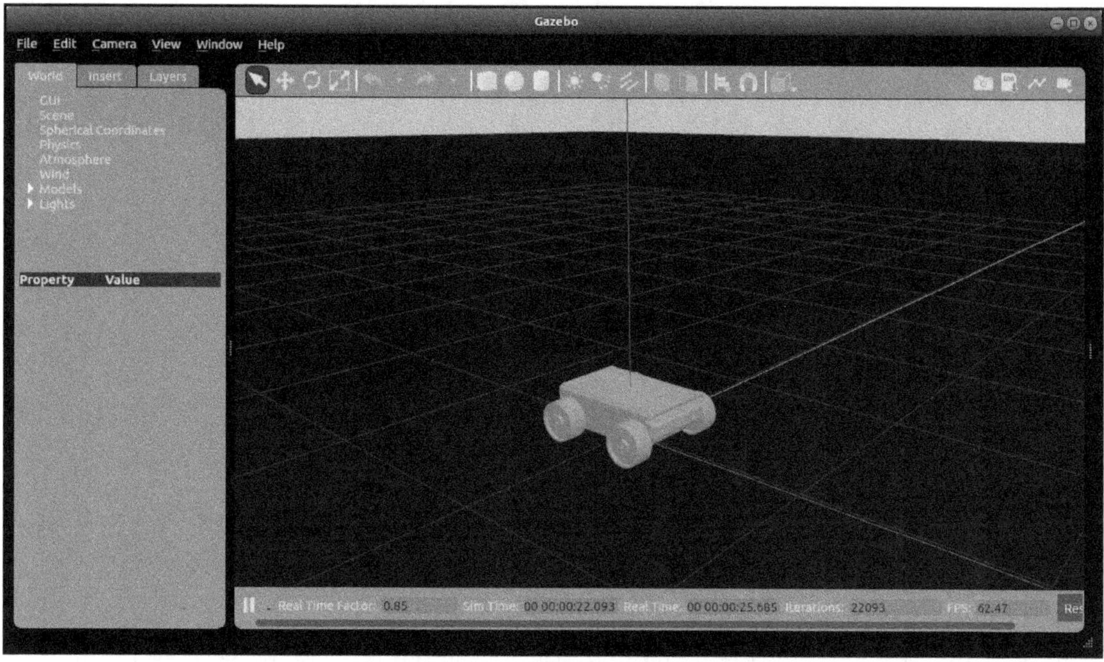

Robot base in Gazebo

5. Try using the `rqt_robot_steering` node and move the robot, as follows:

```
$ rosrun rqt_robot_steering rqt_robot_steering
```

In the window, mention our topic, `/robot_base_controller/cmd_vel`. Now, move the sliders slowly and see how the robot base moves around.

Have a look at the robot URDF file in this book's GitHub repository: `https://github.com/PacktPublishing/ROS-Robotics-Projects-SecondEdition/blob/master/chapter_3_ws/src/robot_description/urdf/robot_base.urdf.xacro` before testing the robot base.

We have completed our robot base. Now, let's learn how to build the robot arm.

Getting started building the robot arm

Now that we built a robot base in URDF and visualized it in Gazebo, let's get to building the robot arm. The robot arm will be built in a similar way to using URDF and there are no differences in terms of its approach. There may be only a few changes that need to be made to the robot base URDF. Let's see how the robot arm is built, step by step.

Robot arm prerequisites

To build a robot arm, we need the following:

- A good set of linkages that move independently with good physical strength
- A good set of actuators that could help us withstand enough payload
- Drive controls

In case you plan to build a real robot arm, you might need similar embedded architecture and electronic controls, real-time communication between the actuators, a power management system, and maybe a good end effector, based on your application requirements. As you know, there's more to consider and this is not in the scope of this book. However, the aim is to emulate a robot arm in ROS so that it works the same way in reality as well. Now, let's look at the robot arm specifications.

Robot arm specifications

Here, we want to build a mobile manipulator. Our robot arm doesn't need to carry payloads of tons. In fact, in reality, such a robot would require heavy machinery drives and mechanical power transmission systems and so it might not be that easy to mount on a movable platform.

Let's be practical and consider common parameters that some of the industrial cobots follow in the market:

- **Type**: 5 **DOF** (short for **degrees of freedom**) robot arm
- **Payload**: Within 3-5 kgs

Now, let's look at the robot arm kinematics.

Robot arm kinematics

Robot arm kinematics is slightly different compared to robot base kinematics. You would need to move five different actuators to five different positions to move the robot arm's tooltip to a required position. The mathematical modeling follows the **Denavit-Hartenberg** (**DH**) method of computing kinematics. Explaining the DH method is out of our scope, so we shall look at the kinematics equation directly.

The arm kinematics equation is defined by a 4 x 4 homogeneous transformation matrix that connects all of the five links with respect to the robot base coordinate system, as shown here:

$$T = \begin{matrix} C_1 C_{234} C_5 + S_1 S_5 & -C_1 C_{234} S_5 + S_1 C_5 & -C_1 S_{234} & C_1(-d_5 S_{234} + a_3 C_{23} + a_2 C_2) \\ C_1 C_{234} C_5 - S_1 S_5 & -S_1 C_{234} S_5 - C_1 C_5 & -S_1 S_{234} & S_1(-d_5 S_{234} + a_3 C_{23} + a_2 C_2) \\ -S_{234} C_5 & S_{234} S_5 & -C_{234} & d_1 - a_2 S_2 - a_3 S_{23} - d_5 C_{234} \\ 0 & 0 & 0 & 1 \end{matrix}$$

Here, we have the following:

$$C_{ijk} = \cos(q_i + qj + qk), S_{ijk} = \sin(q_i + qj + qk),$$

This represents trigonometric equations; q_i is the angle between the normal to rotation axis (usually represented as x_i) and the rotation axis (usually represented as z_i), and is the same for q_j and q_k, respectively. d_i is the distance from the rotation axis (z_i) to the origin (*i-1*) system of axis, and a_i is the shortest distance between two consecutive rotation axes.

From the preceding homogeneous transform, the first 3 x 3 element indicates the rotation of the gripper or tool, the bottom row defines the scale factor, and the tool's pose is given by the remaining elements:

$$Tool_{pose} = \begin{matrix} C_1(-d_5 S_{234} + a_3 C_{23} + a_2 C_2) \\ S_1(-d_5 S_{234} + a_3 C_{23} + a_2 C_2) \\ d_1 - a_2 s_2 - a_3 s_{23} - d_5 C_{234} \end{matrix}$$

Now, let's look at the software parameters.

Software parameters

So, if we consider the arm a black box, the arm gives out a pose based on the commands each actuator receives. The commands may be in the form of position, force/effort, or velocity commands. A simple representation is shown here:

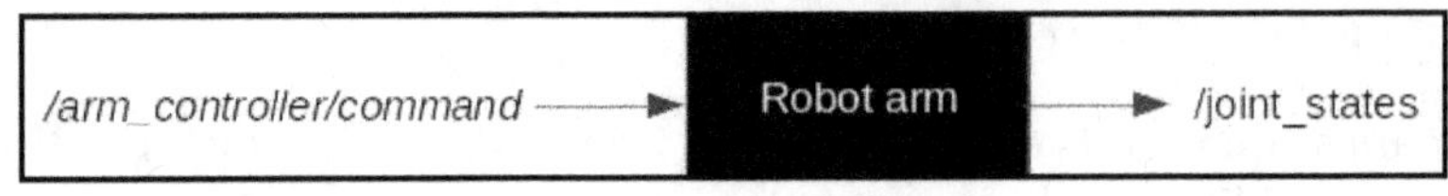

The robot arm as a black box

Let's look into their message representations.

The ROS message format

The `/arm_controller/command` topic, which is used to command or control the robot arm, is of the `trajectory_msgs/JointTrajectory` message format.

This message structure can be found at `http://docs.ros.org/melodic/api/trajectory_msgs/html/msg/JointTrajectory.html`.

ROS controllers

`joint_trajectory_controller` is used for executing joint space trajectories on a list of given joints. The trajectories to the controller are sent using the action interface, `control_msgs::FollowJointTrajectoryAction`, in the `follow_joint_trajectory` namespace of the controller.

More information about this controller is available at the following website: `http://wiki.ros.org/joint_trajectory_controller`.

Now, let's learn how to model the robot arm.

Modeling the robot arm

This is what our robot arm will look like once it's been built in ROS:

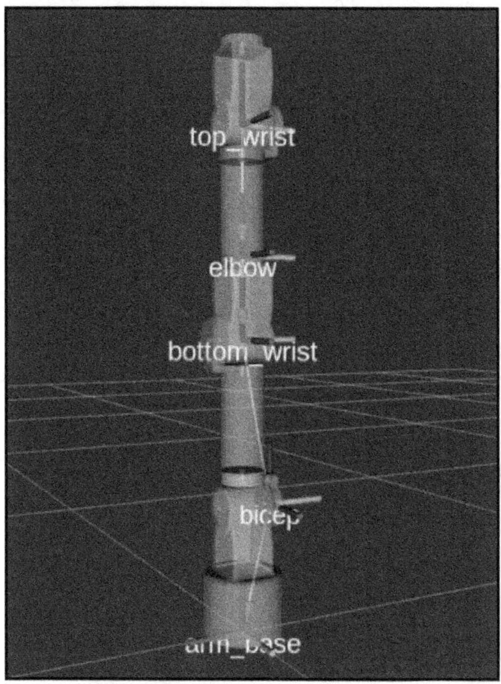

Representing the robot arm in 2D with link representation

Since all of the necessary explanation has already been provided in the *Modeling the robot base* section, and since this section takes the same approach as well, let's get into modeling the robot arm step by step.

Initializing the workspace

To initialize the workspace, we will make use of the same workspace we created for this chapter (`chapter_3_ws`). Download the meshes from the link to the meshes folder you created in the *Building the robot base* section. Now, go to the `urdf` folder and create a file called `robot_arm.urdf.xacro` using the following commands:

```
$ cd ~/chapter3_ws/src/robot_description/urdf/
$ gedit robot_arm.urdf.xacro
```

Initialize the XML version tag and the <robot> tag as follows and begin copying the given XML code into it step by step:

```
<?xml version="1.0"?>
 <robot xmlns:xacro="http://ros.org/wiki/xacro" name="robot_base" >

 </robot>
```

Now that we have initialized the workspace, let's move on to the next step, which is defining the links.

Defining the links

Now, since you're defining the robot model by parts, copy the following code into the <robot> tags (that is, the space between the <robot> tags). Note that we'll define all of the five links directly since they contain coordinate information. Let's look at one of them in detail, the arm base:

```
<link name="arm_base">
    <visual>
      <origin
        xyz="0 0 0"
        rpy="0 0 0" />
      <geometry>
        <mesh filename="package://robot_description/meshes/arm_base.stl" />
      </geometry>
      <material
        name="">
        <color
          rgba="0.79216 0.81961 0.93333 1" />
      </material>
    </visual>
</link>
```

You could find the other links such as bicep, bottom_wrist, elbow, and top_wrist, that have been defined in this file in this book's GitHub repository: https://github.com/PacktPublishing/ROS-Robotics-Projects-SecondEdition/blob/master/chapter_3_ws/src/robot_description/urdf/robot_arm.urdf.xacro.

You could try their more macros for <visual> and <collision> tags with respective parameter definitions, but this sometimes might lead to confusion for beginners. Hence, to keep it simple, we're only defining minimal macros in this book. If you feel you have mastered macros, you are welcome to test your skills by adding more macros.

Defining the joints

The joints we will define for the robot arm are revolute, the joint can move between limits. Since all of the joints in this robot arm move between limits and are common for all, we define them in the same `robot_essentials.xacro` file we created while building the robot base:

```
<xacro:macro name="arm_joint" params="prefix origin">

 <joint name="${prefix}_joint" type="continuous">
   <axis xyz="0 0 1"/>
   <parent link="arm_base"/>
   <child link="${prefix}_joint"/>
   <origin rpy="0 0 0" xyz="${origin}"/>
 </joint>

 </xacro:macro>
```

In the `robot_arm.urdf.xacro` file, we define them as follows:

```
<xacro:arm_joint prefix="shoulder" parent="arm_base" child="bicep"
originxyz="-0.05166 0.0 0.20271" originrpy="0 0 1.5708"/>
 <xacro:arm_joint prefix="bottom_wrist" parent="bicep" child="bottom_wrist"
originxyz="0.0 -0.05194 0.269" originrpy="0 0 0"/>
 <xacro:arm_joint prefix="elbow" parent="bottom_wrist" child="elbow"
originxyz="0.0 0 0.13522" originrpy="0 0 0"/>
 <xacro:arm_joint prefix="top_wrist" parent="elbow" child="top_wrist"
originxyz="0.0 0 0.20994" originrpy="0 0 0"/>
```

Now that you have everything in our file, let's visualize this in `rviz` and see whether it matches our representation. We can do that by using the following commands in a new Terminal:

```
$ initros1
$ cd ~/chapter3_ws/
$ source devel/setup.bash
$ roscd robot_description/urdf/
$ roslaunch urdf_tutorial display.launch model:=robot_arm.urdf.xacro
```

Add the robot model and, in **Global** options, set **Fixed Frame** to `arm_base`. Now, you should see our robot model if everything was successful. You could add the `tf` display and see whether the representation matches the *Representing the robot arm in 2D with link representation* screenshot. You could move the arm links using the sliders that were launched as a result of setting the `gui` argument to `true`.

Simulating the robot arm

The first three steps were initially used to define the arm URDF model that could be understood by ROS. Now that we have a proper robot model that is understood by ROS, we need to add a few more tags to view the model in Gazebo. We will start by defining collisions.

Defining collisions

To visualize the robot in Gazebo, we need to add `<collision>` tags, along with the `<visual>` tags we defined in the `<link>` tag, like we did previously. Hence, just specify the necessary visual and inertial tags on the links:

- For the `arm_base` link, add the following:

```
 <collision>
      <origin
        xyz="0 0 0"
        rpy="0 0 0" />
      <geometry>
       <mesh
filename="package://robot_description/meshes/arm_base.stl" />
      </geometry>
 </collision>

<inertial>
      <origin
        xyz="7.7128E-09 -0.063005 -3.01969999961422E-08"
        rpy="0 0 0" />
      <mass
        value="1.6004" />
      <inertia
        ixx="0.00552196561445819"
        ixy="7.9550614501301E-10"
        ixz="-1.34378458924839E-09"
        iyy="0.00352397447953875"
        iyz="-1.10071809773382E-08"
        izz="0.00553739792746489" />
    </inertial>
```

- For the other links, please have a look at the code for `robot_arm.urdf.xacro` at `https://github.com/PacktPublishing/ROS-Robotics-Projects-SecondEdition/blob/master/chapter_3_ws/src/robot_description/urdf/robot_arm.urdf.xacro`.

Now that the Gazebo properties have been created, let's create the mechanisms.

Defining actuators

Now, we can define the actuator information for all of the links. The actuator macro is defined in the robot_arm_essentials.xacro file, as shown here:

```
<xacro:macro name="arm_transmission" params="prefix ">

 <transmission name="${prefix}_trans" type="SimpleTransmission">
   <type>transmission_interface/SimpleTransmission</type>
   <actuator name="${prefix}_motor">
<hardwareInterface>hardware_interface/PositionJointInterface</hardwareInter
face>
     <mechanicalReduction>1</mechanicalReduction>
   </actuator>
   <joint name="${prefix}_joint">
<hardwareInterface>hardware_interface/PositionJointInterface</hardwareInter
face>
   </joint>
 </transmission>

 </xacro:macro>
```

Call them in the robot arm file as follows:

```
<xacro:arm_transmission prefix="arm_base"/>
<xacro:arm_transmission prefix="shoulder"/>
<xacro:arm_transmission prefix="bottom_wrist"/>
<xacro:arm_transmission prefix="elbow"/>
<xacro:arm_transmission prefix="top_wrist"/>
```

You can find the robot_arm_essentials.xacro file in this book's GitHub repository: https://github.com/PacktPublishing/ROS-Robotics-Projects-SecondEdition/blob/master/chapter_3_ws/src/robot_description/urdf/robot_arm_essentials.xacro.

Now that the mechanisms have been called, let's call the controllers that use them and make our robot dynamic.

Defining ROS_CONTROLLERS

Finally, we can port the plugins that are needed to establish communication between Gazebo and ROS and more. We add one more controller, `joint_state_publisher` into the already created `gazebo_essentials_arm.xacro` file:

```
<Gazebo>
     <plugin name="joint_state_publisher"
filename="libgazebo_ros_joint_state_publisher.so">
       <jointName>arm_base_joint, shoulder_joint, bottom_wrist_joint,
elbow_joint, bottom_wrist_joint</jointName>
     </plugin>
   </Gazebo>
```

The `joint_state_publisher` controller (`http://wiki.ros.org/joint_state_publisher`) is used to publish the robot arm link's state information in space.

You can find the `gazebo_essentials_arm.xacro` file at our repository: `https://github.com/PacktPublishing/ROS-Robotics-Projects-SecondEdition/blob/master/chapter_3_ws/src/robot_description/urdf/gazebo_essentials_arm.xacro`.

Now that we have defined the macros for our robot, along with the Gazebo plugins, let's add them to our robot arm file. This can be done by adding the following two lines to the robot file inside the `<robot>` macro tag:

```
<xacro:include filename="$(find
robot_description)/urdf/robot_arm_essentials.xacro" />
  <xacro:include filename="$(find
robot_description)/urdf/gazebo_essentials_arm.xacro" />
```

Let's create an `arm_control.yaml` file and define the arm controller's configurations:

```
$ cd ~/chapter3_ws/src/robot_description/config/
$ gedit arm_control.yaml'
```

Now, copy the following code into the file:

```
arm_controller:
    type: position_controllers/JointTrajectoryController
    joints:
        - arm_base_joint
        - shoulder_joint
        - bottom_wrist_joint
        - elbow_joint
        - top_wrist_joint
    constraints:
```

```
        goal_time: 0.6
        stopped_velocity_tolerance: 0.05
        hip: {trajectory: 0.1, goal: 0.1}
        shoulder: {trajectory: 0.1, goal: 0.1}
        elbow: {trajectory: 0.1, goal: 0.1}
        wrist: {trajectory: 0.1, goal: 0.1}
      stop_trajectory_duration: 0.5
      state_publish_rate:  25
      action_monitor_rate: 10
  /gazebo_ros_control:
      pid_gains:
        arm_base_joint: {p: 100.0, i: 0.0, d: 0.0}
        shoulder_joint: {p: 100.0, i: 0.0, d: 0.0}
        bottom_wrist_joint: {p: 100.0, i: 0.0, d: 0.0}
        elbow_joint: {p: 100.0, i: 0.0, d: 0.0}
        top_wrist_joint: {p: 100.0, i: 0.0, d: 0.0}
```

Now that we have completed our `robot_base` model, let's test it in Gazebo.

Testing the robot arm

Have a look at the robot URDF file in this book's GitHub repository, `https://github.com/PacktPublishing/ROS-Robotics-Projects-SecondEdition/blob/master/chapter_3_ws/src/robot_description/urdf/robot_arm.urdf.xacro`, into testing the robot base.

Now that we have the complete model of our robot arm, let's get it into action and see how our arm moves. Follow the steps:

1. Create a launch file that would spawn the robot arm and its controllers. Now, go into the `launch` folder and create the following launch file:

```
$ cd ~/chapter3_ws/src/robot_description/launch
$ gedit arm_gazebo_control_xacro.launch
```

Now, copy the following code into it:

```
<?xml version="1.0"?>
<launch>

<param name="robot_description" command="$(find xacro)/xacro --
inorder $(find robot_description)/urdf/robot_arm.urdf.xacro" />
  <include file="$(find gazebo_ros)/launch/empty_world.launch"/>
  <node name="spawn_urdf" pkg="gazebo_ros" type="spawn_model"
args="-param robot_description -urdf -model robot_arm" />
  <rosparam command="load" file="$(find
robot_description)/config/arm_control.yaml" />
```

```
  <node name="arm_controller_spawner" pkg="controller_manager"
type="controller_manager" args="spawn arm_controller"
respawn="false" output="screen"/>
  <rosparam command="load" file="$(find
robot_description)/config/joint_state_controller.yaml" />
  <node name="joint_state_controller_spawner"
pkg="controller_manager" type="controller_manager" args="spawn
joint_state_controller" respawn="false" output="screen"/>
  <node name="robot_state_publisher" pkg="robot_state_publisher"
type="robot_state_publisher" respawn="false" output="screen"/>

</launch>
```

2. Now, you could visualize your robot by running the following command:

```
$ initros1
$ roslaunch robot_description arm_gazebo_control_xacro.launch
```

   You could see the necessary ROS topics by opening another Terminal and running `rostopic list`:

```
$ initros1
$ rostopic list
```

3. Try to move the arm using the following command:

```
$ rostopic pub /arm_controller/command
trajectory_msgs/JointTrajectory '{joint_names: ["arm_base_joint",
"shoulder_joint", "bottom_wrist_joint", "elbow_joint",
"top_wrist_joint"], points: [{positions: [-0.1,
0.210116830848170721, 0.022747275919015486, 0.0024182584123728645,
0.00012406874824844039], time_from_start: [1.0,0.0]}]}'
```

You will learn how to move the arm without using publishing values in the upcoming chapters. We'll learn how to set up the mobile manipulator in the next section.

Putting things together

You have now created a robot base and robot arm successfully and have simulated them in Gazebo. Now we're just one step away from getting our mobile manipulator model.

Modeling the mobile manipulator

The use of `xacro` will help us connect both with ease. Before creating the final URDF, let's understand something that we saw in the *Successful Gazebo launch* screenshot. The arm needs to connect to the base, and that's our goal. Hence, all you need to do now is connect the robot arm's `arm_base` link to the robot base's `base_link`. Create a file called `mobile_manipulator.urdf.xacro` and copy the following code into it:

```
<?xml version="1.0"?>

<robot xmlns:xacro="http://ros.org/wiki/xacro" name="robot_base" >

<xacro:include filename="$(find
robot_description)/urdf/robot_base.urdf.xacro" />
<xacro:include filename="$(find
robot_description)/urdf/robot_arm.urdf.xacro" />

<xacro:arm_joint prefix="arm_base_link" parent="base_link" child="arm_base"
originxyz="0.0 0.0 0.1" originrpy="0 0 0"/>

</robot>
```

You could view the model the same way you viewed in `rviz` for the robot base and robot arm using the following command:

```
$ cd ~/chapter_3_ws/
$ source devel/setup.bash
$ roslaunch urdf_tutorial display.launch model:=mobile_manipulator.urdf
```

The `rviz` display will look as follows:

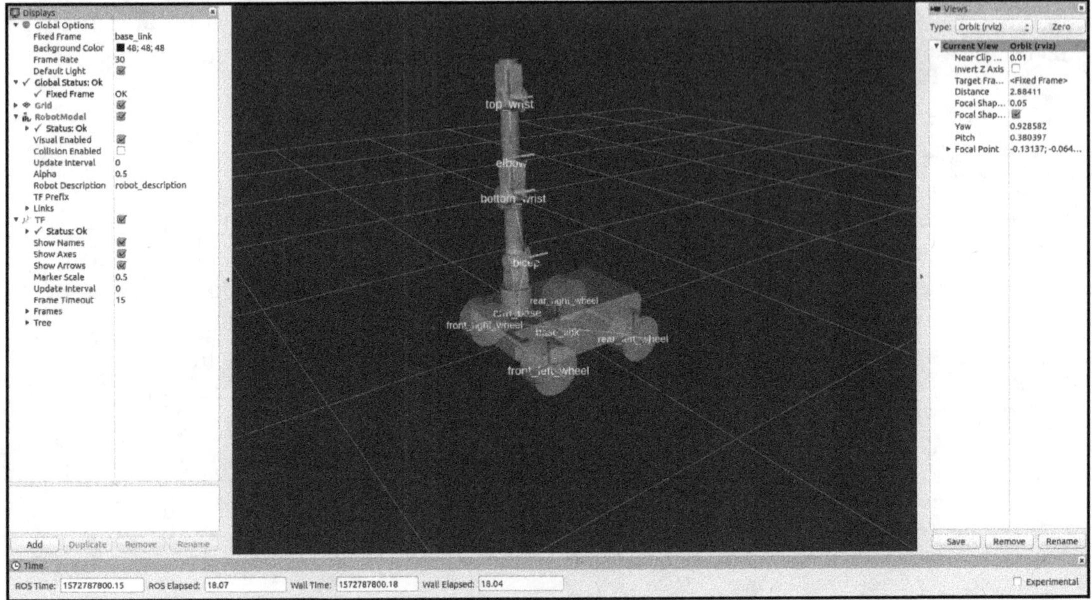

The rviz model view of the mobile manipulator

Now, let's simulate the model.

Simulating and testing the mobile manipulator

Let's create the following launch file:

```
$ cd ~/chapter3_ws/src/robot_description/launch
$ gedit mobile_manipulator_gazebo_control_xacro.launch
```

Now, copy the following launch file:

```
<?xml version="1.0"?>
<launch>

<param name="robot_description" command="$(find xacro)/xacro --inorder
$(find robot_description)/urdf/mobile_manipulator.urdf" />
  <include file="$(find gazebo_ros)/launch/empty_world.launch"/>
  <node name="spawn_urdf" pkg="gazebo_ros" type="spawn_model" args="-param
robot_description -urdf -model mobile_manipulator" />
  <rosparam command="load" file="$(find
```

```
robot_description)/config/arm_control.yaml" />
  <node name="arm_controller_spawner" pkg="controller_manager"
type="controller_manager" args="spawn arm_controller" respawn="false"
output="screen"/>
  <rosparam command="load" file="$(find
robot_description)/config/joint_state_controller.yaml" />
  <node name="joint_state_controller_spawner" pkg="controller_manager"
type="controller_manager" args="spawn joint_state_controller"
respawn="false" output="screen"/>
  <rosparam command="load" file="$(find
robot_description)/config/control.yaml" />
  <node name="base_controller_spawner" pkg="controller_manager"
type="spawner" args="robot_base_joint_publisher
robot_base_velocity_controller"/>
  <node name="robot_state_publisher" pkg="robot_state_publisher"
type="robot_state_publisher" respawn="false" output="screen"/>

</launch>
```

Now, you can visualize your robot by running the following command:

```
$ initros1
$ cd ~/chapter_3_ws
$ source devel/setup.bash
$ roslaunch robot_description mobile_manipulator_gazebo_xacro.launch
```

You can now try moving the base and arm, as shown in their respective sections. Some might be happy with the way the robot has been imported and is moving, but some might not be. Obviously, they're shaking, they're jerky, and they don't intend to look great in motion. This is simply because the controllers haven't been tuned properly.

Summary

In this chapter, we understood how a robot can be defined, modeled, and simulated in ROS with the help of Gazebo. We defined the robot's physical characteristics through plugins that were available in Gazebo. We began with creating a robot base, created a five DOF robot arm, and finally, put them together to create a mobile manipulator. We simulated the robot in Gazebo to understand how a robot application could be defined in ROS. This chapter helped us understand the use of robots in industries.

With this understanding, let's move on to the next chapter, where we will learn how the robot can handle complex tasks. Then, we will learn how to create a robot application in ROS and simulate it in Gazebo.

4
Handling Complex Robot Tasks Using State Machines

Robots are machines that need to understand the environment in order to perform any action the application demands. They sense the environment through a variety of sensors to the very last detail, after which they compute the logic through their compute systems and convert them into necessary control actions. To be more practical, while computing such logic, they're also supposed to take care of other factors that would affect their actions.

In this chapter, we shall look at how state machines are helpful in robotics and ROS. You will understand the fundamentals of state machines, be introduced to tools that handle tasks with feedback mechanisms, such as `actionlib`, and be introduced to ROS state machine packages called `executive_smach`. You will be introduced to writing examples in `actionlib` and state machines, which will help us to build our industrial application in the next chapter using the robot we created in the previous chapter.

In this chapter, the following topics will be covered:

- Introduction to ROS actions
- Waiter robot analogy
- Introduction to state machines
- Introduction to SMACH
- Getting started with SMACH examples

Technical requirements

Let's look into the technical requirements for this chapter:

- ROS Melodic Morenia on Ubuntu 18.04, with Gazebo 9 (preinstalled)
- ROS packages, that is, `actionlib`, `smach`, `smach_ros`, and `smach_viewer`
- Timelines and test platform:
    - **Estimated learning time**: On average, 100 minutes
    - **Project build time (inclusive of compile and run time)**: On average, 45 minutes
    - **Project test platform**: HP Pavilion laptop (Intel® Core™ i7-4510U CPU @ 2.00 GHz × 4 with 8 GB Memory and 64-bit OS, GNOME-3.28.2)

The code for this chapter is available at `https://github.com/PacktPublishing/ROS-Robotics-Projects-SecondEdition/tree/master/chapter_4_ws/src`.

Let's begin this chapter by understanding what ROS actions are.

Introduction to ROS actions

Let's start this section with an example. Let's assume there's a restaurant that uses robots as its waiters to help serve food to customers. Let's say that once the customer has taken their seat, they call the waiter by pressing a simple button on their table. The waiter robot understands the call and navigates to the table, takes the order from the customer, goes to the kitchen to place the order to the chef, and delivers the food to the customer once food is ready. Here, the robot tasks are to navigate to the customer's place, take the order, go to the kitchen, and bring food back to the customer.

A typical approach would be to define a script with multiple functions for individual tasks, put them together through a series of condition or case statements, and run the application. Well, the application might work as expected sometimes, but might not be the same always. Looking at some practical constraints, what if the robot was obstructed by a few people on its course, and what if the robot's battery goes down and so the robot either doesn't deliver the food or take the order on time and the customer leaves the restaurant in dismay? These robot behaviors can't always be achieved with scripts due to redundancy and the complexity of coding but can be sorted out by using state machines.

Considering the preceding waiter robot example, after taking the order from the customer, the robot would need to navigate to various locations. While navigating, the robot navigation may be disturbed by people walking around. Initially, the obstacle avoidance algorithm may help the robot avoid people, but over a certain period of time, there are chances that the robot still might not have reached the goal. This might possibly be due to cases where the robot is stuck in a loop trying to avoid people. In this case, the robot is asked to traverse to the goal without any feedback mechanism, because of which it got stuck in a loop trying to reach the destination. This is time-consuming and the food might not be delivered to the customer on time. To avoid such situations, we can use an ROS concept called **actions** (`http://wiki.ros.org/actionlib`). ROS actions are a type of ROS implementation for achieving time-consuming or goal-oriented behavior. They're more flexible and powerful to use with complex systems with just a little more effort. We'll look at how they work in the upcoming sections.

The client-server concept

ROS actions follow a client-server concept, as shown in the following diagram. The client is responsible for sending control signals, whereas the server is responsible for listening to those control signals and providing the necessary feedback:

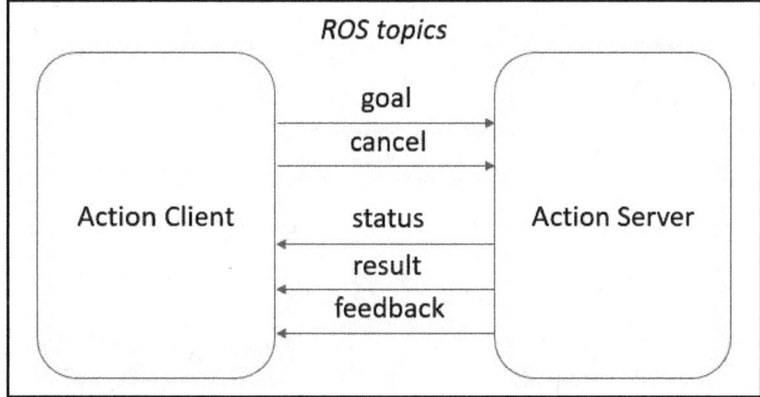

ROS actions concept

The client node and the server node communicate via the ROS action protocol, which is built on top of ROS messages. The client node will send a goal or a list of goals or cancel all the goals, whereas a server application would return the status of that goal, the result upon the completion of a goal, and feedback as periodic information about the goal being achieved or cancelled. Let's look at this in more detail by looking at an example.

An actionlib example – robot arm client

Let's go back to the previous chapter and run our robot arm Gazebo example that we built:

```
$ roslaunch robot_description mobile_manipulator_gazebo_xacro.launch
```

Now, in another Terminal, run `$ rostopic list`. Do you see anything interesting? Look at the following screenshot:

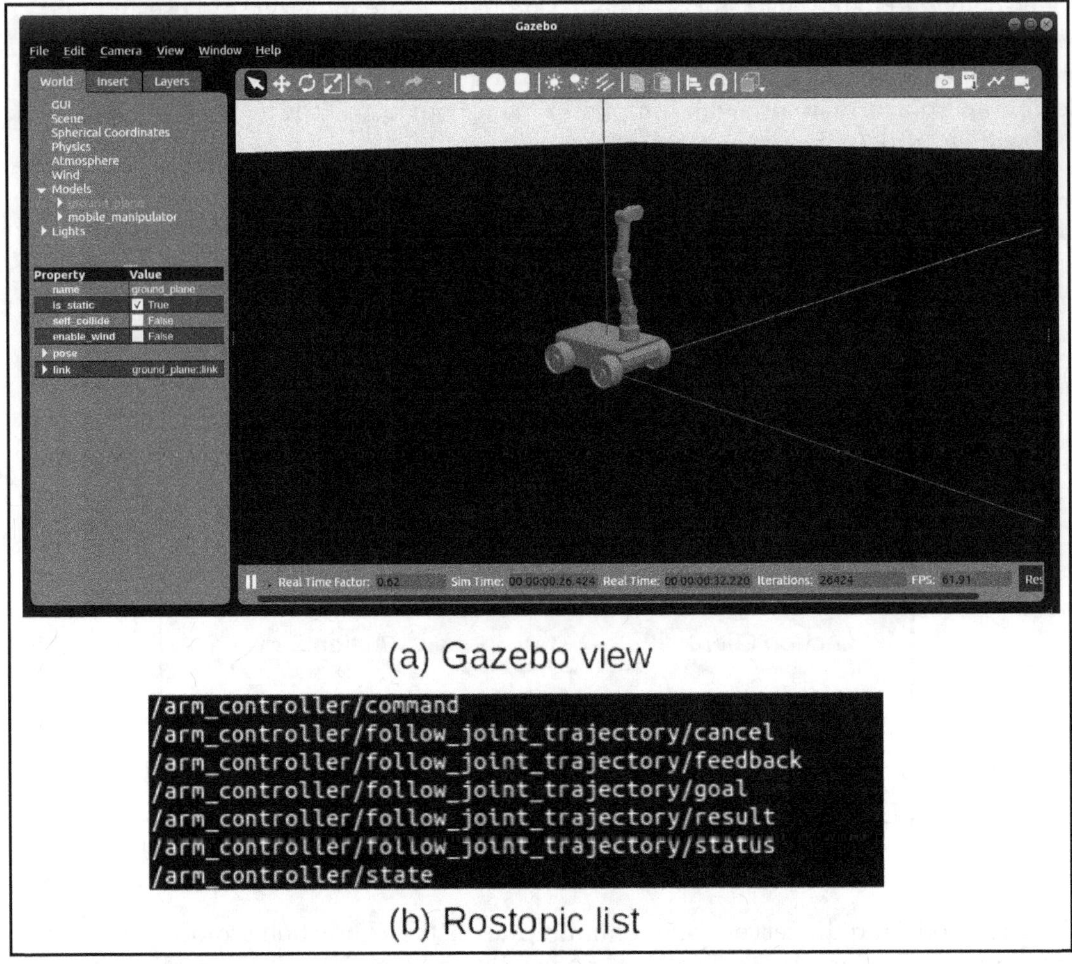

Robot arm Gazebo ROS topics

Do you see a list of topics that end with `/goal`, `/cancel`, `/result`, `/status`, and `/feedback`? That's an ROS action implementation.

joint_trajectory_controller is used for executing joint space trajectories on a list of given joints. The trajectories to the controller are sent by means of the control_msgs::FollowJointTrajectoryAction action interface in the follow_joint_trajectory namespace of the controller. FollowJointTracjectoryAction is a result of position_controllers/JointTrajectoryController, which we called as the arm_controller plugin.

You can find more information about this at http://wiki.ros.org/joint_trajectory_controller. The FollowJointTrajectoryAction definition can be found here: http://docs.ros.org/api/control_msgs/html/action/FollowJointTrajectory.html.

As a result of the arm_controller plugin call, the FollowJointTrajectory action server is already implemented as part of the Gazebo node. You can see that the topics ending with /result, /status, and /feedback are published by the Gazebo node (alias, the robot) and the topics ending with /goal and /cancel are being subscribed to by the robot. Hence, to move the robot arm, we need to send a goal to the action server that's been implemented. Let's learn how to send goals to our robot arm via an action client implementation. Have a look at the following function for an action client implementation:

- Before getting into the code, let's create the chapter_4_ws workspace and arm client package using the following commands:

```
$ initros1
$ mkdir -p /chapter_4_ws/src
$ cd ~/chapter_4_ws/src
$ catkin_init_workspace
$ catkin_create_pkg arm_client
$ cd ~/chapter_4_ws/src/arm_client
```

- You can find the whole code here: https://github.com/PacktPublishing/ROS-Robotics-Projects-SecondEdition/blob/master/chapter_4_ws/src/battery_simulator/src/arm_action_client.py. Download the file and place it into the folder, give root permissions using the $ chmod +x command, and compile the workspace using the $ catkin_make command.

Let's break down the code into chunks and try to understand it. The following lines of code are the necessary import statements for using ROS functions, the action library, and ROS messages:

```
import rospy
import actionlib
from std_msgs.msg import Float64
```

```
from trajectory_msgs.msg import JointTrajectoryPoint
from control_msgs.msg import JointTrajectoryAction, JointTrajectoryGoal,
FollowJointTrajectoryAction, FollowJointTrajectoryGoal
```

In our main program, we initialize our node and client and send the goal. We're using `SimpleActionClient` in our example. We call the server name, `arm_controller/follow_joint_trajectory`, which is a result of the `ros_controller` plugin and its message type. We then wait for a response from the server. Once the response is received, the `move_joint` function is executed:

```
if __name__ == '__main__':
    rospy.init_node('joint_position_tester')
    client =
actionlib.SimpleActionClient('arm_controller/follow_joint_trajectory',
FollowJointTrajectoryAction)
    client.wait_for_server()
    move_joint([-0.1, 0.210116830848170721, 0.022747275919015486,
0.0024182584123728645, 0.00012406874824844039])
```

The `move_joint()` function accepts the angle values of each joint and sends them as a trajectory to the robot arm. As you may recall from the previous chapter, we published some information to the `/arm_controller/follow_joint_trajectory/command` topic. We need to pass the same information via the `FollowJointTrajectoryGoal()` message, which needs `joint_names`, points (that is, joint values), and the time to move the arm to a specified trajectory:

```
def move_joint(angles):
    goal = FollowJointTrajectoryGoal()
    goal.trajectory.joint_names = ['arm_base_joint',
'shoulder_joint','bottom_wrist_joint' ,'elbow_joint', 'top_wrist_joint']
    point = JointTrajectoryPoint()
    point.positions = angles
    point.time_from_start = rospy.Duration(3)
    goal.trajectory.points.append(point)
```

The following line of code is used to send a goal via the client. This function will send the goal until the goal is complete. You can also define a timeout here and wait until the timeout is exceeded:

```
client.send_goal_and_wait(goal)
```

The preceding example of the `SimpleActionClient` implementation supports only one goal at a time. This is an easy to use wrapper for the user.

You can test the preceding code using the following steps:

1. In one Terminal, open the robot Gazebo file using the following command:

```
$ cd ~/chapter_3_ws/
$ source devel/setup.bash
$ roslaunch robot_description
mobile_manipulator_gazebo_xacro.launch
```

2. In another Terminal, run the action client file to see the robot arm move. You can do this with the following command:

```
$ cd ~/chapter_4_ws/
$ source devel/setup.bash
$ rosrun arm_client arm_action_client.py
```

You should see that the arm moves to the given position in the code.

You can find more information about this implementation here: `https://docs.ros.org/melodic/api/actionlib/html/classactionlib_1_1SimpleActionClient.html#_details`.

Now, let's look at another example.

An actionlib example – battery simulator server-client

In the previous example, you wrote a client that sends goals to the server, which was already implemented as a part of the Gazebo plugin. In this example, let's try to implement ROS actions from scratch. To create an ROS action, follow these steps:

1. Create a package and create a folder called action inside it.
2. Create an action file that has the goal, result, and feedback.
3. Modify the package files and compile the package.
4. Define a server.
5. Define a client.

Let's consider an example of a battery simulator. When you power up a robot, the robot runs on a battery supply and eventually loses charge over a certain amount of time, depending on the battery type. Let's simulate the same scenario using a server-client implementation. The simulator will be our server and will be initiated when the robot is powered up. We will use a client to change the state of the battery, either as charging or discharging, based on which the battery server charges or discharges. Hence, let's get on with the steps for implementing our own ROS action.

Creating a package and a folder action inside it

In our created workspace, `chapter_4_ws/src`, create a package called `battery_simulator` using the following command:

```
$ cd ~/chapter_4_ws/src
$ catkin_create_pkg battery_simulator actionlib std_msgs
```

Now, let's go into our workspace and create a folder called `action`:

```
$ cd ~/chapter_4_ws/src/battery_simulator/
$ mkdir action
```

The `action` folder is where we will define our action file in the next step.

Creating an action file that has the goal, result, and feedback

Let's create an action file named `battery_sim.action`:

```
$ gedit battery_sim.action
```

Add the following content to it:

```
#goal
bool charge_state

#result
string battery_status
---
#feedback
float32 battery_percentage
```

Here, you can see that the goal, result, and feedback are defined in a hierarchical order, separated by three dashes `---`.

Modifying the package files and compiling the package

To make our package understand action definitions and aid in their usage, we need to modify our `package.xml` and `CMakeLists.txt` files. In the `package.xml` file, ensure the following lines of code are available to call the action package. If not, please add them:

```
<build_depend>actionlib_msgs</build_depend>
<exec_depend>actionlib_msgs</exec_depend>
```

In the `CMakeLists.txt` file, add the action files to the `add_action_files()` call:

```
add_action_files(
DIRECTORY action
FILES battery_sim.action
)
```

They may be commented. You can either uncomment and add them or simply copy and paste these lines into your `CMakeLists.txt` file.

Indicate the dependencies in the `generate_messages()` call:

```
generate_messages(
DEPENDENCIES
actionlib_msgs
std_msgs
)
```

Indicate the dependencies in the `catkin_package()` call:

```
catkin_package(
CATKIN_DEPENDS
actionlib_msgs
)
```

Now that the package has been configured, let's compile the workspace:

```
$ cd ~/chapter_4_ws
$ catkin_make
```

You should see the following list of files:

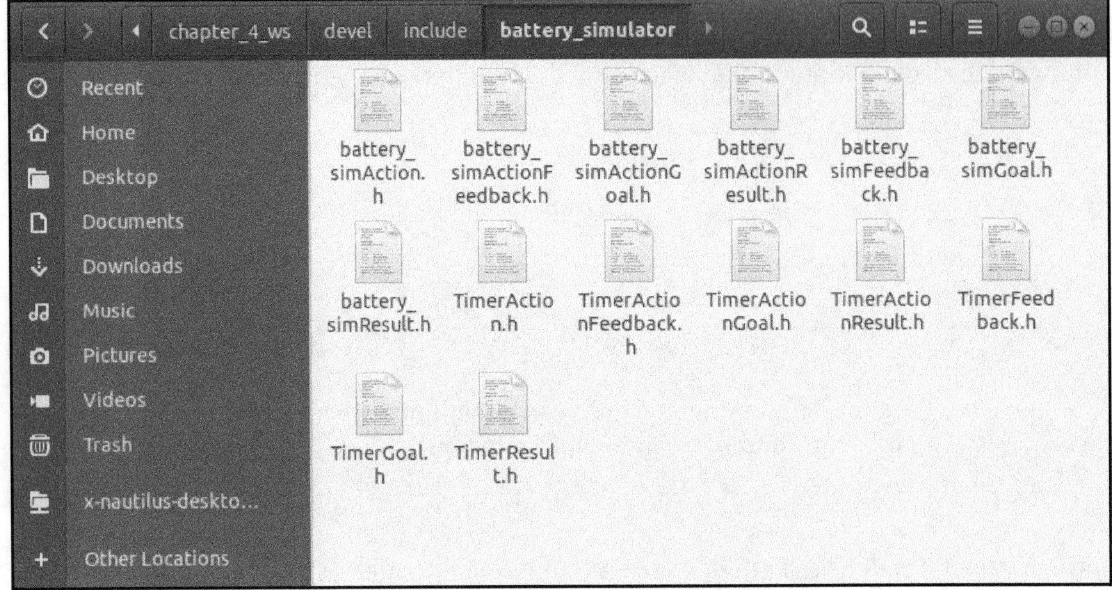

Compilation of action messages and the list of action files

We will use these files in our code as a message structure. Now, let's define the server and client.

Defining a server

In this section, we will define the server. Let's look at the code for this. The necessary import statements will be called upon.

You can have a look at the code in our repository here: `https://github.com/PacktPublishing/ROS-Robotics-Projects-SecondEdition/blob/master/chapter_4_ws/src/battery_simulator/src/battery_sim_server.py`.

Note that we enable multiprocessing and call the necessary `actionlib` messages from our `battery_simulator` package:

```
#! /usr/bin/env python

import time
import rospy
from multiprocessing import Process
from std_msgs.msg import Int32, Bool
import actionlib
from battery_simulator.msg import battery_simAction, battery_simGoal,
battery_simResult, battery_simFeedback
```

In the main function, we initialize the node and server's `battery_simulator` and start the server. To keep the server up and running, we use `rospy.spin()`:

```
if __name__ == '__main__':
rospy.init_node('BatterySimServer')
server = actionlib.SimpleActionServer('battery_simulator',
battery_simAction, goalFun, False)
server.start()
rospy.spin()
```

The server calls upon the `goalFun()` function, which starts the `batterySim()` function in parallel. We assume that the goal from the client is Boolean. If the goal that's received is `0`, then it means that the battery is in a discharging state and if it is `1`, the battery is in a charging state. For ease of use, we will set ROS parameters so that we can enable or disable charging:

```
def goalFun(goal):
  rate = rospy.Rate(2)
  process = Process(target = batterySim)
  process.start()
  time.sleep(1)
  if goal.charge_state == 0:
  rospy.set_param("/MyRobot/BatteryStatus",goal.charge_state)
  elif goal.charge_state == 1:
  rospy.set_param("/MyRobot/BatteryStatus",goal.charge_state)
```

The `batterySim()` function checks for the parameter value and runs the battery's incremented or decremented code based on the `charge_state` goal it has received:

```
def batterySim():
  battery_level = 100
  result = battery_simResult()
  while not rospy.is_shutdown():
  if rospy.has_param("/MyRobot/BatteryStatus"):
    time.sleep(1)
    param = rospy.get_param("/MyRobot/BatteryStatus")
    if param == 1:
      if battery_level == 100:
        result.battery_status = "Full"
        server.set_succeeded(result)
        print "Setting result!!!"
        break
      else:
        print "Charging...currently, "
        battery_level += 1
        print battery_level
        time.sleep(4)
    elif param == 0:
      print "Discharging...currently, "
      battery_level -= 1
      print battery_level
      time.sleep(2)
```

If the charging is full, then the result is set as successful using the `server.set_succeeded(result)` function. Now, let's define the client.

Defining a client

Now that we have defined a server, we need to give suitable goals to the server using a client, which you should be familiar by now after our *An actionlib example – robot arm client* section. Let's create a `battery_sim_client.py` file and copy the following code into it:

```
#! /usr/bin/env python

import sys
import rospy
from std_msgs.msg import Int32, Bool
import actionlib
from battery_simulator.msg import battery_simAction, battery_simGoal,
battery_simResult

def battery_state(charge_condition):
```

```
    goal = battery_simGoal()
    goal.charge_state = charge_condition
    client.send_goal(goal)

  if __name__ == '__main__':
    rospy.init_node('BatterySimclient')
    client = actionlib.SimpleActionClient('battery_simulator',
  battery_simAction)
    client.wait_for_server()
    battery_state(int(sys.argv[1]))
    client.wait_for_result()
```

We receive the charge state as an argument and send the goal to the server accordingly while charging or discharging. We define the server to that we need to send goals to in our client declaration (in the main function). We wait for its response and send the necessary goal.

Now that we have our very own ROS action implementation, let's see `battery_simulator` in action:

- Start `$ roscore` in the first Terminal.
- In the second Terminal, start the server using the following command:

```
$ cd ~/chapter_4_ws/
$ source devel/setup.bash
$ rosrun battery_simulator battery_sim_server.py
```

- In the third Terminal, start the client using the following command:

```
$ cd ~/chapter_4_ws/
$ source ~/chapter_4_ws/devel/setup.bash
$ roslaunch battery_simulator battery_sim_client.py 0
```

You can see the battery information being printed out as a result of ROS logging. Change the state to `1` and run the node again. You can see that the battery is now charging.

Now, let's cover the waiter robot analogy based on the example we have already discussed.

Waiter robot analogy

Let's continue with our waiter robot analogy to understand state machines better. Have a look at the following setup:

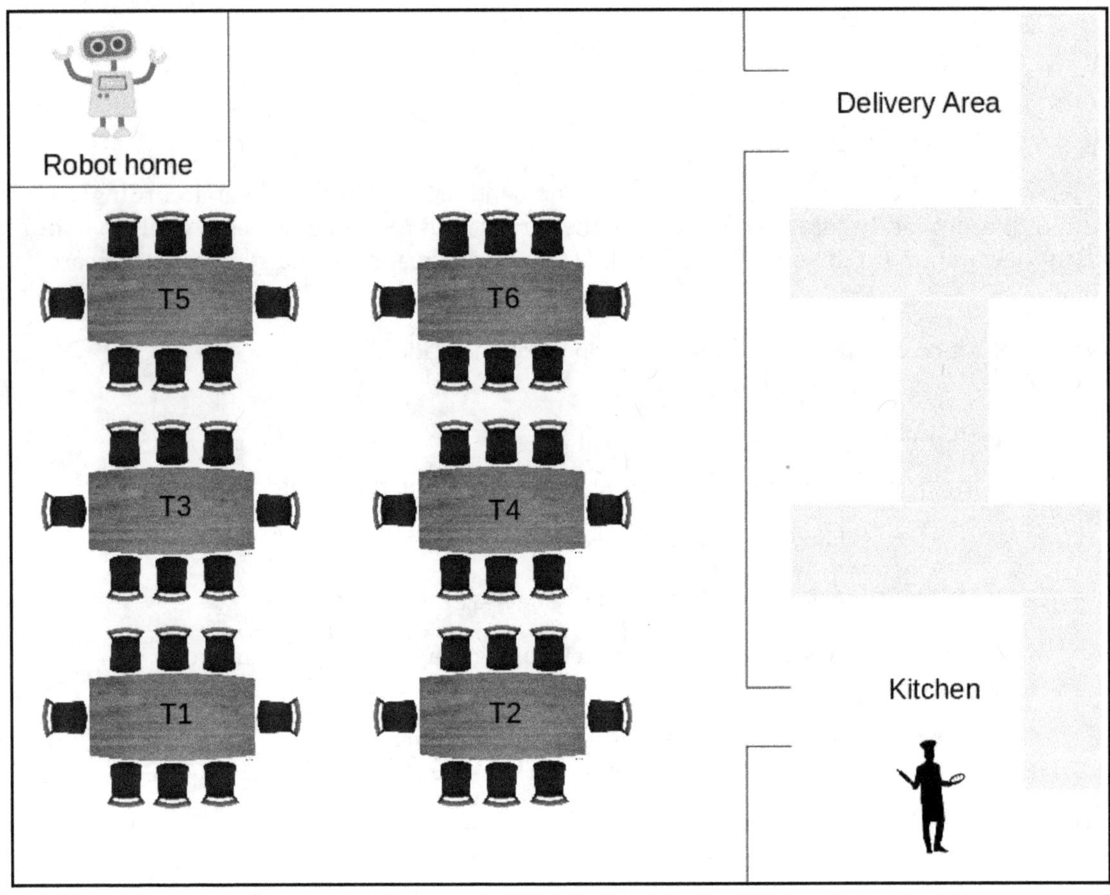

Waiter robot analogy

Let's try to get in-depth and list the possible tasks that need to be carried out by the robot:

- Navigate to the tables (T1, T2, ..., and T6, as shown in the preceding diagram).
- Take the order from the customer.
- Go to the kitchen (and confirm with the chef if necessary).
- Bring the food to the customer (from the delivery area, as shown in the preceding diagram).

The robot can navigate around the restaurant autonomously and reach customer locations based on the table the customer is seated at. In this process, the robot has the necessary information such as the locations of the table, delivery, kitchen, storeroom, and charging area through the map it has created. Once the robot has reached the table, let's assume that the robot takes the order from the customer through a touch-based interaction system combined with voice-based interactions.

Once the order is received, the robot sends it to the kitchen through a central management system. Once the kitchen confirms the order reception, the robot goes to the delivery location and waits for the food to be delivered. If the order is not confirmed by the kitchen, the robot goes to the kitchen and confirms the order with the chef. Once the food is ready, the robot takes the food from the delivery location and delivers it to the specific customer table. Once it is delivered, the robot goes to its standoff position or gets charged. The pseudocode for this may look as follows:

```
table_list = (table_1, table_2,....)
robot_list = (robot_1, robot_2,....)
locations = (table_location, delivery_area, kitchen, store_room,
charging_area, standoff)

customer_call = for HIGH in [table_list], return table_list(i) #return
table number
customer_location = customer_call(table_list)

main ():
 while customer_call is "True"
   navigate_robot(robot_1, customer_location)
   display_order()
   accept_order()
   wait(60) # wait for a min for order acceptance from kitchen
   if accept_order() is true
      navigate_robot(robot_1, delivery_area)
   else
      navigate_robot(robot_1, kitchen)
      inform_chef()
      wait(10)
   wait_for_delivery_status()
```

```
    pick_up_food()
    navigate_robot(robot_1, customer_location)
    deliver_food()
```

This pseudocode allows us to do all of the tasks in sequence based on specific function calls. What if we wanted to do more than just the sequence? For instance, if the robot needs to keep track of its battery usage and perform necessary tasks, the preceding `main()` function can get inside another condition block that monitors the battery state and makes the robot navigate to `charging_area` in case the battery state is less than the threshold mentioned.

But what if the robot actually goes into a low battery state while trying to deliver food? It may go into the charging station with the food instead of going to the customer table. Instead, the robot can predict its battery state before attempting delivery and charge while the food is being cooked or—in the worst case scenario—it can call another robot that is free and with sufficient charge to deliver the food. Writing scripts for such characteristics becomes complex and can give the developer nightmares. Instead, what we would need is a simple method to implement such complex behavior in a standard manner and have the ability to diagnose, debug, and run tasks in parallel. Such an implementation is done through state machines.

The next section will help us get a better understanding of state machines.

Introduction to state machines

State machines are graphical representations of a problem, decomposed into chunks of smaller operations or processes. These smaller operations then communicate with each other either in series or in parallel and in a certain order to accomplish the problem at hand. State machines are fundamental concepts in computer science and are used to solve complex systems by spending more time solving the problem visually than coding.

Let's see how our waiter robot analogy is represented via a state machine:

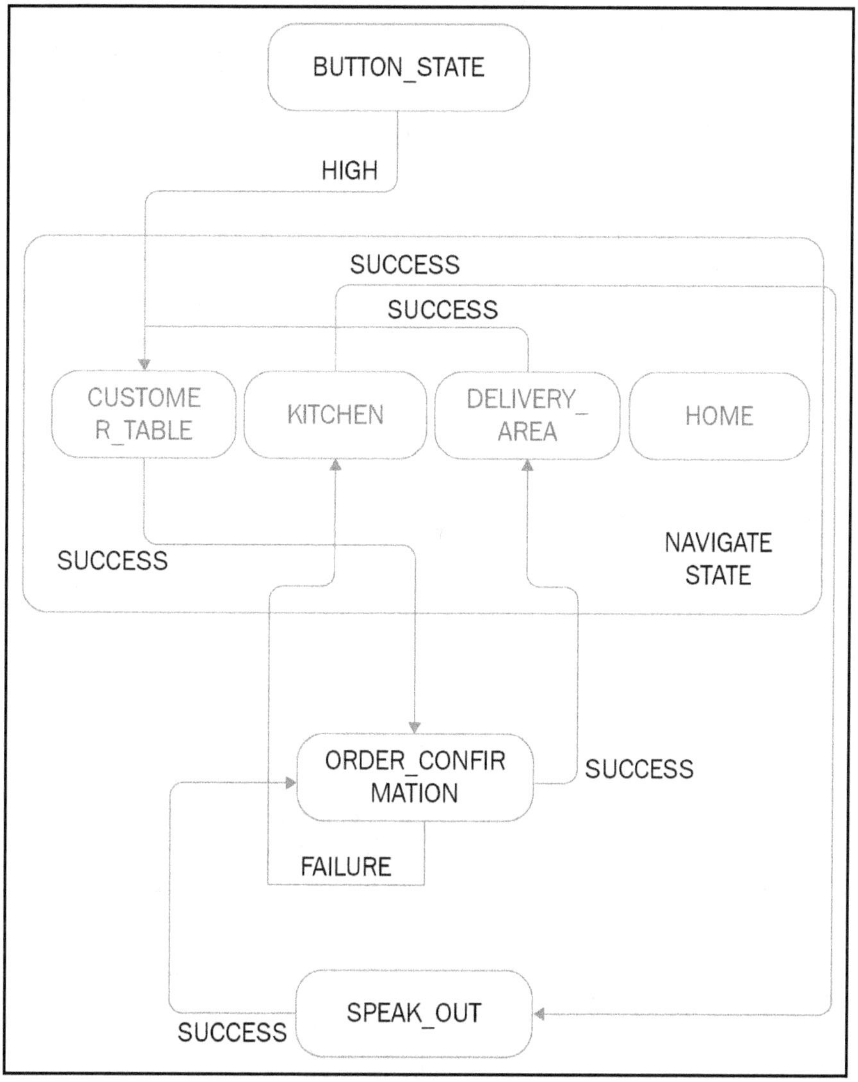

State machine representation of our analogy

The circled components are robot states. These states are the ones in which the robot is performing certain actions to accomplish that specific state. The lines that connect these circles are called edges. They represent the transition between states. The values or descriptions on these lines are the outcomes that represent whether the state is complete, incomplete, or in progress.

This is just a simple representation to help you understand state machines and is not the complete representation of the robot application. When the customer has pressed the button, there is a transition to a state called `NAVIGATION_STATE`. As you can see, `NAVIGATION_STATE` has a few other states internally, which are defined as `CUSTOMER_TABLE`, `KITCHEN`, `DELIVERY_AREA`, and `HOME`. You shall see how simplified they are and how they are defined in the upcoming section. Once the robot reaches `CUSTOMER_TABLE`, the robot displays the order, takes the order (not displayed in the preceding diagram), and waits for the chef to confirm the order (the `ORDER_CONFIRMATION` state). If `SUCCESS`, the robot navigates to `DELIVERY_AREA`; otherwise, it navigates to `KITCHEN` and voices out the order confirmation request though the `SPEAK_OUT` state. If `SUCCESS`, it goes to `DELIVERY_AREA`, gets the order, and delivers the same to `CUSTOMER_TABLE`.

In the next section, we will cover SMACH.

Introduction to SMACH

SMACH (pronounced as smash) is a Python-based library that helps us handle complex robot behaviors. It is a standalone Python tool that can be used to build hierarchical or concurrent state machines. While its core is ROS-independent, wrappers are written in the `smach_ros` package (`http://wiki.ros.org/smach_ros?distro=melodic`) to support integration with ROS systems via actions, services, and topics. SMACH is best useful when you're able to describe robot behaviors and actions more explicitly (like in our waiter robot analogy-state machine diagram). SMACH is a simple and wonderful tool to use and define state machines and is quite powerful when used in combination with ROS actions. The resultant combination can help us build more sophisticated systems. Let's look at some of its concepts in detail to help understand SMACH in more detail.

SMACH concepts

The underlying core, SMACH, is lightweight with good logging and utility functions. It has two fundamental interfaces:

- **States**: States are what you want the robot to exhibit as a behavior or simply execute an action. They have potential outcomes as a result of the execution of actions. In ROS, the behavior or actions are defined in the `execute()` function, which runs until it returns an identified outcome.

- **Containers**: Containers are a collection of states. They implement an execution policy that can be hierarchical, execute more than one state at a time (concurrent), or execute the state for a certain period of time (iterative):

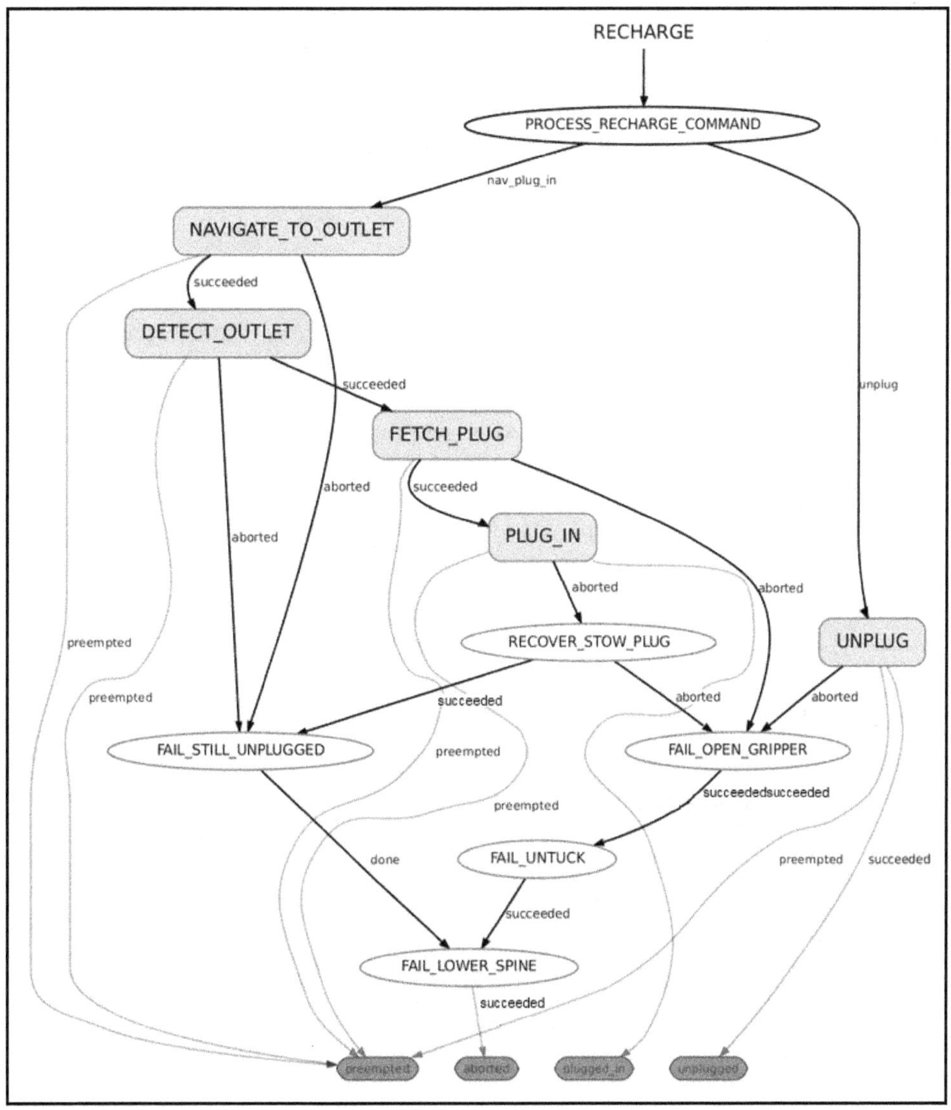

A simple state machine in ROS (source: http://wiki.ros.org/smach/Tutorials/Getting%20Started. Image from ros.org. Licensed under Creative Commons CC-BY-3.0: https://creativecommons.org/licenses/by/3.0/us/legalcode)

Let's consider the preceding diagram and look into SMACH concepts in a bit more detail.

Outcome

The outcome is a potential result of a state that the state returns after the execution of an action or behavior. From the preceding diagram, the outcomes are `nav_plug_in`, unplug, succeeded, aborted, preempted, and done. The results of states can be different and can do different things to different states. For instance, it may help in the transition to another state or may terminate the state. Hence, the outcome is irrelevant from the perspective of state. In ROS, outcomes are initialized along with the state initialization to ensure consistency and are returned after the `execute()` block (we shall look into them in the upcoming examples).

User data

States or containers may need inputs from the users or previous states (in an autonomous system) to undergo a transition. The information can be a sensor reading from the environment or a prioritization call based on the time taken for a task. They may also need to return specific information to other states or containers to allow the execution of those states. This information is described as input and output keys. Input keys are what a state may need for execution, so a state cannot manipulate what it receives (such as sensor information). Output keys are what the state returns as output to other states (such as prioritization calls) and can be manipulated. User data helps prevent errors and aids in debugging.

Preemption

Preemption is a state of interruption so that we can draw immediate attention to something other than what action is being executed. In SMACH, the `State` class has an implementation for handling preempt requests between the states and containers. This built-in behavior helps in smoothly transitioning to, say, a termination state either from the user or an executive.

Introspection

The state machine we design or create in `smach_ros` can be visualized for debugging or analysis through a tool called SMACH viewer. The following diagram shows the use of SMACH viewer with the preceding PR2 robot example (the text and numbers in this image are intentionally illegible):

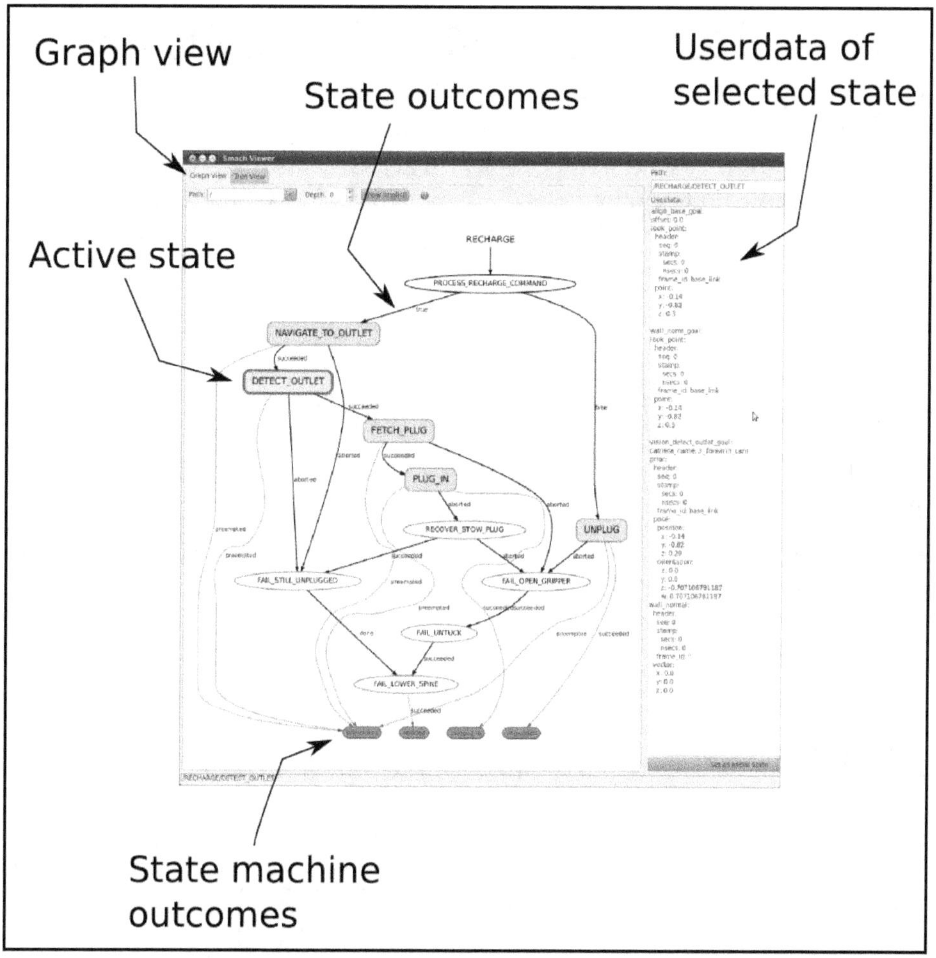

Graph view

State outcomes

Userdata of selected state

Active state

State machine outcomes

SMACH viewer representation (source: http://wiki.ros.org/smach_viewer. Image from ros.org. Licensed under Creative Commons CC-BY-3.0: https://creativecommons.org/licenses/by/3.0/us/legalcode)

The robot is currently in the `DETECT_OUTLET` state, which is indicated in green.

To view these states in the SMACH viewer, you must define an introspection server in your SMACH code.

We now have a basic understanding of SMACH. Now, let's look at a few examples based on SMACH.

Getting started with SMACH examples

The best way to learn about SMACH is through examples. Let's look at some simple examples of how to get started with SMACH so that we can create our state machine for our waiter robot analogy. Note that this book will act as a starting point for SMACH ROS and not all the examples and methodologies used in SMACH will be explained due to our limited scope. Hence, I suggest that you have a look at the official tutorials link, `http://wiki.ros.org/smach/Tutorials`, which has a very good and comprehensive list of tutorials.

Installing and using SMACH-ROS

The installation of SMACH is pretty straightforward and it can be installed directly using the following command:

```
$ sudo apt-get install ros-melodic-smach ros-melodic-smach-ros ros-melodic-executive-smach ros-melodic-smach-viewer
```

Let's look at a very simple example to explain this concept.

Simple example

In this example, we have four states: **A**, **B**, **C**, and **D**. This example aims to transition from one state to another on reception of an output. The end result looks like this:

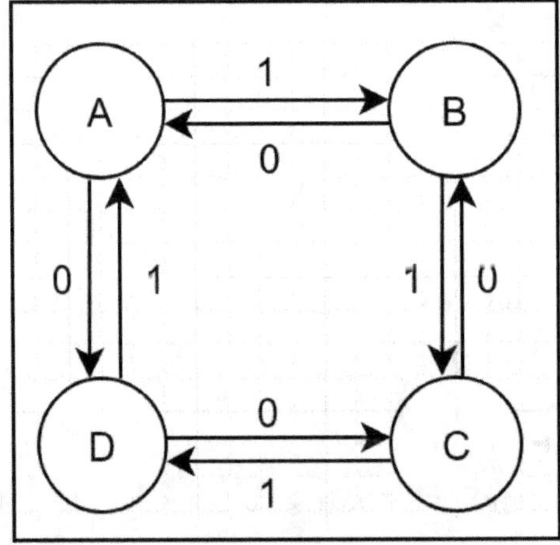

Simple transition between states

Let's look at the code in detail. The complete code is available in this book's GitHub repository: `https://github.com/PacktPublishing/ROS-Robotics-Projects-SecondEdition/blob/master/chapter_4_ws/src/smach_example/simple_fsm.py`.

Let's look into the code in detail. We will bring in the following import statements, along with the Python shebang file:

```
#!/usr/bin/env python
import rospy
from smach import State,StateMachine
from time import sleep
import smach_ros
```

Let's define the states individually in such a way that they receive an input from the user:

```
class A(State):
    def __init__(self):
        State.__init__(self, outcomes=['1','0'], input_keys=['input'],
output_keys=[''])

    def execute(self, userdata):
        sleep(1)
        if userdata.input == 1:
        return '1'
        else:
        return '0'
```

Here, we define the state, `A`, which has the outcomes `1` and `0` as a result of state accomplishment. We define the input required for the state via `input_keys`. We keep `output_keys` empty as we're not going to output any user data. We execute the state actions in the `execute()` function. For this example, we simply return `1` if the user input is `1` and return `0` if the user input is `0`. We define other states in the same way.

In the main function, we initialize the node and the state machine and assign user data to the state machine:

```
rospy.init_node('test_fsm', anonymous=True)
sm = StateMachine(outcomes=['success'])
sm.userdata.sm_input = 1
```

With the help of Python's `with()` function, we add the preceding states and define the transitions:

```
with sm:
    StateMachine.add('A', A(), transitions={'1':'B','0':'D'},
remapping={'input':'sm_input','output':''})
    StateMachine.add('B', B(), transitions={'1':'C','0':'A'},
remapping={'input':'sm_input','output':''})
    StateMachine.add('C', C(), transitions={'1':'D','0':'B'},
remapping={'input':'sm_input','output':''})
    StateMachine.add('D', D(), transitions={'1':'A','0':'C'},
remapping={'input':'sm_input','output':''})
```

For state `A`, we define the state name as `A`, call the `A()` class we defined, and remap the input to the one we had defined in the main function. Then, we define the state transition based on the state outcome. In this case, if our outcome is `1`, then we call state `B` and if the outcome is `0`, we call state `D`. We can add other states in a similar fashion.

To check our state transitions, we use the `smach_viewer` package from `smach_ros`. To view the states in `smach_viewer`, we need to call `IntrospectionServer()` with a server name, state machine, and root name for the states to connect to. We then start `IntrospectionServer`:

```
sis = smach_ros.IntrospectionServer('server_name', sm, '/SM_ROOT')
sis.start()
```

Finally, we execute our state machine in a loop:

```
sm.execute()
rospy.spin()
sis.stop()
```

We stop the introspection server using the `stop()` function once the transition is complete. You can run this example as follows. The following diagram shows the `smach_viewer` view of our example. The individual text is for reference and does not need to be read. This is what the output will look like:

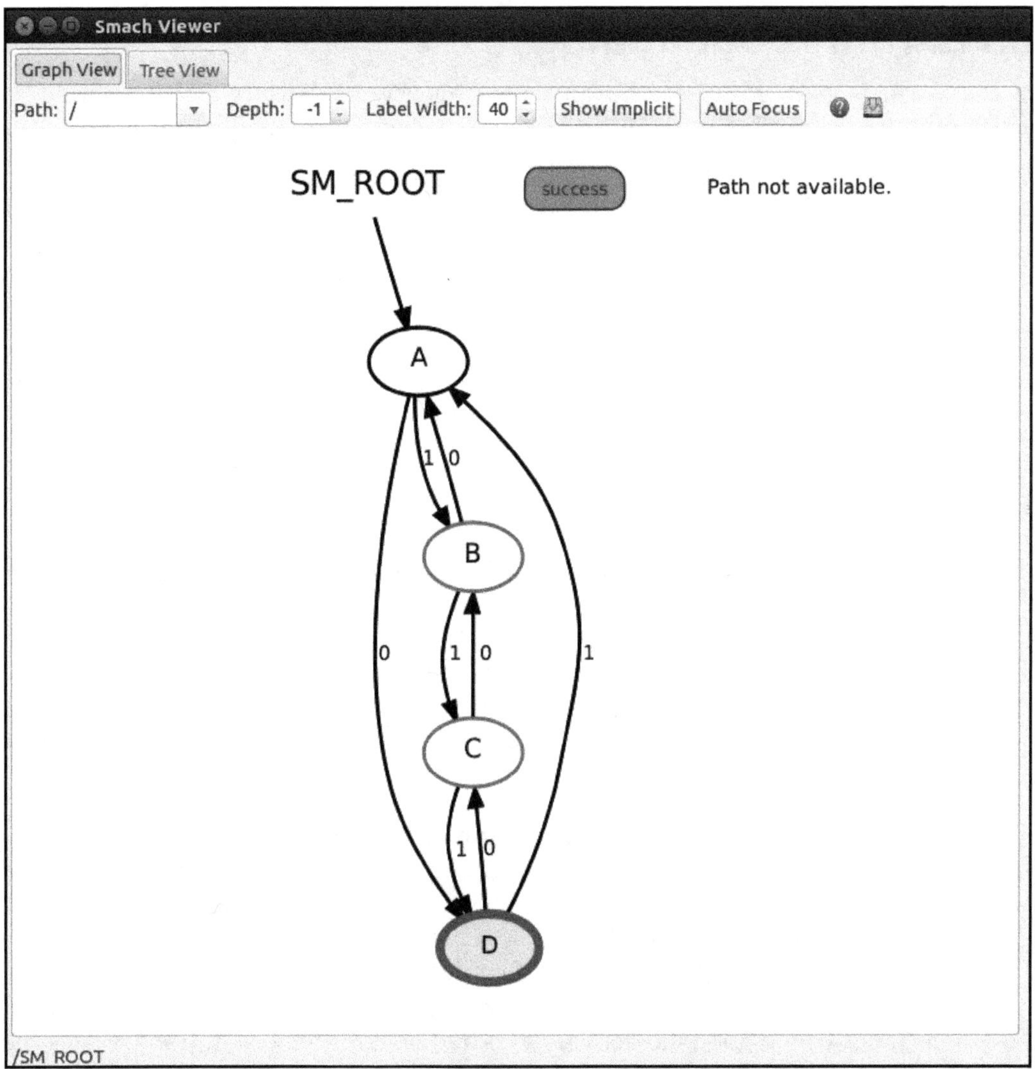

The smach_viewer view of our example

Here, you can see that the green color shows the state, which is currently active. You can also see the state transition between each states. If you change `sm_input` to `0` in the main function, you will see the transition changes.

Restaurant robot analogy

Now that we have had a glimpse at how state machines work, let's try creating a state machine for our waiter robot analogy. To be more practical, the robot has to carry out the following tasks:

- Power on the robot
- Check for a customer call
- Navigate to the table based on call
- Take the order from the customer
- Go to the delivery area or kitchen if the order has been confirmed or has failed
- Deliver food to the customer

Now, we can go about defining our analogy:

- `STATES_MACHINE`: The robot needs to be powered ON and initialized, so we need to have a `POWER_ON` state. Once the robot is powered ON, we need to check whether any customer has placed any orders. We do this via `BUTTON_STATE`. Once the robot has received a call from the customer, the robot has to navigate to the customer table, which is why we define the `GO_TO_CUSTOMER_TABLE` state. Since the robot has more such navigation goals, we define the respective states, that is, `GOT_TO_DELIVERY_AREA`, `GO_TO_KITCHEN`, and `GO_TO_HOME`, via action implementation.

  The robot has to accept confirmation from the customer once the customer decides to place the order to the kitchen. Let's imagine that the customer interacts with the robot via a touch screen and hits the place order button once the order is placed. This needs a `ORDER_CONFIRMATION` process state. The `ORDER_CONFIRMATION` state not just confirms the user order, but also places the order to the kitchen. If the order is confirmed from the kitchen, say, by the chef, the robot would go to the delivery area and wait for the order. If the order is not confirmed, the robot would go to the kitchen and inform the chef in person. This can be done as a voice out using the `SPEAK_OUT` state. Once the chef confirms this, the robot goes to the delivery area.

- `STATES`: Now that we have the state machine, let's define our states. The robot needs to power ON and initialize itself. We do this with our `PowerOnRobot()` class. Once the state is called upon, it (for our example) virtually powers ON and sets the state to `succeeded`:

```
class PowerOnRobot(State):
    def __init__(self):
```

```
            State.__init__(self, outcomes=['succeeded'])

        def execute(self, userdata):
            rospy.loginfo("Powering ON robot...")
            time.sleep(2)
            return 'succeeded'
```

Then, we need to read the button state and go to the specific table location. In this case, we simplify the case by assuming there's only one table and navigating to that location. In reality, you would have more tables and the same can be defined by a dictionary of table labels and poses. So, in our example, if the button state is high, the state is set to `succeeded` or `aborted`:

```
class ButtonState(State):
    def __init__(self, button_state):
        State.__init__(self, outcomes=['succeeded','aborted','preempted'])
        self.button_state=button_state

    def execute(self, userdata):
        if self.button_state == 1:
            return 'succeeded'
        else:
            return 'aborted'
```

The robot communicates with the user through a touch screen, as per our example, and accepts the order. If the order is confirmed from the user's side, a hidden subprocess happens (that is not defined in this analogy), and that process tries to place the order to the chef. If the chef accepts the order, then we set the state as succeeded; otherwise, the state is aborted. We receive the chef's confirmation as user input:

```
class OrderConfirmation(State):
    def __init__(self, user_confirmation):
        State.__init__(self, outcomes=['succeeded','aborted','preempted'])
        self.user_confirmation=user_confirmation

    def execute(self, userdata):
        time.sleep(2)
        if self.user_confirmation == 1:
            time.sleep(2)
            rospy.loginfo("Confirmation order...")
            time.sleep(2)
            rospy.loginfo("Order confirmed...")
            return 'succeeded'
        else:
            return 'preempted'
```

If the order is not confirmed by the chef, the robot needs to voice out. We do this using the `SpeakOut()` class, where the robot speaks out to confirm the order and sets the state accordingly:

```
class SpeakOut(State):
def __init__(self,chef_confirmation):
State.__init__(self, outcomes=['succeeded','aborted','preempted'])
self.chef_confirmation=chef_confirmation

def execute(self, userdata):
sleep(1)
rospy.loginfo ("Please confirm the order")
sleep(5)
if self.chef_confirmation == 1:
return 'succeeded'
else:
return 'aborted'
```

Finally, we need to move the robot between places. This can be defined as a specific state but, for simplicity, we've defined them as several states. However, all the states have a specific state in common that makes the robot move, that is, `move_base_state`. We make use of the ROS actions we saw previously in this case. The robot has a `move_base` server that accepts the goal in a proper message format. We define a client call for such a state through the following code:

```
move_base_state = SimpleActionState('move_base', MoveBaseAction,
goal=nav_goal, result_cb=self.move_base_result_cb,
exec_timeout=rospy.Duration(5.0), server_wait_timeout=rospy.Duration(10.0))
```

We give our goals as follows:

```
quaternions = list()
euler_angles = (pi/2, pi, 3*pi/2, 0)
for angle in euler_angles:
q_angle = quaternion_from_euler(0, 0, angle, axes='sxyz')
q = Quaternion(*q_angle)
quaternions.append(q)

# Create a list to hold the waypoint poses
self.waypoints = list()
self.waypoints.append(Pose(Point(0.0, 0.0, 0.0), quaternions[3]))
self.waypoints.append(Pose(Point(-1.0, -1.5, 0.0), quaternions[0]))
self.waypoints.append(Pose(Point(1.5, 1.0, 0.0), quaternions[1]))
self.waypoints.append(Pose(Point(2.0, -2.0, 0.0), quaternions[1]))
room_locations = (('table', self.waypoints[0]),
('delivery_area', self.waypoints[1]),
('kitchen', self.waypoints[2]),
```

```
('home', self.waypoints[3]))
self.room_locations = OrderedDict(room_locations)
nav_states = {}

for room in self.room_locations.iterkeys():
nav_goal = MoveBaseGoal()
nav_goal.target_pose.header.frame_id = 'map'
nav_goal.target_pose.pose = self.room_locations[room]
move_base_state = SimpleActionState('move_base', MoveBaseAction,
goal=nav_goal, result_cb=self.move_base_result_cb,
exec_timeout=rospy.Duration(5.0),
server_wait_timeout=rospy.Duration(10.0))
nav_states[room] = move_base_state
```

The entire working code is available in this book's GitHub repository: `https://github.com/PacktPublishing/ROS-Robotics-Projects-SecondEdition/blob/master/chapter_4_ws/src/smach_example/waiter_robot_anology.py`. You should see an output like this once everything is up and running. The following diagram shows the robot analogy state machine; the individual text is for reference and does not need to be read:

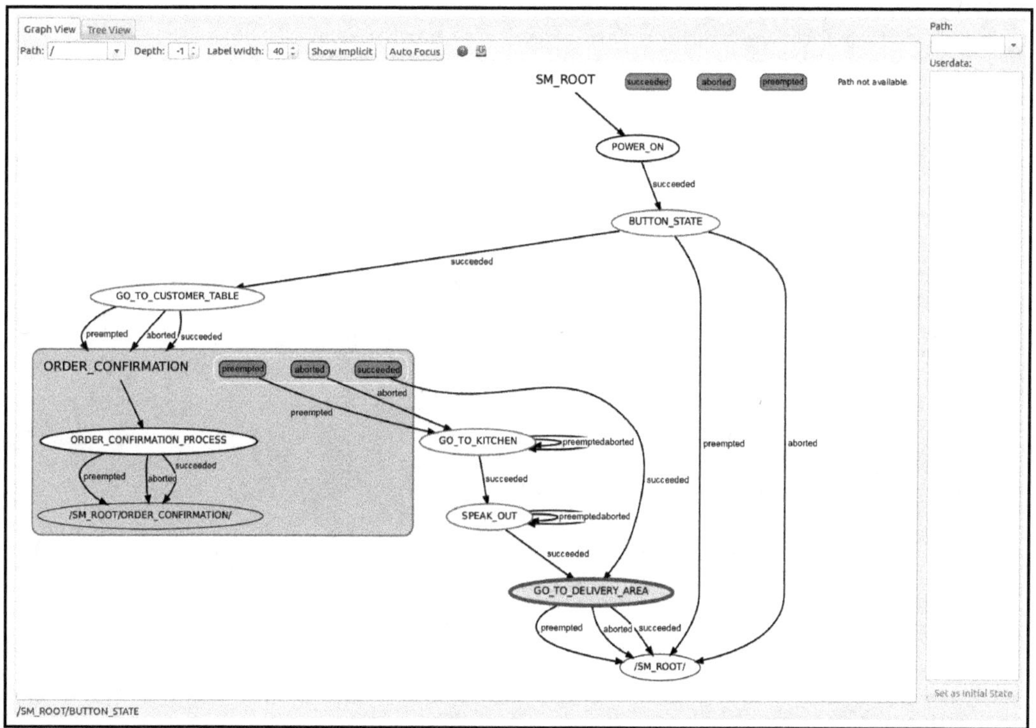

Robot analogy state machine

The preceding is a view of `smach_viewer`. You can see the respective outcomes, that is, succeeded, aborted, or preempted, from each and every state and their transitions. The green color shows the current state the robot is in. Also, as you may have noticed, `ORDER_CONFIRMATION` is a substate and is represented in gray.

Summary

In this chapter, a simple analogy was taken and split into smaller chunks and we saw how to use ROS to do the same. Initially, a feedback-based messaging system was introduced and showed us how effective it is compared to topics and services. Later, we learned how to create such messaging mechanisms. Then, we learned how the robot can handle complex tasks and perform the same by defining the smaller chunks as states and run these states in sequences, concurrently, iteratively, and in a nested manner. We shall make use of the knowledge we acquired in this chapter and the previous one to learn how to actually implement a robot application.

In the next chapter, we will learn how to build an industrial application using state machines and the robot we created in the previous chapter.

5
Building an Industrial Application

By now, you should be familiar with building your own robots with ROS and ensuring the robot can handle complex tasks using state machines. With this understanding, let's put all of the concepts we learned about in the previous chapters together and develop a use case. This chapter will help us to effectively learn how to make use of ROS for producing proof of concepts. You will be introduced to a real-time use case where robots are researched and used. At the end of this chapter, there is a section on improvements that will help you overcome some limitations that the robot faces while performing its tasks as a part of the application.

In this chapter, the following topics will be covered:

- Application use case—robot home delivery
- Making our robot base intelligent
- Making our robot arm intelligent
- Simulating the application
- Improvements to the robot

Technical requirements

Let's look into the technical requirements for this chapter:

- ROS Melodic Morenia on Ubuntu 18.04, with Gazebo 9 (preinstalled)
- ROS packages: `moveit` and `slam-gmapping`
- Timelines and test platform:
    - **Estimated learning time**: On average, 120 minutes
    - **Project build time (inclusive of compile and run time)**: On average, 90 minutes
    - **Project test platform**: HP Pavilion laptop (Intel® Core™ i7-4510U CPU @ 2.00 GHz × 4 with 8 GB Memory and 64-bit OS, GNOME-3.28.2)

The code for this chapter is available at `https://github.com/PacktPublishing/ROS-Robotics-Projects-SecondEdition/tree/master/chapter_5_ws/src`.

Let's begin by understanding the application use case we will be working on in this chapter.

Application use case – robot home delivery

In the last two decades, there has been huge growth in the e-commerce sector, driven by big e-commerce companies such as Amazon and Alibaba, across the world. This has led to retailers around the globe looking to expand their operations to include online shopping. This has brought opportunities as well as challenges for delivery service companies. As demand increases among competitors, the delivery service companies have scrambled to offer product delivery as early as possible. These delivery times may vary from a few hours to a couple of days. This has created a huge demand for delivery agents as well.

The use of robots in this sector could have huge potential and could enhance business prospects. For instance, robots could be used in packaging or sorting retail items, delivering items within warehouses, or even delivering the items to our houses. This way, they could be more productive and deliver at least twice as fast as a human could in terms of time and the number of deliveries.

Let's consider this as our use case and use the mobile manipulator we created in `Chapter 3`, *Building an Industrial Mobile Manipulator*, to act as a delivery agent. Generally, the products are shipped from the retailers to the specific city and then they're split and sent to different streets that have a common delivery office for every street or group of streets. We shall simulate that delivery office scenario after the products have arrived from the retailer in Gazebo and ROS. Our environment shall look as follows:

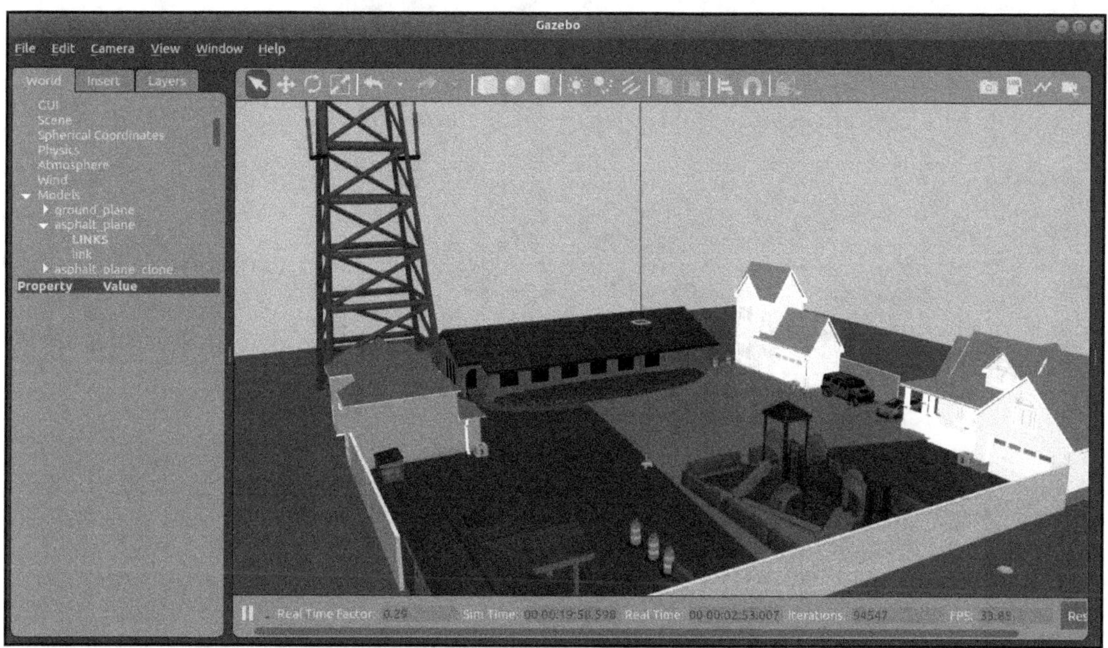

Gazebo environment

Here, the robot would be located in the post office. The robot is supposed to deliver items to the three houses in the colony and are numbered 1, 2, and 3, respectively:

From top left to bottom right—post office (robot location), houses 1, 2, and 3

Once the items are received by the post office, the office shall know of the list of products that need to be delivered to specific locations. This information is shared with the robot through a central material movement application software. Furthermore, the products are placed in the delivery area that the robot has access to. The robot then receives the list and updates itself with the delivery locations, along with the respective products, and plans its delivery.

In this case, the robot needs to know the products, know the locations, and pick and deliver the items to and from the necessary locations. If we break the application into chunks, we should make the robot base autonomous, then make our arm pick and place items intelligently, and finally put all of these into a series of actions using state machines. Now, let's set up our environment in Gazebo.

Setting up the environment in Gazebo

The environment we talked about earlier is saved as a Gazebo world file called `postoffice.world`. You can visualize this world by copying this file into your workspace and opening Gazebo (assuming you open it from the same Terminal where the file is downloaded or copied) with this world file using the following command:

```
$ gazebo postoffice.world
```

Gazebo might take a while to load all of the necessary models from the internet, due to which it may take time to launch. Hence, provide around 10 minutes of settling time once the preceding command has been entered.

Since we're going to concentrate on the application from now on, let's set the environment for this chapter:

1. Create a new workspace and add the robot files we created in `Chapter 3`, *Building an Industrial Mobile Manipulator*, to it:

```
$ initros1
$ mkdir -p chapter_5_ws/src
$ cd ~/chapter_5_ws/src
$ catkin_init_workspace
$ cp -r ~/chapter_3_ws/src/robot_description ~/chapter_5_ws/src/
$ cd ~/chapter_5_ws
$ catkin_make
```

2. Now, let's create a folder called `worlds` in the `robot_description` package and save our world file there:

```
$ cd ~/chapter_5_ws/src/robot_description/
$ mkdir worlds
```

   Copy the downloaded world into the preceding folder.

3. Now, let's modify the world file in our `mobile_manipulator_gazebo_xacro.launch` file to add the preceding world file as an argument:

```
<include file="$(find gazebo_ros)/launch/empty_world.launch">
  <arg name="world_name"
value="home/robot/chapter_5_ws/src/robot_description/worlds/postoff
ice.world"/>
</include>
```

Try launching the robot. You will see that the robot is positioned in our new environment. The robot is currently positioned at the (0, 0) coordinates of the world file. If we wish to change the position of the robot, we need to add the following additional arguments to the `<node>` tag in the `mobile_manipulator_gazebo_xacro.launch` file:

```
<node name="spawn_urdf" pkg="gazebo_ros" type="spawn_model" args="-param
robot_description -urdf -x 1 -y 2 -z 1 -model mobile_manipulator" />
```

The preceding code would set the robot position (`x, y, z`) to (`1, 2, 1`). Now that the environment is set, let's learn how to modify the robot.

Making our robot base intelligent

Let's begin by making our robot base autonomous or intelligent. Our robot base needs to understand the environment and move around the environment with ease. We shall do this using **Simultaneous Localization And Mapping** (**SLAM**). SLAM has been a research problem for almost a decade but ROS has good open source packages that can help us map and localize the robot in an environment. We shall add the necessary sensors to the robot, configure the available ROS packages, and make the robot maneuver autonomously in the environment. Let's start by adding the laser sensor.

Adding a laser sensor

To make the robot understand the environment, we shall make use of a laser scanner sensor. A laser scanner sensor uses a laser light as a source (more information can be found at `https://en.wikipedia.org/wiki/Lidar`) and time of flight principle for distance calculation (more information can be found at `https://en.wikipedia.org/wiki/Time-of-flight_camera`). This laser light source is made to rotate at a certain rate and, as a result, a two-dimensional scan of the environment is achieved. The scan trace range of the laser scanner varies for different types of scanners, from 90 degrees to 360 degrees. The two-dimensional output of the sensor purely depends on the sensor positioning on the robot. There are chances that the robot might not be able to identify a table or a truck as expected.

In Gazebo, there is a plugin for the laser scanner sensor that we would make use of in our robot. To simulate the plugin, let's create a simple geometry that would virtually indicate a laser scanner sensor:

- Add the following lines of code to our robot file, `mobile_manipulator.urdf`, which we created in `Chapter 3`, *Building an Industrial Mobile Manipulator*:

```
<link name="laser_link">
  <collision>
   <origin xyz="0 0 0" rpy="0 0 0"/>
   <geometry>
    <box size="0.1 0.1 0.1"/>
   </geometry>
  </collision>
  <visual>
   <origin xyz="0 0 0" rpy="0 0 0"/>
    <geometry>
     <box size="0.05 0.05 0.05"/>
    </geometry>
  </visual>
  <inertial>
    <mass value="1e-5" />
    <origin xyz="0 0 0" rpy="0 0 0"/>
    <inertia ixx="1e-6" ixy="0" ixz="0" iyy="1e-6" iyz="0"
izz="1e-6" />
  </inertial>
</link>
```

Here, we have created a simple square block that resembles a laser scanner and connected it to our robot's `base_link`.

- Let's call the Gazebo plugin for the laser scanner:

```
<gazebo reference="laser_link">
<sensor type="ray" name="laser">
<pose>0 0 0 0 0 0</pose>
<visualize>true</visualize>
<update_rate>40</update_rate>
<ray>
    <scan>
        <horizontal>
        <samples>720</samples>
        <resolution>1</resolution>
        <min_angle>-1.570796</min_angle>
        <max_angle>1.570796</max_angle>
        </horizontal>
    </scan>
```

```
        <range>
            <min>0.10</min>
            <max>30.0</max>
            <resolution>0.01</resolution>
        </range>
    </ray>
    <plugin name="gazebo_ros_head_hokuyo_controller"
    filename="libgazebo_ros_laser.so">
      <topicName>/scan</topicName>
      <frameName>laser_link</frameName>
    </plugin>
    </sensor>
    </gazebo>
```

If the `<visualize>` tag is set to `true`, we will be able to see a blue ray of light with a 180-degree sweep. We assume that our laser scan sweep has a 180-degree sweep, that is, it ranges between -90 to +90 degrees; hence, we indicate the sweep in the `<min_angle>` and `<max_angle>` tags in radians. The sensor resolution we assume is 10 centimeters and ranges to 30 meters with an accuracy of 1 centimeter. We also mention the topic name so that we can read the value and frame of reference we defined in the preceding URDF to the `<plugin>` tag.

Now, launch the robot in Gazebo once again. You should see the following output:

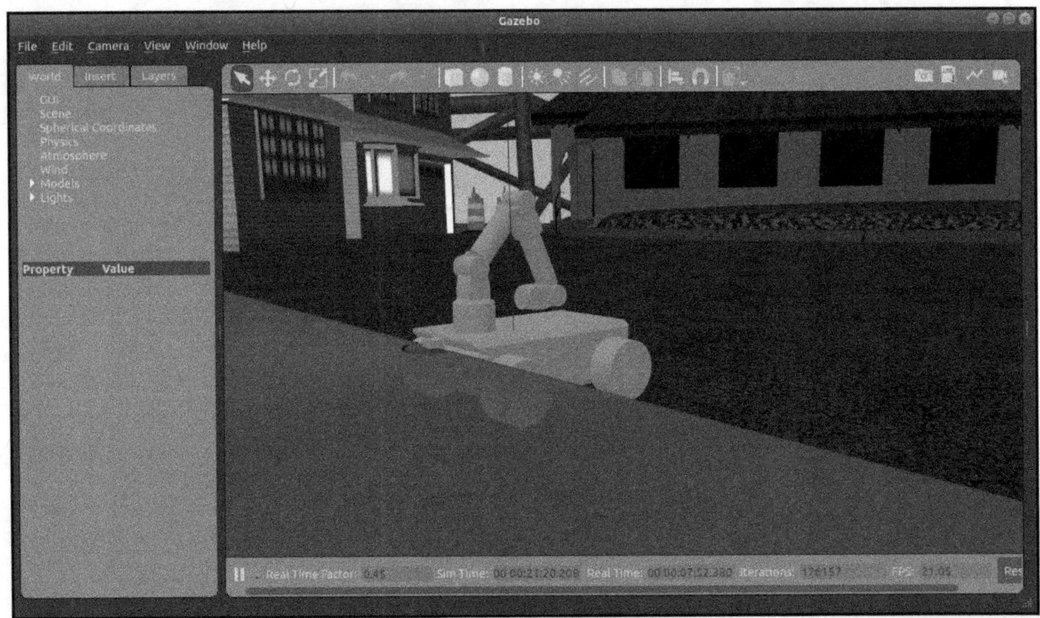

Gazebo view of the robot with a laser scanner

You could try and simulate different laser scanner values with respect to the sensors available on the market, such as Hokuyo UST10 or 20LX or a rplidar with a 360-degree scan sweep. Now, let's configure the navigation stack.

Configuring the navigation stack

Now that a laser scanner has been added to the robot base, we shall configure the `move_base` server, which will help the robot navigate autonomously with ROS packages. What is `move_base`? Let's consider that the robot moves autonomously in an environment through a known map. The map consists of information, such as obstructions, walls, or tables (called costs in robotics), that are known to us. This known information is treated as global information.

While the robot tries to move around with this global information, the robot might be introduced to sudden dynamic changes in the environment, such as changes in the position of a chair or dynamic movement such as a person walking. These changes are treated as local information. In simple terms, global information is with respect to the environment and local information is with respect to the robot. `move_base` is a complicated (in terms of code) yet simple (in terms of understanding) node that helps in linking together global and local information to accomplish a navigation task. `move_base` is an ROS action implementation. When given a goal, the robot base would try to reach that goal. A simple representation of `move_base` is shown in the following diagram:

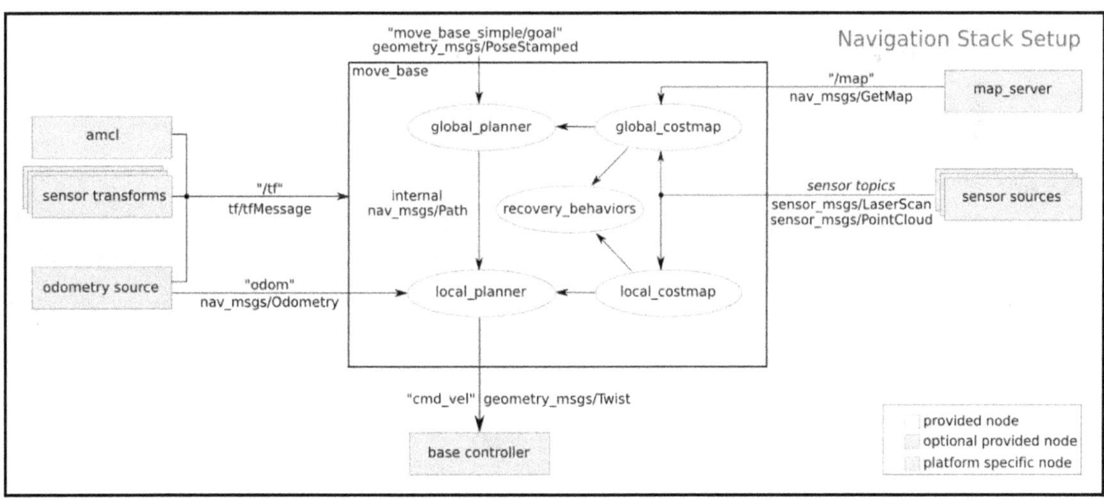

move_base implementation (source: http://wiki.ros.org/move_base. Image from ros.org. Licensed under Creative Commons CC-BY-3.0: https://creativecommons.org/licenses/by/3.0/us/legalcode)

From the preceding diagram, it is evident that if the robot base receives a goal command, the `move_base` server understands the goal and gives out a series of command velocities to direct the robot toward the goal. For this, the robot base would need to be defined with certain parameters that are defined in the following YAML files:

- `costmap_common_params.yaml`
- `global_costmap_params.yaml`
- `local_costmap_params.yaml`
- `base_local_planner.yaml`

Now, let's follow these steps to configure the navigation stack:

1. Let's create a navigation package and add these files to them:

```
$ cd ~/chapter_5_ws/src
$ catkin_create_pkg navigation
$ cd navigation
$ mkdir config
$ cd config
$ gedit costmap_common_params.yaml
```

2. Copy the following code into the `costmap_common_params.yaml` file:

```
footprint: [[0.70, 0.65], [0.70, -0.65], [-0.70, -0.65], [-0.75,
0.65]]
observation_sources: laser_scan_sensor
laser_scan_sensor:
 sensor_frame: laser_link
 data_type: LaserScan
 topic: scan
 marking: true
 clearing: true
```

   Since our robot is of rectangular geometry, we have defined its footprint as coordinate limits. Since we are using a laser scanner, we provide the necessary information.

3. Now, save the file, close it, and create another file and name it `global_costmap_params.yaml`:

```
$ gedit global_costmap_params.yaml
```

Copy the following code into it:

```
global_costmap:
 global_frame: map
 robot_base_frame: base_link
 static_map: true
```

In general, our frame of reference for movement is with respect to the world. In the case of the robot, it is with respect to the environment that is defined as a map. It is the robot's `base_link` that geometrically moves with respect to the map.

4. Now, save the file, close it, and create another file and name it `local_costmap_params.yaml`:

```
$ gedit local_costmap_params.yaml
```

Copy the following code into it:

```
local_costmap:
 global_frame: odom
 robot_base_frame: base_link
 rolling_window: true
```

As we mentioned earlier, local information is with respect to the robot. Hence, our `global_frame` of reference is the robot's `odom` frame. The local cost in the environment is updated only within a certain footprint, which could be defined as a `rolling_window` parameter.

Now that we have the necessary information, let's go about mapping and localizing our robot in the given environment.

Mapping the environment

To move the robot autonomously in the environment, the robot needs to know the environment. This could be done using a mapping technique such as `gmapping` or `karto`. Both of these techniques are available as ROS packages and both would need encoder information (say, `odom`, in our case), laser scan data, and transform information. Gmapping uses particle filters and is open source while Karto is Apache-licensed. However, you could still use the open source wrapper for testing and educational purposes in Karto. For our application, let's use `gmapping` to map the environment.

We shall install `gmapping` SLAM using the following command:

```
$ sudo apt-get install ros-melodic-slam-gmapping
```

Now that the mapping technique has been installed, let's look at how to localize the robot on the map we've acquired.

Localizing the robot base

Once the map of the environment is available, the robot is ready to be localized. We make use of the `amcl` package (`http://wiki.ros.org/amcl`) in ROS, which helps us with autonomous navigation. `amcl` uses a particle filter to determine the position of the robot against a known map. The particle filter provides a list of poses based on the robot's sensor information and certain assumptions (you can find more information here: `http://wiki.ros.org/amcl`). As the robot moves, the pose cloud (the list of poses) converges toward the robot, stating the robot's positional assumption. If the poses are beyond a certain range, they are neglected automatically.

Since we use a differential drive robot, we shall make use of the already available `amcl_diff.launch` file as a template for our robot. To simulate our robot's navigation stack, let's create a file called `mobile_manipulator_move_base.launch`, all our `move_base` with the necessary parameters, and include the `amcl` node. Our launch file will look as follows:

```
<launch>
<node name="map_server" pkg="map_server" type="map_server"
args="/home/robot/test.yaml"/>

<include file="$(find navigation)/launch/amcl_diff.launch"/>

<node pkg="move_base" type="move_base" respawn="false" name="move_base"
output="screen">
 <rosparam file="$(find navigation)/config/costmap_common_params.yaml"
command="load" ns="global_costmap" />
 <rosparam file="$(find navigation)/config/costmap_common_params.yaml"
command="load" ns="local_costmap" />
 <rosparam file="$(find navigation)/config/local_costmap_params.yaml"
command="load" />
 <rosparam file="$(find navigation)/config/global_costmap_params.yaml"
command="load" />
 <rosparam file="$(find navigation)/config/base_local_planner_params.yaml"
command="load" />
 <remap from="cmd_vel" to="/robot_base_velocity_controller/cmd_vel"/>
</node>
</launch>
```

Now that the robot base navigation has been set, let's learn how to set the arm in motion.

Making our robot arm intelligent

As you may recall, we control the robot arm by publishing values to a rostopic, which we did in `Chapter 3`, *Building an Industrial Mobile Manipulator*. Later, as a result of the `follow_joint_trajectory` plugin, we made use of the action server implementation and wrote a client that could move the robot to the desired position. These implementations follow forward kinematics where, knowing the link lengths and limits, you give a rotation value for each and every joint. As a result, the arm reaches a certain pose in the environment. What if you knew the pose of an object in the environment and you wanted to move the robot arm to that pose? This is called inverse kinematics and this is what we aim to achieve in this section through a specialized software called Moveit. Let's look at the basics of Moveit.

Introduction to Moveit

Moveit is a sophisticated piece of software written on top of ROS for achieving inverse kinematics, motion or path planning, 3D perception of the environment, collision checking, and so on. It is the primary source of functionality for manipulation in ROS. Moveit understands a robot arm configuration (geometry and link information) through `urdf` and ROS message definitions and utilizes the **ROS visualizing** (**RViz**) tool to perform manipulation.

Moveit is used in more than 100 robot arms and you can find more information about those robots here: `https://moveit.ros.org/robots/`. Moveit has lots of advanced features and is used by many industrial robots as well. Covering all the Moveit concepts is out of the scope of this book, so we shall only look at what we need to move and control our robot arm.

Let's begin by looking at the architecture of Moveit:

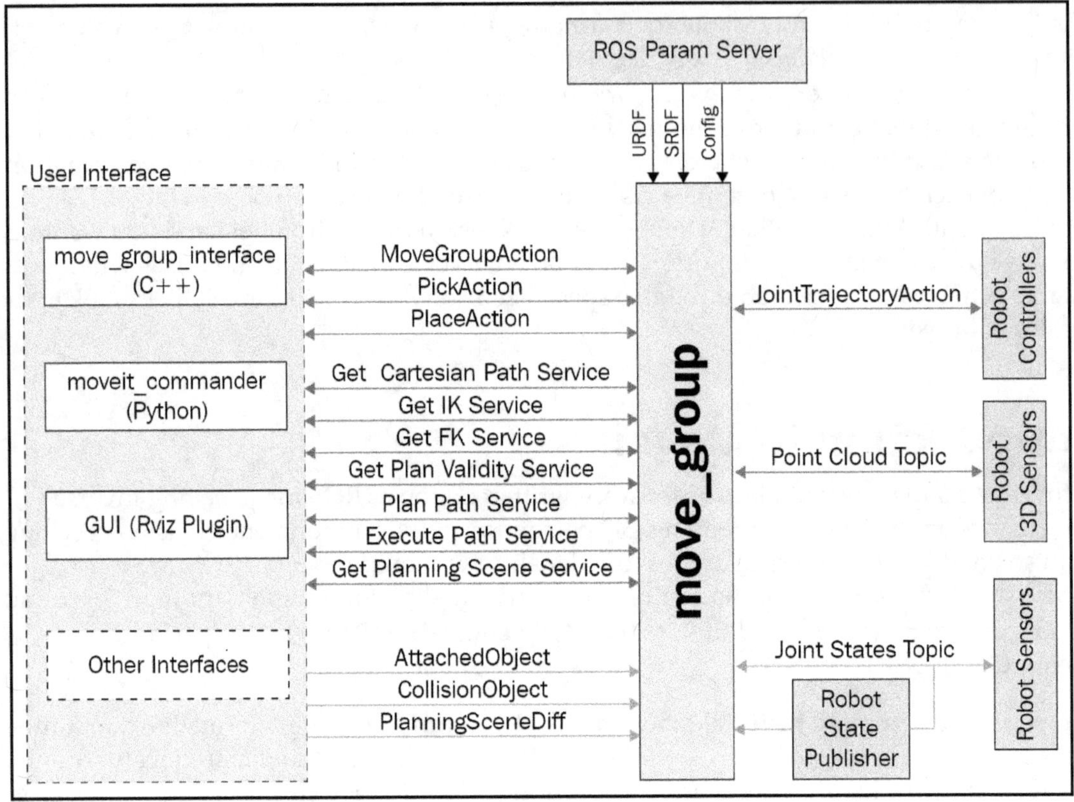

Moveit architecture

Here, we have the most important component, that is, the `move_group` node, which is responsible for putting all other components together to provide the user with necessary actions and service calls to use. The user could interface by using a simple scripting interface (for beginners) called the `moveit_commander` interface, a C++ wrapper called `move_group_interface`, a Python interface written on top of `move_it_commander`, or the GUI interface using an RViz plugin.

The `move_group` node would need the robot information that is defined through URDF, as well as configuration files. Moveit understands the robot in a format called SRDF that Moveit converts into URDF while setting up the robot arm. Also, the `move_group` node understands the robot arm's joint states and talks back via the `FollowJointTrajectoryAction` client interface.

For more information on Moveit concepts, have a look at `https://moveit.ros.org/documentation/concepts/`.

Now, let's learn how to install and configure Moveit for our mobile robot.

Installing and configuring Moveit for our mobile robot

Installing and configuring Moveit is a multistep process. Let's begin by learning how to install it.

Installing Moveit

To install Moveit, we'll use some pre-built binaries:

- Enter the following command in a Terminal:

```
$ sudo apt install ros-melodic-moveit
```

- You might also want to install additional Moveit functionalities, so use the following command as well:

```
$ sudo apt-get install ros-melodic-moveit-setup-assistant
$ sudo apt-get install ros-melodic-moveit-simple-controller-manager
$ sudo apt-get install ros-melodic-moveit-fake-controller-manager
```

Once they're all installed, we can begin configuring our robot using a Moveit setup assistant wizard.

Configuring the Moveit setup assistant wizard

This wizard is very useful, particularly because it helps us save time. Some of the things that we can do with this wizard are as follows:

- Define collision zones for our robot arm
- Set custom poses

- Choose the necessary kinematics library
- Define ROS controllers
- Create the necessary simulation files

Let's invoke the setup assistant by using the following command:

```
$ initros1
$ roslaunch moveit_setup_assistant setup_assistant.launch
```

You should see the window shown here:

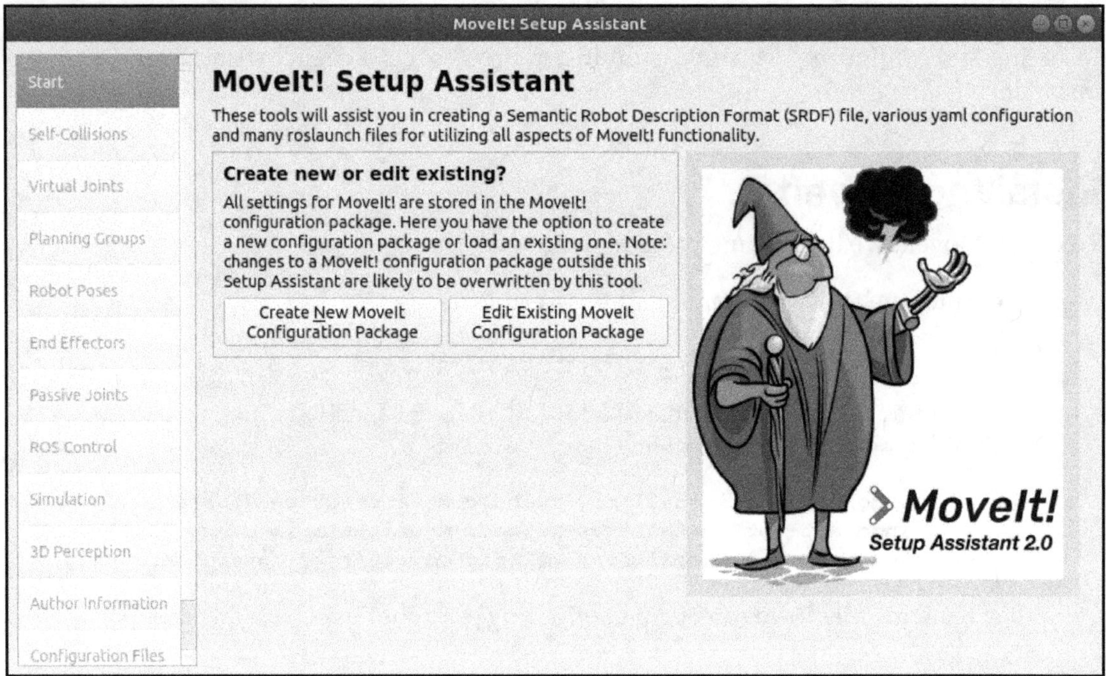

The Moveit setup wizard

Now, let's look into the configuration steps, one by one.

Loading the robot model

Let's configure our robot in Moveit by selecting the respective robot URDF. We do that by clicking **Create New Moveit Configuration Package**, loading our robot URDF, `mobile_manipulator.urdf`, and selecting **Load Files**. You should see a success message, along with our robot in the right-hand pane:

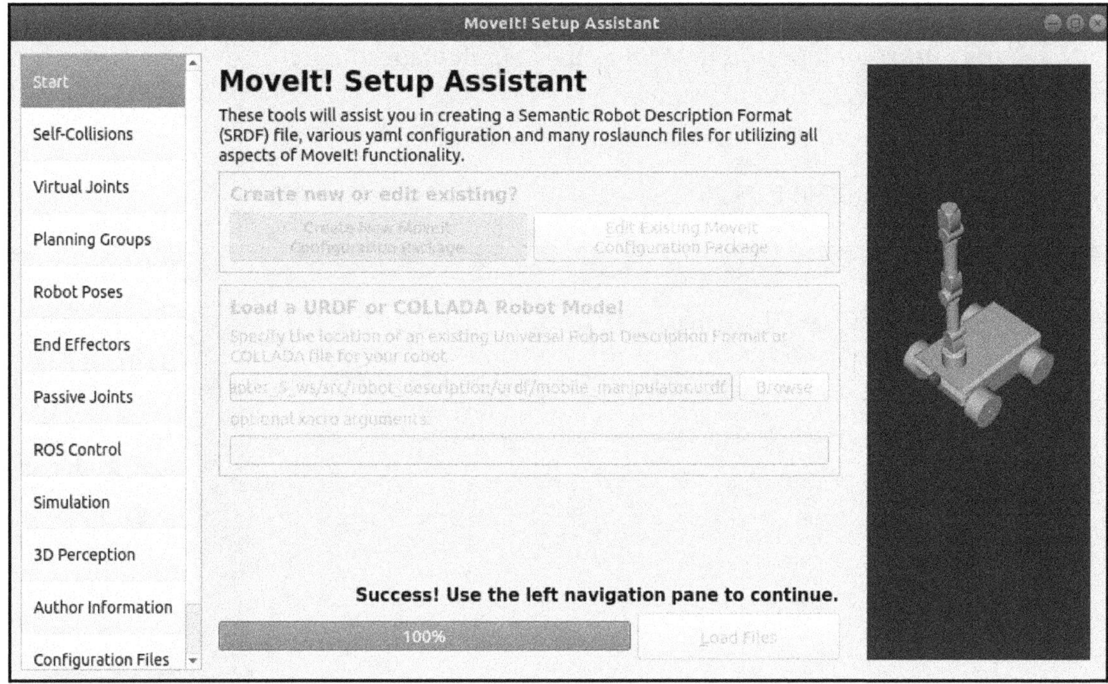

Moveit success

Now, let's set up the components on the left-hand pane.

Setting up self-collisions

Click on **Self-Collisions** on the left pane and select **Generate Collision Matrix**. Here, you can set the sampling density high if you wish to move the robot arm in a more confined space. This may increase the planning time for the robot to execute a trajectory and may sometimes fail execution due to a collision assumption. However, we can avoid this failure by manually defining and checking the collisions for each and every link and disabling or enabling them accordingly. Let's not assume we have a virtual joint for our robot.

Setting up planning groups

Let's set up planning groups by following these steps:

1. In **Planning Groups**, add our robot arm group by selecting **Add Group**.
2. Name our group `arm`.

3. Select **Kinematic Solver** as `kdl_kinematics_plugin/KDLKinematicsPlugin`. Set the resolution and timeout as the default values.
4. Select `RRTStar` as our **Planner**.
5. Now, add our robot arm joints and click **Save**.

Your final window should look as follows:

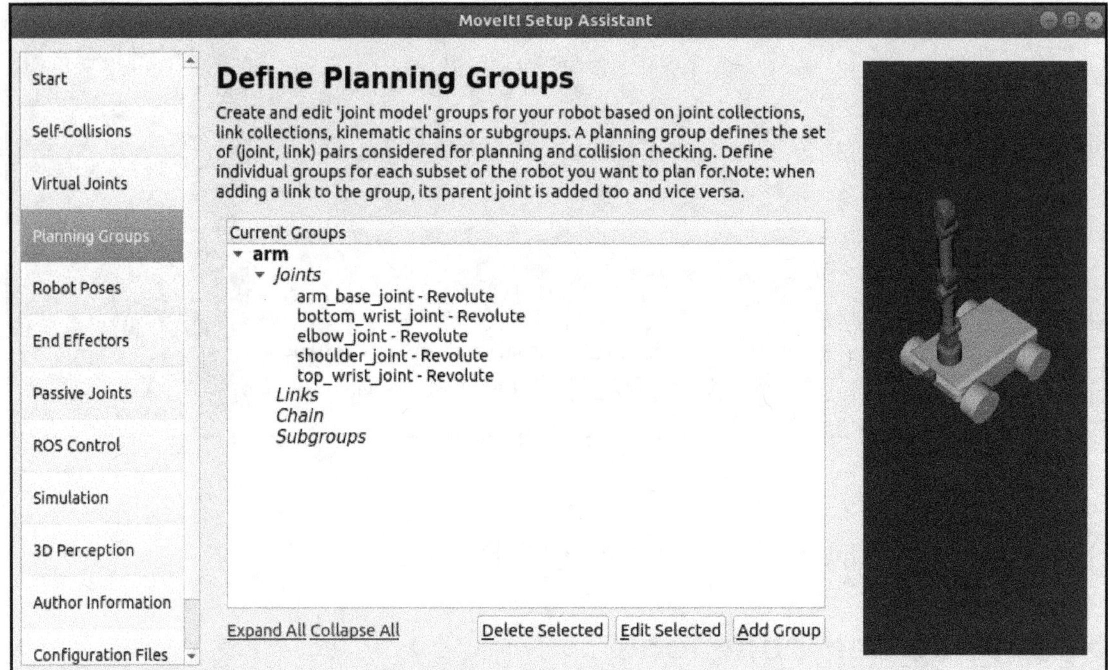

Moveit planning groups

Once the arm group has been set, we can set the poses for the arm.

Setting up arm poses

Now, let's define the robot poses. Click **Add Pose** and add the following poses in the following format (`Posename : arm_base_joint, shoulder_joint, bottom_wrist_joint, elbow_joint, top_wrist_joint`):

- **Straight**: 0.0, 0.0, 0.0, 0.0, 0.0
- **Home**: 1.5708, 0.7116, 1.9960, 0.0, 1.9660:

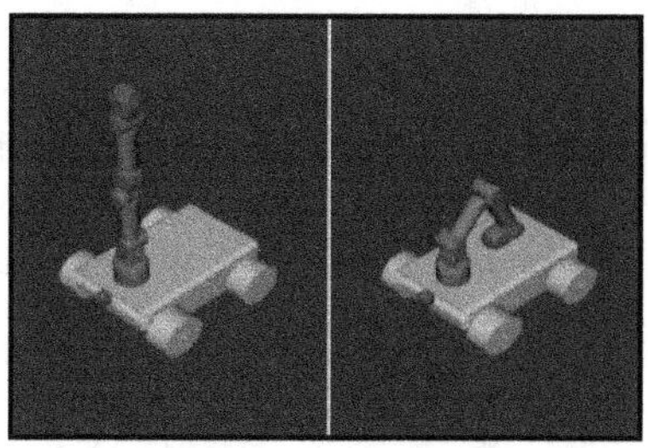

Robot arm poses

We don't have an end effector, so we can skip this step.

Setting up passive joints

Now, let's define the **Passive Joints**—those whose joint states are not expected to be published:

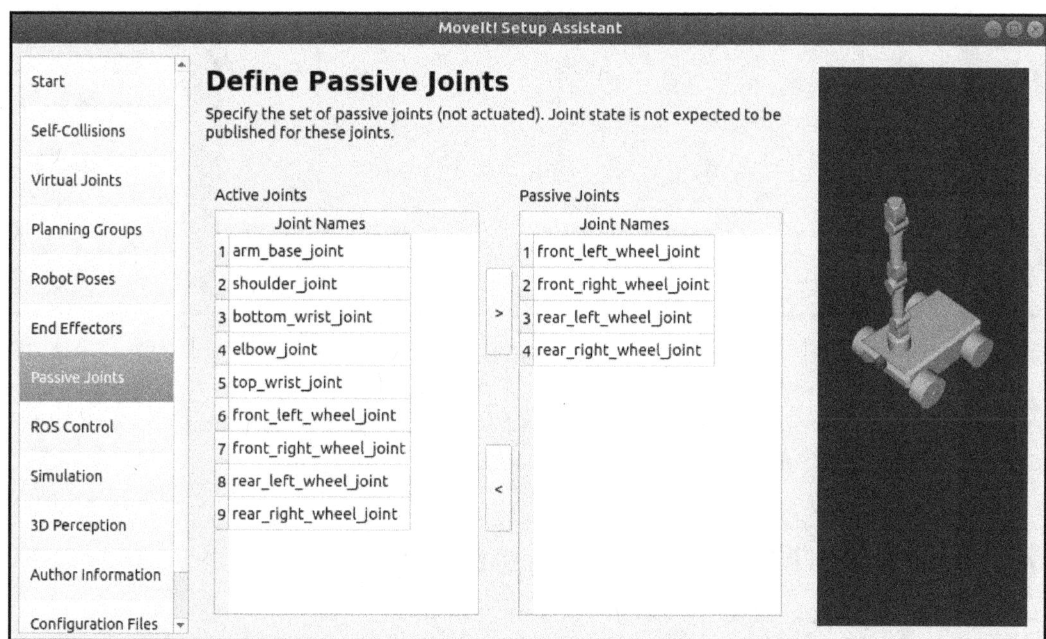

Moveit passive joints

Now, it's time to check the ROS controllers we set up with the robot URDF.

Setting up ROS controllers

Now, we need to connect our robot with Moveit for manipulation through the ROS controllers we defined. Click **Auto Add FollowJointsTrajectory Controllers For Each Planning Group**. You should see the controller being automatically ported in, as shown here:

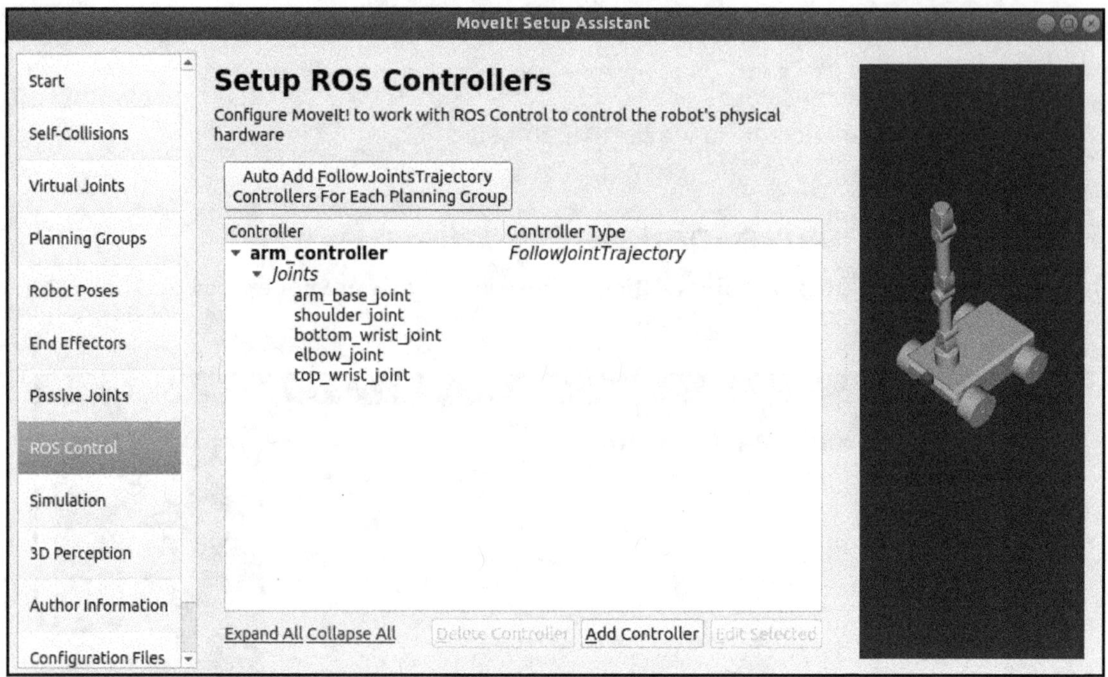

Moveit Setup ROS Controllers

The `FollowJointTrajectory` plugin we had called upon in our plugin is shown in the preceding screenshot. Now, let's finalize the `Moveitconfig` package.

Finalizing the Moveitconfig package

The next step will autogenerate a URDF for simulation:

1. In case you made any changes, these changes will be highlighted in green. We can skip this step as we didn't change anything.
2. We don't need to define a 3D sensor, so skip this step as well.
3. Add any appropriate information in the **Author Information** tab.

4. The final step is the **Configuration Files**, where you will see a list of files that have been generated. The window is as follows:

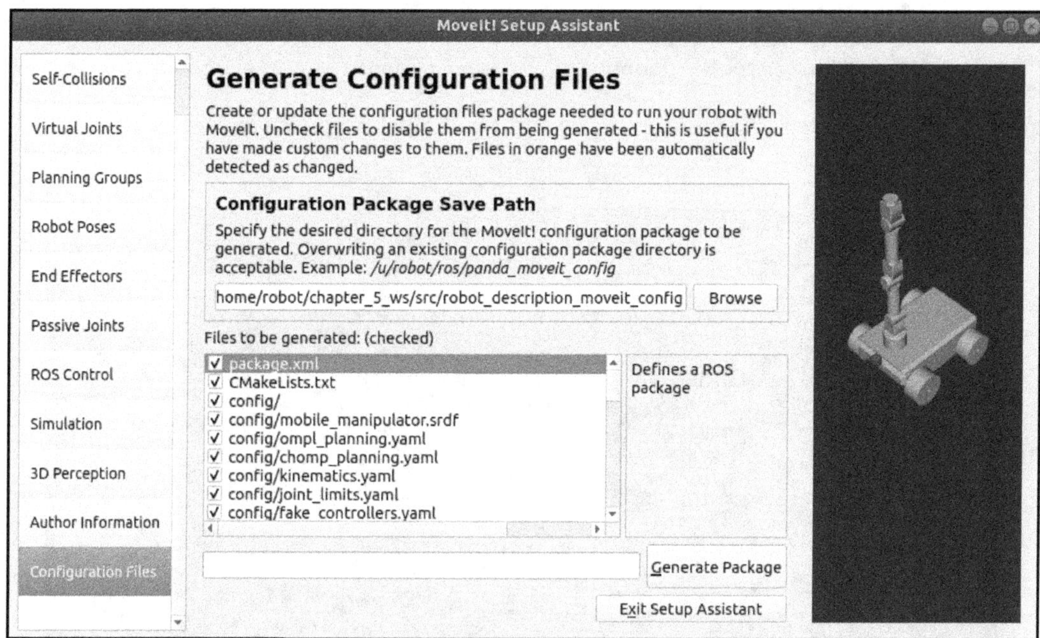

Configuration files

5. Give a configuration name such as `robot_description_moveit_config` in`chapter_5_ws/src`, click on **Generate Package**, and exit the setup assistant.

Now, let's control the robot arm using Moveit.

Controlling the robot arm using Moveit

Once Moveit has been configured, we can test our robot arm manipulation using the GUI interface (RViz plugin):

1. Launch the mobile manipulator in Gazebo:

```
$ initros1
$ source devel/setup.bash
$ roslaunch robot_description
mobile_manipulator_gazebo_xacro.launch
```

2. In a new Terminal, open the `move_group.launch` file that was auto-generated by the Moveit setup assistant wizard:

```
$ initros1
$ source devel/setup.bash
$ roslaunch robot_description_moveit_config move_group.launch
```

Your Terminal's output would be similar to what's shown in the following screenshot:

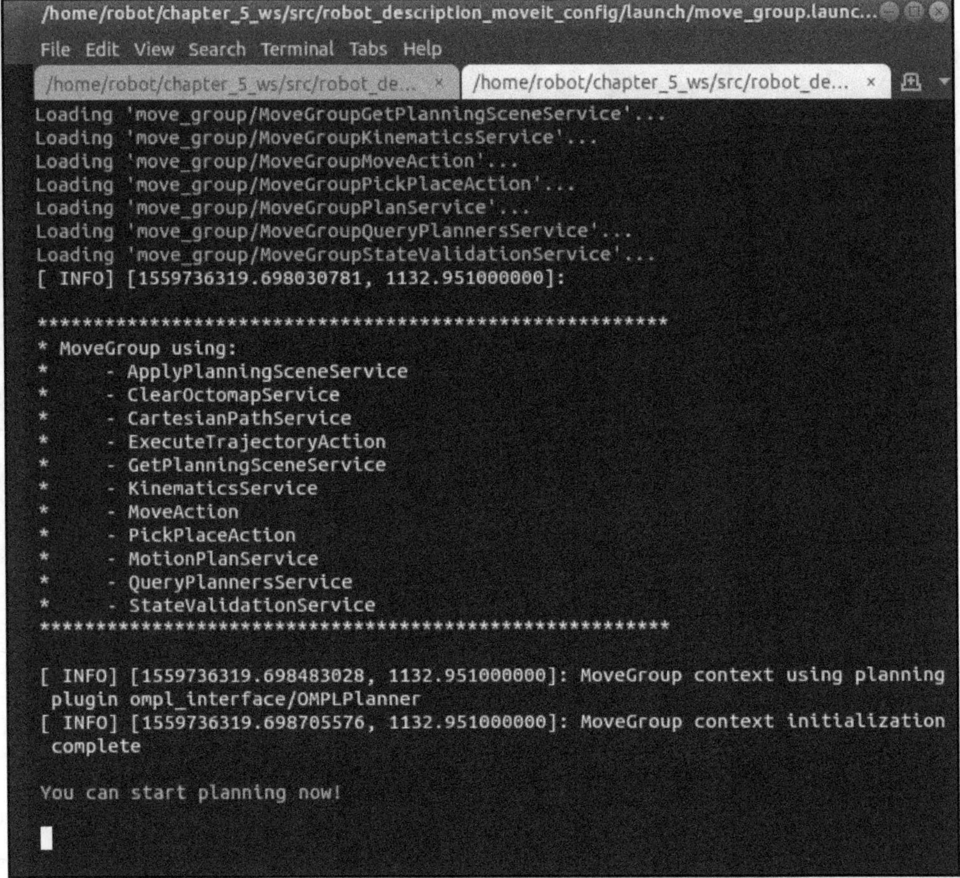

move_group.launch

3. Now, let's open RViz to control the robot's motion:

```
$ initros1
$ source devel/setup.bash
$ roslaunch robot_description_moveit_config movit_rviz.launch
config:=True
```

You should see the following window:

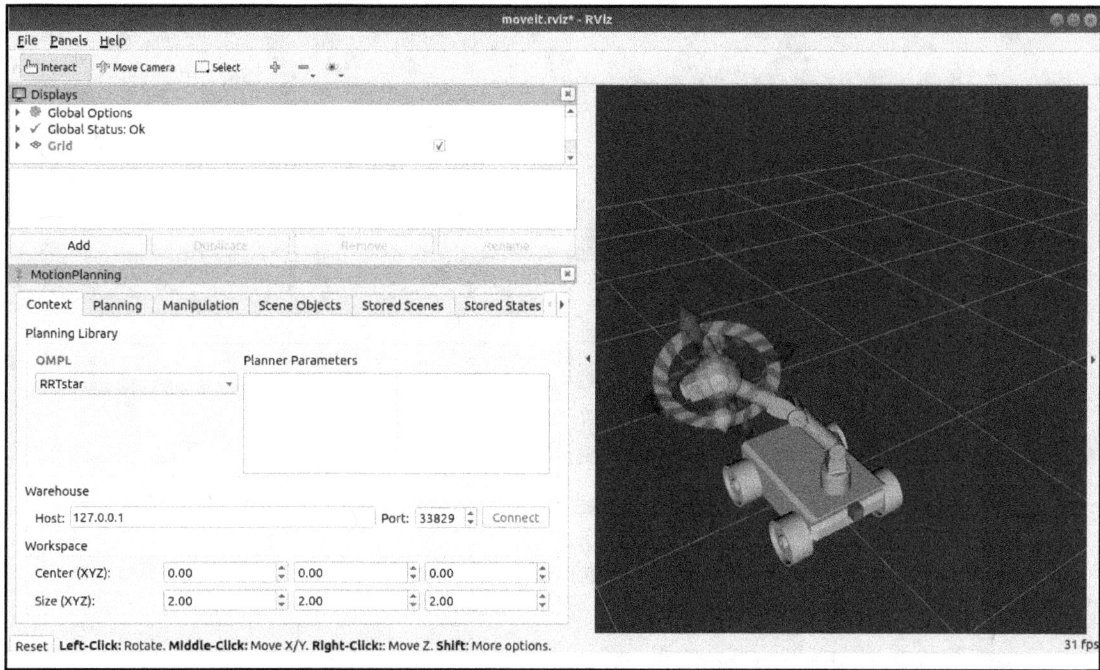

RViz Moveit launch

4. Go to the **Planning** tab, select **home** in **Goal State**, and click on **Plan**. You should see a visual of the robot arm planning (moving) to the target position.

5. To actually make the arm move, select **Execute**:

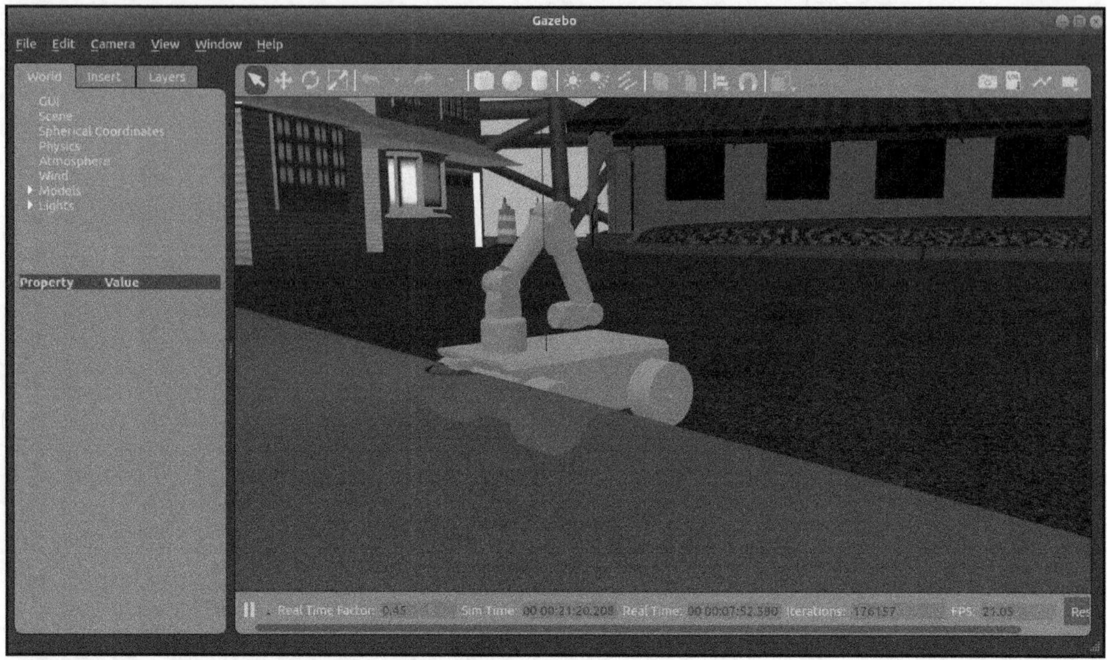

Gazebo arm to home pose

You should see that the arm moved to the selected position in Gazebo, as shown in the preceding screenshot.

In case the arm failed to plan the path, try to replan and execute once again using the buttons in the **Planning** tab.

Now, let's learn how to simulate the application.

Simulating the application

Now, everything is set for the final demonstration. We have a robot base that can move autonomously in the environment and the arm can move to any location that we define it to go to. So, to run our application, we need to perform the following steps:

1. Map and save the environment.
2. Choose the points on the environment.
3. Add the points to our library.
4. Define the respective states.
5. Complete the state machine.

Let's look at these steps in detail.

Mapping and saving the environment

Follow these steps to map and save the environment:

1. The first step is to make the robot understand the environment. For this, we use an open source mapping technique we saw in the *Making our robot arm intelligent* section. Once you have the robot with the Gazebo environment launched, in a new Terminal, start the `gmapping` node using the following command:

   ```
   $ initros1
   $ rosrun gmapping slam_gmapping
   ```

2. Now, open the RViz file in our repository to view the map that is being created by the robot using the following command:

   ```
   $ rosrun rviz rviz -d navigation.rviz
   ```

3. Now, move the robot around the environment using the `teleop` node command:

   ```
   $ rosrun teleop_twist_keyboard teleop_twist_keyboard.py
   cmd_vel:=/robot_base_velocity_controller/cmd_vel
   ```

   You should see the map being created in RViz.

4. Now, save the map by using the following command in a new Terminal:

   ```
   $ rosrun map_server map_saver -f postoffice
   ```

You should see a couple of files of the `.yaml` and `.pgm` format being created at the location that your Terminal is pointed to.

Choosing the points on the environment

Now that the map has been created, let's save the house goals as individual points. Follow these steps:

1. Load the map file we created earlier in the `mobile_manipulator_navigation.launch` file, which we created in the *Making our robot base intelligent* section. Replace the following line in the launch file with the map file path:

    ```
    <node name="map_server" pkg="map_server" type="map_server"
    args="/home/robot/chapter_5_ws/postoffice.yaml"/>
    ```

Please note that you should only load the `.yaml` file. Also, ensure you have `map_server` installed using the following command: `sudo apt-get install ros-melodic-map-server`.

2. Now, with the Gazebo node running, open the navigation launch file in another Terminal:

    ```
    $ source devel/setup.bash
    $ roslaunch robot_description mobile_manipulator_move_base.launch
    ```

3. Open the RViz file in our repository to view the map that is being created by the robot using the following command:

    ```
    $ rosrun rviz rviz -d navigation.rviz
    ```

4. Choose the **Publish Point** tool on RViz and move intuitively toward the houses on the map. This might be tricky the first time you try to identify the house, but you will get used to this quickly. Once you click on the point in the location, you will be able to see a list of three values shown in RViz at the bottom, next to the **Reset** button. Make a note of these points for all of the destinations.

Adding the points to our library

Now, let's save the points that we made note of as a dictionary here:

```
postoffice = [x,y,z,qz]
house1 = [x,y,z,qz]
house2 = [x,y,z,qz]
house3 = [x,y,z,qz]
```

The points are in the following format: translational pose (`x`, `y,z`) and rotation component `qz`.

Completing the state machine

The code for the preceding steps can be found at the following GitHub page: `https://github.com/PacktPublishing/ROS-Robotics-Projects-SecondEdition/tree/master/chapter_5_ws/src`. Once you run the code, you should see that the robot starts from the post office by picking up one package, places it on robot, and delivers it to the houses individually.

Now, let's look at the improvements we can make to the robot.

Improvements to the robot

When working with robots, it can sometimes be hectic and tedious to get them to work just the way we want them to. Let's look into some improvements that we could consider to make our application successful enough in certain aspects:

- **Use of a high-end CPU- or GPU-based compute**: The preceding application should work just fine in general. But since robot algorithms are mostly based on probabilistic approaches, most of the solutions are assumption-based. Hence, there are possibilities for errors or the application to not work as expected. At times, if you have a lower-spec computer, there are chances that Gazebo might crash and might cause you trouble when you view the complete application.

- **Tuning the navigation stack for better localization**: There are some best practices that can help us tune certain parameters in the algorithm. This helps us get the application up and running successfully in most cases. Take a look at the navigation tuning guide by Kaiyu Zhen (`http://wiki.ros.org/navigation/Tutorials/Navigation%20Tuning%20Guide`) and the document available at `https://github.com/zkytony/ROSNavigationGuide/blob/master/main.pdf` to learn how to tune the robot base parameters.
- **Making the application interesting by finding the package directly**: As for our application, instead of providing fixed poses for the arm to pick the object of interest, try to use `find_2d_package`, which was explained by Lentin Joseph in the first edition of this book.
- **Sensor fusion to improve accuracy**: In case you want to improve the odometry accuracy you receive from the mobile base, try sensor fusion. You could add an IMU plugin and combine its information with your wheel encoder, that is, `/odom`. The IMU plugin example can be found here: `http://gazebosim.org/tutorials?tut=ros_gzplugins#IMUsensor(GazeboRosImuSensor)`. Once IMU is defined, you can make use of a filter such as the Kalman filter to combine the IMU and encoder value. The result would be a steady pose value without any possible spikes in values due to the hardware constraints or environment parameters. Check out the following ROS packages for that: `http://wiki.ros.org/robot_pose_ekf` and `http://wiki.ros.org/robot_localization`.
- **Improved mapping options**: Getting an accurate map can also be challenging. Try other open source mapping packages such as Karto (`http://wiki.ros.org/slam_karto`) and Google cartographer (`https://google-cartographer-ros.readthedocs.io/en/latest/`) SLAM for mapping the environment.

Summary

In this chapter, you learned how to make your mobile robot intelligent. You were introduced to sensor plugins and how to configure the navigation stack for autonomous localization. You were also introduced to inverse kinematics concepts for controlling the robot arm. In the end, both of the preceding concepts were combined to create an application that can deliver packages in a neighborhood for better understanding. In this chapter, we were introduced to an actual industrial application where we learned how to use ROS-based robots. We acquired certain useful skills while simulating the application to learn how the robot is capable of handling intricate tasks using state machines.

In the next chapter, you will learn how to work with multiple robots at the same time for a robotics application.

6
Multi-Robot Collaboration

In the previous chapters, we have seen how to build a robot and simulate it in ROS. We also learned how to make our robot base autonomous and our robot arm intelligent enough to move to a pose in the environment. We also covered the waiter robot analogy (defined in `Chapter 4`, *Handling Complex Robot Tasks Using State Machines*), where we used multiple robots to collectively communicate between each other and serve customers, and using more robots to deliver products to houses (an application we defined in the previous chapter).

In this chapter, we will learn how to bring multiple robots into a simulation environment and establish communication between them. You will be introduced to a few methods and workarounds in ROS regarding how to differentiate between multiple robots and communicate effectively between them. You will also see the issues that arise due to these workarounds and learn how to overcome them using the `multimaster-fkie ros` package. In this chapter, we will learn about swarm robotics and then look at the difficulties of using a single ROS master. Then, we will install and set up the `multimaster_fkie` package.

In this chapter, we will cover the following topics:

- Understanding the swarm robotics application
- Swarm robot classification
- Multiple robot communication in ROS
- Introduction to the multimaster concept
- Installing and setting up the `multimaster_fkie` package
- A multi-robot use case

Technical requirements

Let's look into the technical requirements for this chapter:

- ROS Melodic Morenia on Ubuntu 18.04, with Gazebo 9 (preinstalled)
- ROS packages: The `multimaster_fkie` package
- Timelines and test platform:
    - **Estimated learning time**: On average, 90 mins
    - **Project build time (inclusive of compile and run time)**: On average, 60 mins
    - **Project test platform**: HP Pavilion laptop (Intel® Core™ i7-4510U CPU @ 2.00GHz × 4 with 8 GB Memory and 64-bit OS, GNOME-3.28.2)

The code for this chapter is available at `https://github.com/PacktPublishing/ROS-Robotics-Projects-SecondEdition/tree/master/chapter_6_ws/src`.

Let's begin this chapter by understanding swarm robotics applications.

Understanding the swarm robotics application

When we speak about using more than one or two robots for an application, the application is usually termed a swarm robotics application. Swarm robotics is the study of how a collection of robots are used to perform a complex task. They are inspired by biological species that work in groups, such as a collection of bees, a flock of birds, or a group of ants.

All of these creatures work collectively and carry out tasks such as building the beehive, collecting food, or building ant nests, respectively. If you consider ants, they have a load-carrying capacity of 50-100 times their own weight. Now, imagine a group of such ants lifting even more than what they could lift individually. This is how swarm robotics works too. Imagine that our robot arm only has a designed payload of 5 kg and needs lift a 15-20 kg object. We could achieve the same using five such robot arms. Swarm robots are those that have the following characteristics:

- They work together as a group and have a leader or a master who is responsible for leading the robots to perform the tasks.

- They handle applications effectively so that, in case of one of the robots in the group, the remaining robots ensure that the application is still up and running smoothly.
- They are simple robots with simple characteristics and only limited sensors and actuators.

With this simple understanding, let's look into the advantages and limitations of swarm robotics.

The advantages of swarm robotics are as follows:

- They can work in parallel to complete a complex task.
- They are scalable as they can handle and split the task accordingly in their group fails, as in the case of the removal or addition of robots.
- Since simpler robots are used to perform tasks instead of sophisticated robots, they are highly efficient.

The limitations of swarm robotics are as follows:

- The communication architecture between the robots is complex.
- Sometimes, it is difficult for the robot to detect other robots and this could cause interference.
- Robots are already costly, so imagine the application cost when using more than one such robot.

Now that we have a basic idea of swarm robotics, let's look at the classification of swarm robots.

Swarm robot classification

Swarm robots are classified into two types:

- Homogeneous swarm robots
- Heterogeneous swarm robots

Homogeneous swarm robots are those that are a collection of robots of the same kind and type, such as a group of mobile robots (as shown in the following photograph). They replicate a biological species' behavior to perform a task.

The following photograph shows a group of robots that are trying to maneuver over rough terrain:

Homogeneous swarm robots (source: https://en.wikipedia.org/wiki/S-bot_mobile_robot#/media/File:Sbot_mobile_robot_passing_gap.jpeg. The image was taken by Francesco Mondada and Michael Bonani. Licensed under Creative Commons CC-BY-SA 3.0: https://creativecommons.org/licenses/by-sa/3.0/legalcode)

Heterogeneous swarm robots are those that are a mere collection of robots working together to perform an application's code. That's it? Yes, although, to be a bit more precise, they are a collection of robots of *varied kinds*. This is quite the trend in robotics nowadays. In areas such as geographical inspections, a drone and mobile robot combination is used to perform mapping and the analysis of a certain environment. In manufacturing industries, mobile robots and static robot arms are used to work in parallel for machine tending applications such as loading or unloading workpieces. In warehouses, drones and automated guided vehicles are used for the inspection and loading of materials onto racks.

There are applications called multi-robot system applications that shouldn't be confused with swarm robotics. The difference here is that the robots would communicate with other robots in the system autonomously, but work independently. If one of the robots in the system breaks down, the other robots may not be able to carry out the task of the specific broken-down robot and hence would bring the application to a standstill unless the problem was looked into.

Let's look at how multiple robots can communicate in ROS.

Multiple robot communication in ROS

The ROS system is a distributed computing environment that can run several nodes—not just on a single machine, but on multiple machines that communicate with each other, provided they are on the same network. This is advantageous in robotics applications as certain sensors demand sophisticated machines.

For instance, if we have a mobile robot that understands the environment through its sensors, such as ultrasonic sensors and cameras, it may need a simple processor that communicates serially with the microcontroller that the ultrasonic sensor is connected to. But a sensor such as a camera may need a more sophisticated processor for processing camera information. Instead of using high-end compute hardware, which is generally costly and sometimes bigger in form, we could make use of cloud services such as AWS or Google Cloud for processing camera data.

All you would need for this system is a good wireless connection. Hence, your ultrasonic sensor node would run in your simple compute that sits on the robot and the camera image processing node would run on the cloud and communicate with the robot's compute over wireless communication (however, the camera node would sit on the robot's simple compute). Let's learn how to establish communication between multiple robots or machines with ROS.

Single roscore and common networks

The simplest method to establish communication between different machines in the same network is through network configurations. Let's consider the example here:

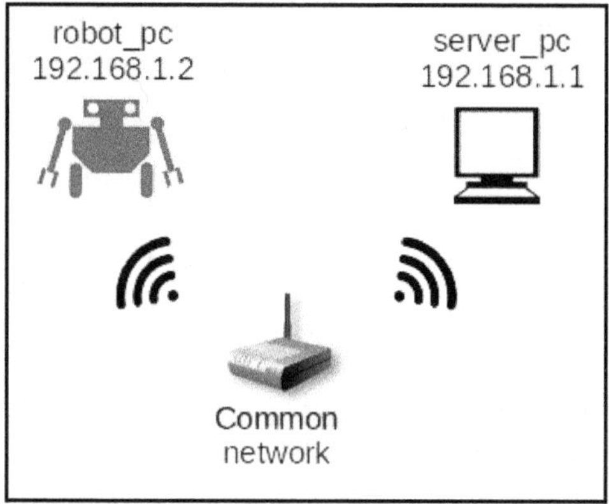

Communicating between different machines in the same network

Let's assume this example is an Industry 4.0 use case. The robot has a compute that is responsible for its control and mobility operations and the robot shares its status and health to the server compute, which analyzes the robot's health and helps predict the possible failure of an event or the robot itself. Both the robot and server computes are in the same network.

For ROS to understand and communicate between both of them, let's assume that the server compute is the master compute. Hence, the roscore would run on the server compute and the robot compute would provide any necessary topic information to this ROS master. To achieve this, you would need to set up a hostname and IP for each compute individually to help differentiate both and help them communicate with each other. Assuming that the server compute is `192.168.1.1` and the robot compute is `192.168.1.2`, and roscore is run on the server compute and the robot compute to connect to it, set up the following environment variables in each compute system:

In the server compute, use the following commands:

```
$ export ROS_MASTER_URI=http://192.168.1.1:11311
$ export ROS_IP=http://192.168.1.1
```

In the robot compute, use the following commands:

```
$ export ROS_MASTER_URI=http://192.168.1.1:11311
$ export ROS_IP=http://192.168.1.2
```

So, what did we do here? We set `ROS_MASTER_URI` to the server compute's IP and are connecting other computes (such as our robot compute) so that they connect to that specific ROS master. Also, to help us differentiate between the computes, we set explicit names for the computes through the `ROS_IP` environment variable.

You could also use `ROS_HOSTNAME` instead of `ROS_IP`, with the robot's name defined as an entry in `/etc/hosts`. Refer to the example at `http://www.faqs.org/docs/securing/chap9sec95.html` to learn how to set up robot names.

You could copy the preceding environment variable entries into their respective bash files to avoid recalling the variables every time a new Terminal is opened.

There is a possibility of timing issues and topic synchronization from occurring, and you may sometimes see a TF warning about extrapolation in the future. These are usually a result of a possible mismatch in system times across the computes.

You can verify this by using `ntpdate`, which can be installed using the following command:

```
$ sudo apt install ntpdate
```

Run the following command to test the date of the other compute:

```
$ ntpdate -q 192.168.1.2
```

In the case of discrepancies, you could install `chrony` using the following command:

```
$ sudo apt install chrony
```

You can edit the configuration file of the robot compute to get its time from the server compute by making the following change in `/etc/chrony/chrony.conf`:

```
$ server 192.168.1.1 minpoll 0 maxpoll 5 maxdelay .05
```

Take a look at `https://chrony.tuxfamily.org/manual.html` for more information.

Issues with a common network

Now, we know how to communicate between nodes in different machines that are in the same network. But what if the nodes in different machines have the same topic names? Consider the example shown in the following diagram:

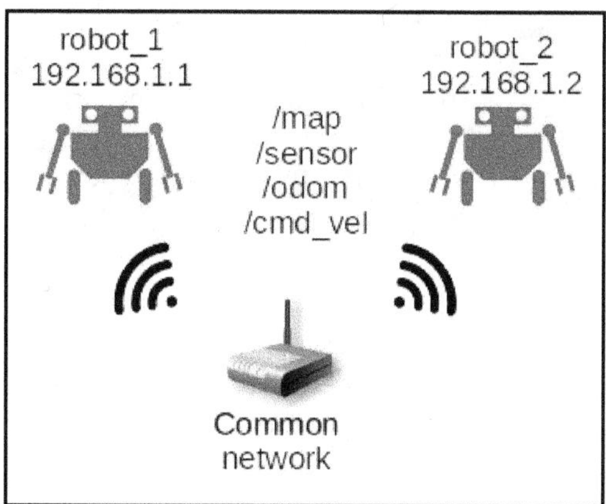

Communication between the same robots in the same network

In a mobile robot, a `move_base` node would need sensor and map information to give out trajectory points as `cmd_vel` commands. If you're planning to use another mobile robot in your application, the mobile robots may have the same topics for communication since both are in the same network and, as a result, this would cause chaos. Both robots may try to follow the same commands and do the same operation instead of performing tasks individually. This is one of the major issues of using a common ROS network. There is no way the robot understands individual controls as they have the same topic name in the network. We'll learn how to overcome this scenario in the upcoming section.

Using groups/namespaces

If you have worked intensely with ROS online tutorials, you will have kind of figured out how to solve this issue. Let's consider the `turtlesim` example for this. Launch the `turtlesim` node using the following commands.

In one Terminal, run the following commands:

```
$ initros1
$ roscore
```

In another Terminal, run the following commands:

```
$ initros1
$ rosrun turtlesim turtlesim_node
```

Check out the `rostopic` list to see `cmd_vel` and the pose of `turtle1`. Now, let's spawn another turtle into the GUI using the following command:

```
$ rosservice call /spawn 3.0 3.0 0.0 turtle2
```

Well, what just happened? You spawned in another turtle with the same topic names, but with an added prefix, `/turtle2`. This is the namespace technique in ROS that's used to spawn multiple robots. Now, our example will be refined, as shown here:

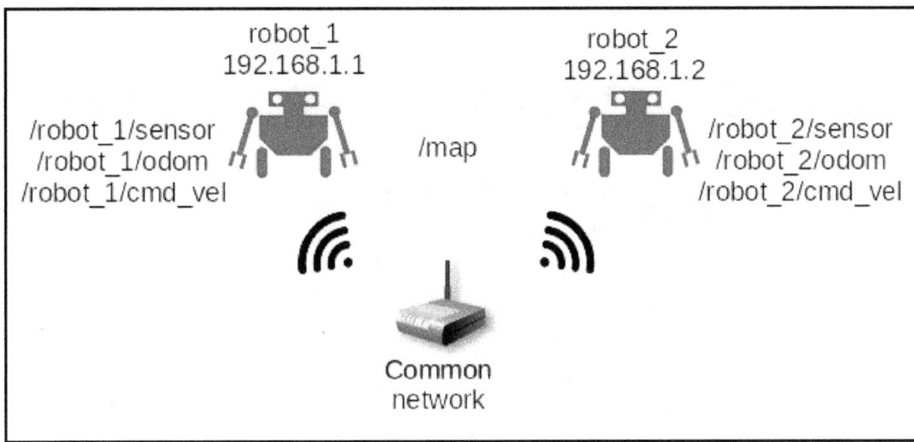

Modified communication between the same robots in the same network

As you can see, each robot has its own topics prefixed with the robot names, respectively. There could be some common topics among the robots too, such as the `/map` topic, since both robots are on the same environment. We'll learn how to use this namespace technique for the robot we created in `Chapter 3`, *Building an Industrial Mobile Manipulator,* in the upcoming section.

Example – multi-robot spawn using groups/namespaces

Let's consider the mobile base we created in `Chapter 3`, *Building an Industrial Mobile Manipulator,* for this example. The workspace is available in GitHub at the following link: `https://github.com/PacktPublishing/ROS-Robotics-Projects-SecondEdition/tree/master/chapter_3_ws/src/robot_description`. Download it in to a new workspace and then compile the workspace. The robot base is launched using the following commands:

```
$ initros1
$ roslaunch robot_description base_gazebo_control.xacro.launch
```

You should then see the gazebo model, as shown here:

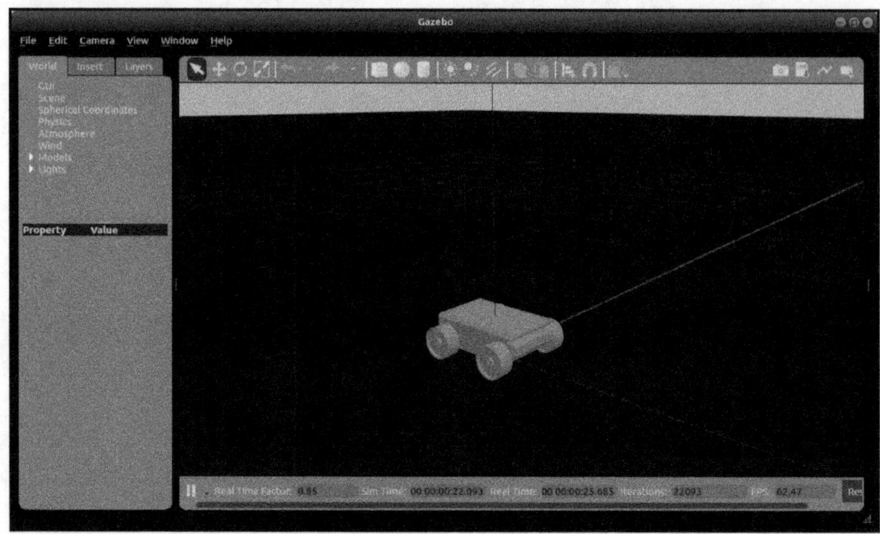

Gazebo view of the robot base model

Now, our intent is to spawn another robot into the gazebo, so let's name the robots `robot1` and `robot2`. To the preceding `base_gazebo_control.xacro.launch` file, we are launching the empty gazebo world, loaded with our robot and controller configurations, into the ROS server and loading the controller node. We need to do the same, but for more than one robot. For this, we would launch the robot under a namespace group tag. We need to ensure that the robot transforms are differentiated using the `tf_prefix` parameter and a `<robot_name>` prefix while loading the robot's URDF.

Finally, we have to differentiate each robot's controllers using the `ns` argument while loading the controller and the `--namespace` argument while launching the controller node. These changes can be applied to `robot1` like so:

```
<group ns="/robot1">
<param name="tf_prefix" value="robot1" />
 <rosparam file="$(find robot_description)/config/control.yaml"
command="load" ns="/robot1" />
 <param name="/robot1/robot_description" command="$(find xacro)/xacro --
inorder $(find robot_description)/urdf/robot_base.urdf.xacro nsp:=robot1"/>
     <node name="urdf_spawner_1" pkg="gazebo_ros" type="spawn_model"
       args="-x -1.0 -y 0.0 -z 1.0 -unpause -urdf -model robot1 -param
robot_description " respawn="false" output="screen">
     </node>
```

```
        <node pkg="robot_state_publisher" type="robot_state_publisher"
    name="robot_state_publisher_1">
          <param name="publish_frequency" type="double" value="30.0" />
        </node>

        <node name="robot1_controller_spawner" pkg="controller_manager"
    type="spawner"
          args="--namespace=/robot1
          robot_base_joint_publisher robot_base_velocity_controller
          --shutdown-timeout 3">
        </node>
    </group>
```

To launch two robots, simply copy and paste this whole block of code twice and replace `robot1` with `robot2` in the second copied block. You could go about creating as many blocks as you like. For simplicity, we moved this block of code into another launch file called `multiple_robot_base.launch`. The full code for launching two such robots is available on GitHub (`https://github.com/PacktPublishing/ROS-Robotics-Projects-SecondEdition/blob/master/chapter_6_ws/src/robot_description/launch/multiple_robot_base.launch`). The main launch file is available on GitHub under the `robotbase_simulation.launch`, which starts the gazebo empty world and the `multiple_robot_base` launch file.

Your gazebo view will look as follows:

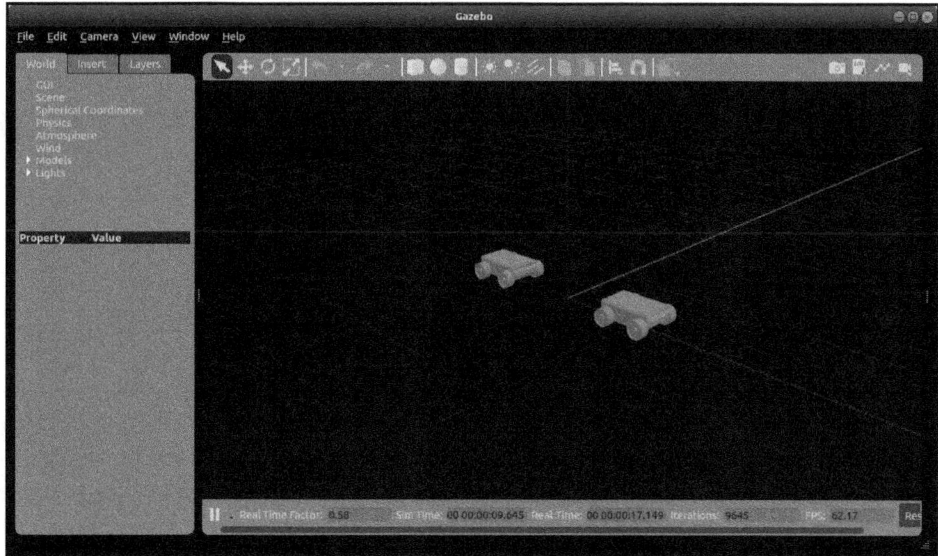

Gazebo view of multiple robots

The following are the `rostopic` lists:

```
robot@robot-pc:~/chapter_6_ws$ initros1
robot@robot-pc:~/chapter_6_ws$ rostopic list
/clock
/gazebo/link_states
/gazebo/model_states
/gazebo/parameter_descriptions
/gazebo/parameter_updates
/gazebo/set_link_state
/gazebo/set_model_state
/robot1/joint_states
/robot1/robot_base_velocity_controller/cmd_vel
/robot1/robot_base_velocity_controller/odom
/robot1/robot_base_velocity_controller/parameter_descriptions
/robot1/robot_base_velocity_controller/parameter_updates
/robot2/joint_states
/robot2/robot_base_velocity_controller/cmd_vel
/robot2/robot_base_velocity_controller/odom
/robot2/robot_base_velocity_controller/parameter_descriptions
/robot2/robot_base_velocity_controller/parameter_updates
/rosout
/rosout_agg
/tf
/tf_static
robot@robot-pc:~/chapter_6_ws$
```

rostopic list of multiple robots

Note that this is just a simple representation of a group tag and is not the only way to represent groups. You could enhance the same using suitable arguments and loops to increase your robot count.

Now that we have launched multiple robots using namespaces, let's look at its limitations.

Issues with using groups/namespaces

While they do solve the purpose of communicating between robots of the same type, there are some concerns when it comes to using this technique. They may be quite challenging to set when your robot is loaded with features, where you have to try to provide namespaces for almost all of the launch files that come along with the robot. This could be time-consuming and, at times, intimidating during setup and may lead to chaos if they're not set properly as it doesn't provide any diagnostics or exceptions in the case of issues.

Let's say that you plan to buy a robot base such as Husky or TurtleBot; to use them together, you need to set up the wrapper files (necessary namespace launch files) for each robot. The reason you would have chosen to go for a robot base is to ensure they are ready to use and that you don't need to set the packages up specifically for your application. However, you would end up creating additional files on top of what is available, which would add to more development time. Also, through this method, your `rostopic` list would be loaded with a cumbersome list of topics with specific robot prefixes. What if you want only selected topics from the robot to be controlled and hence displayed to the main master? What if you need to know which robots are connected or disconnected in the network?

All of these questions will be answered in the upcoming section.

Introduction to the multimaster concept

To understand the multimaster concept, let's consider an industry use case, as shown in the following diagram:

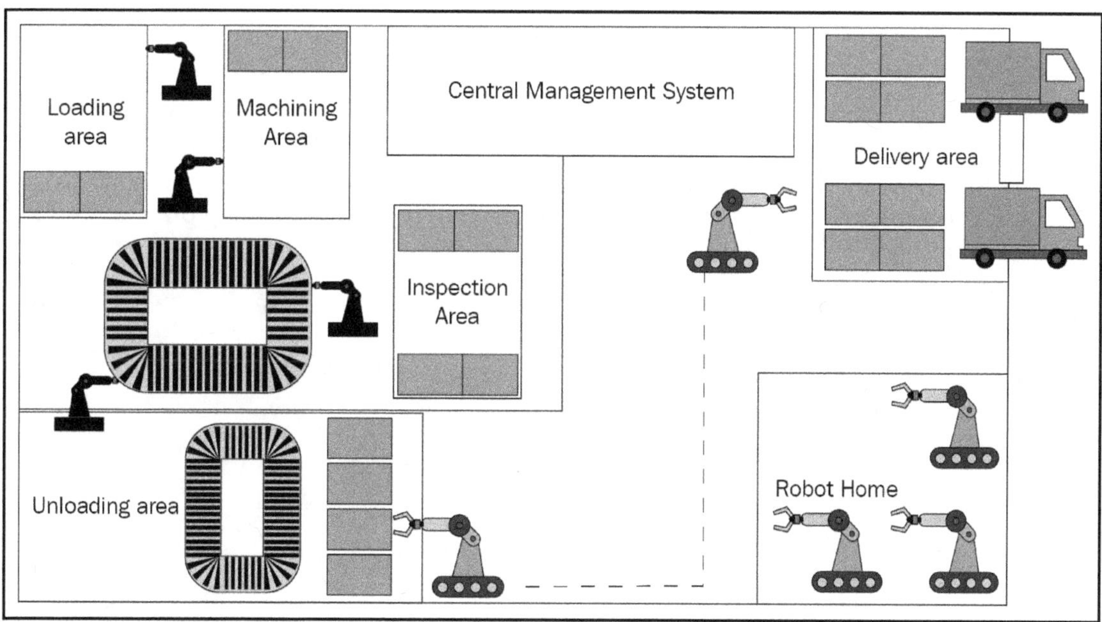

Industry robot use case

Imagine that we have robots such as our robot manipulator (which we created in `Chapter 3`, *Building an Industrial Mobile Manipulator*) working toward loading and unloading goods and we have at least five such robots doing this. Also, imagine that there are a few robot arms that are working along with machines for tending applications, such as loading onto the machining area and then onto the conveyor and unloading for delivery. Finally, there is a central system that monitors all of these robots' tasks and health. For such a setup, there are multiple local networks that would connect to a common network (which is where the central system sits).

There is a mechanism defined in ROS that can help us exchange information between ROS subsystems using common ROS-tools. The solution is provided as a ROS package called `multimaster_fkie`.

Introduction to the multimaster_fkie package

The `multimaster_fkie` package has two major nodes:

- The `master_discovery` node
- The `master_sync` node

The `master_discovery` node helps us detect other ROS masters in the network and identify changes, if any, in the network and share them with the other ROS masters. The `master_sync` node helps other ROS masters to synchronize their topics and services with the local ROS master. This effectively helps the ROS masters in the common network to only share necessary topic information with the common network, keeping other abundant topic information in their local network.

Let's look at some examples to understand how to set up the `multimaster_fkie` package.

More information about `multimaster_fkie` can be found at `http://wiki.ros.org/multimaster_fkie`.

Installing the multimaster_fkie package

Let's install the `multimaster_fkie` package from the source:

1. For this, we need to install `pip` on our system. You can use the following command to do so:

```
$ sudo apt install python-pip
```

2. Once `pip` is installed, clone the multimaster repository into your workspace and build the package using the following commands:

```
$ cd chapter_6_ws/src
$ git clone https://github.com/fkie/multimaster_fkie.git
multimaster
$ rosdep update
$ rosdep install -i --as-root pip:false --reinstall --from-paths
multimaster
```

You will be prompted to accept a few dependency installations after running the preceding command; ensure you accept all the dependency installations. Once the dependencies have been installed, you can build the workspace.

3. Let's use the `catkin_make_isolated` command this time to build the workspace. This command helps build the packages individually and not as a whole:

```
$ cd ..
$ catkin_build_isolated
```

Now that the installation has been set, let's try using this package.

Setting up the multimaster_fkie package

To set up the `multimaster_fkie` package, it is best to run this example in two separate systems. The setup is a three-step process:

1. Setting up hostnames and IPs.
2. Checking and enabling the multicast feature.
3. Testing the setup.

Let's look at each of these steps next.

Setting up hostnames and IPs

Let's assume that we have two systems with us and let's name them `pc1` and `pc2`, respectively. Now, follow these steps to set up the systems:

1. In `pc1`, go to the `/etc/hosts` file and make the following changes:

```
127.0.0.1 localhost
127.0.0.1 pc1

192.168.43.135 pc1
192.168.43.220 pc2
```

2. Now, in `pc2`, go to the `/etc/hosts` file and make the following changes:

```
127.0.0.1 localhost
127.0.0.1 pc1

192.168.43.135 pc1
192.168.43.220 pc2
```

You need `sudo` permissions to edit these files since they're in the system folder.

3. As you may recall, you would have added `ROS_MASTER_URI` and `ROS_IP` for the previous examples. Let's do the same for each computer with their respective IPs and `ROS_MASTER_URIs` in their respective bash scripts.
4. Open the bash script using `$ sudo gedit ~/.bashrc`:
    - In the `pc1` bash script, use the following code:

```
export ROS_MASTER_URI=http://192.168.43.135
export ROS_IP=192.168.43.135
```

    - In the `pc2` bash script, use the following code:

```
export ROS_MASTER_URI=http://192.168.43.220
export ROS_IP=192.168.43.220
```

Let's move on to the next section to check and enable the multicast feature.

Checking and enabling the multicast feature

To initiate synchronization between multiple roscores, you need to check whether the multicast feature is enabled on each computer. They're generally disabled in Ubuntu. To check whether they're enabled, use the following command:

```
$ cat /proc/sys/net/ipv4/icmp_echo_ignore_broadcasts
```

If the value is `0`, this means that the multicast feature is enabled. If the value is `1`, reset the value to `0` using the following command:

```
$ sudo sh -c "echo 0 >/proc/sys/net/ipv4/icmp_echo_ignore_broadcasts"
```

Once they're enabled, check for the multicast IP address using the following command:

```
$ netstat -g
```

The standard IP is usually `220.0.0.x`, so ping that IP in each computer to see whether the communication is happening between the computers. In my case, my multicast IP was `220.0.0.251`, and so `220.0.0.251` is the ping in both computers to check for connectivity:

```
$ ping 220.0.0.251
```

Now that the system is set up, let's test the setup.

Testing the setup

In each Terminal, begin roscores individually:

```
$ initros1
$ roscore
```

Now, go to the respective `multimaster_fkie` package folder and run the following command in each computer:

```
$ rosrun fkie_master_discovery master_discovery _mcast_group:=220.0.0.251
```

Here, `_mcast_group` is the multicast IP address we acquired using the `netsat` command.

If the network was set up properly, you should see the `master_discovery` node identifying the roscores in each computer, as shown here:

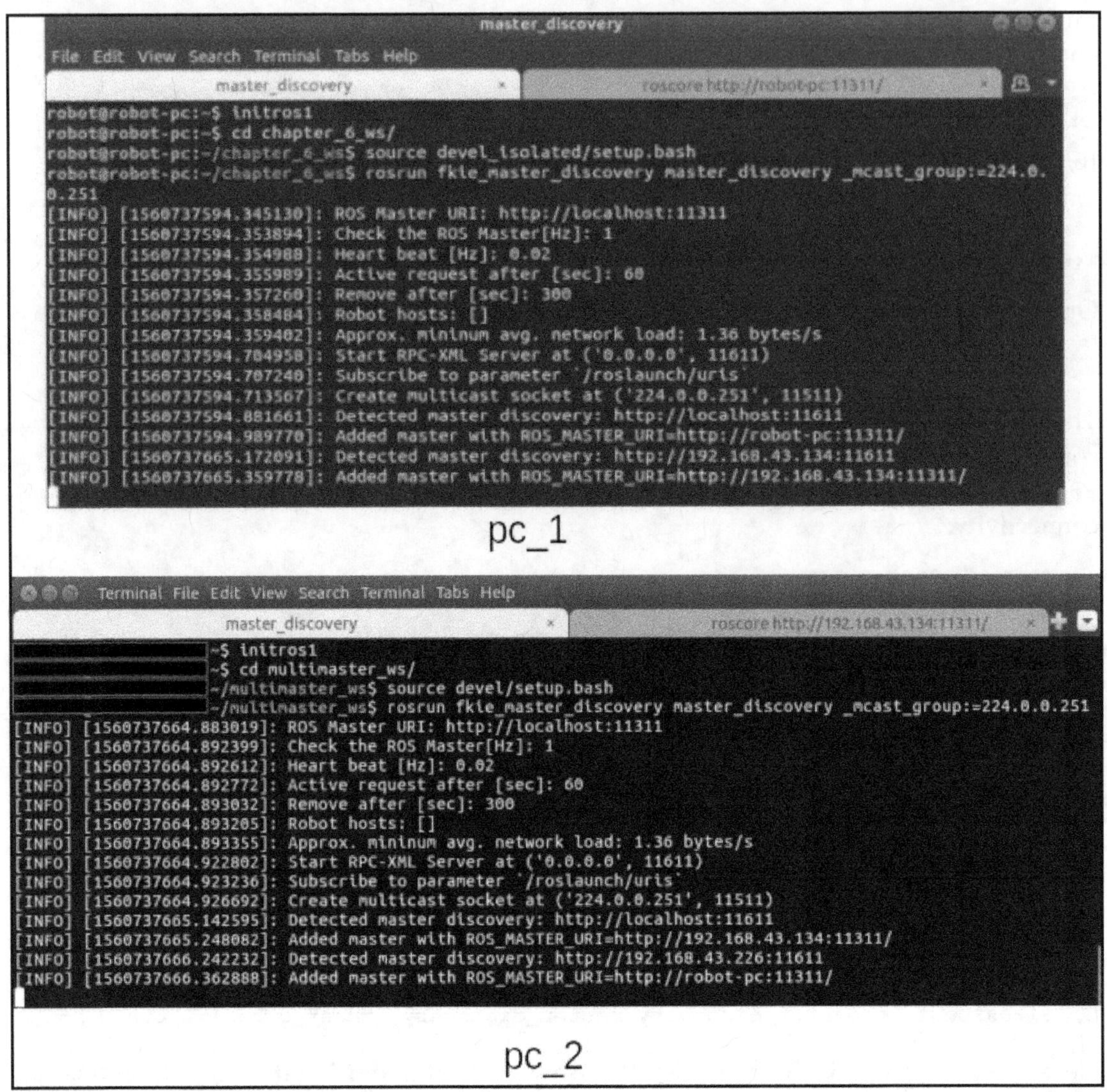

master_discovery nodes in pc1 and pc2

Now, to synchronize between the topics of each ROS master, run the `master_sync` node in each computer. As we can see, the `multimaster_fkie` package is set. Let's try the same with an example to understand this better.

A multi-robot use case

Let's try to launch the robot simulation from one PC and control the robots with a teleop on another PC. We shall use the same robot we set up in the *Example – multi-robot spawn using groups/namespaces* section. Assuming there is no change in the preceding setup, let's run the multi-robot launch on one PC:

1. In `pc1`, run the following commands:

```
$ initros1
$ roslaunch robot_description robotbase_simulation.launch
```

2. Then, start the `master_discovery` node in another Terminal:

```
$ initros1
$ source devel_isolated/setup.bash
$ rosrun fkie_master_discovery master_discovery
_mcast_group:=224.0.0.251
```

3. In another Terminal, start the `master_sync` node to synchronize all of the topics:

```
$ initros1
$ source devel_isolated/setup.bash
$ rosrun fkie_master_sync master_sync
```

   Now, run the necessary commands in `pc2`. In `pc2`, assuming you have started roscore in one Terminal, run the `master_discovery` and `master_sync` nodes in each Terminal, one by one.

4. In one Terminal, run the following commands:

```
$ initros1
$ rosrun fkie_master_discovery master_discovery
_mcast_group:=224.0.0.251
```

5. In another Terminal, run the following commands:

```
$ initros1
$ rosrun fkie_master_sync master_sync
```

Once the topics have been synchronized, you should see a window, as shown here:

```
[INFO] [1560785525.819599, 43.163000]: ROS masters obtained from '/master_discov
ery/list_masters': ['192.168.43.226']
[INFO] [1560785555.932963, 49.699000]: ROS masters obtained from '/master_discov
ery/list_masters': ['192.168.43.226']
[INFO] [1560785573.082508, 53.391000]: [192.168.43.134] ignore_nodes: ['/node_ma
nager', '/master_sync', '/rosout', '/node_manager_daemon', '/zeroconf', '/master
_discovery']
[INFO] [1560785573.127443, 53.400000]: [192.168.43.134] sync_nodes: []
[INFO] [1560785573.158611, 53.405000]: [192.168.43.134] ignore_topics: ['/rosout
', '/rosout_agg']
[INFO] [1560785573.186362, 53.410000]: [192.168.43.134] sync_topics: []
[INFO] [1560785573.204387, 53.412000]: [192.168.43.134] ignore_services: ['/*get
_loggers', '/*set_logger_level']
[INFO] [1560785573.227225, 53.415000]: [192.168.43.134] sync_services: []
[INFO] [1560785573.248637, 53.418000]: [192.168.43.134] ignore_type: ['bond/Stat
us']
[INFO] [1560785573.265998, 53.420000]: [192.168.43.134] ignore_publishers: []
[INFO] [1560785573.280129, 53.422000]: [192.168.43.134] ignore_subscribers: []
[INFO] [1560785575.303136, 53.832000]: SyncThread[192.168.43.134] Requesting rem
ote state from 'http://192.168.43.134:11611'
[INFO] [1560785576.329207, 54.051000]: SyncThread[192.168.43.134] Applying remot
e state...
[INFO] [1560785576.361536, 54.054000]: SyncThread[192.168.43.134] remote state a
pplied.
[INFO] [1560785586.045449, 56.192000]: ROS masters obtained from '/master_discov
ery/list_masters': ['192.168.43.134', '192.168.43.226']
[INFO] [1560785598.853209, 59.030000]: SyncThread[192.168.43.134] Requesting rem
ote state from 'http://192.168.43.134:11611'
[INFO] [1560785598.891440, 59.033000]: SyncThread[192.168.43.134] Applying remot
e state...
[INFO] [1560785598.907966, 59.035000]: SyncThread[192.168.43.134] remote state a
pplied.
[INFO] [1560785616.135324, 62.554000]: ROS masters obtained from '/master_discov
ery/list_masters': ['192.168.43.134', '192.168.43.226']
```

The multimaster_fkie topic sync

6. Now, run `$ rostopic list`. You should see the `/robot1` and `/robot2` topics simultaneously. Try moving the robots using the following command with appropriate topic names. You should see the robot moving:

```
$ rosrun rqt_robot_steering rqt_robot_steering
```

A simple representation of the robot's movement in the simulation is shown here:

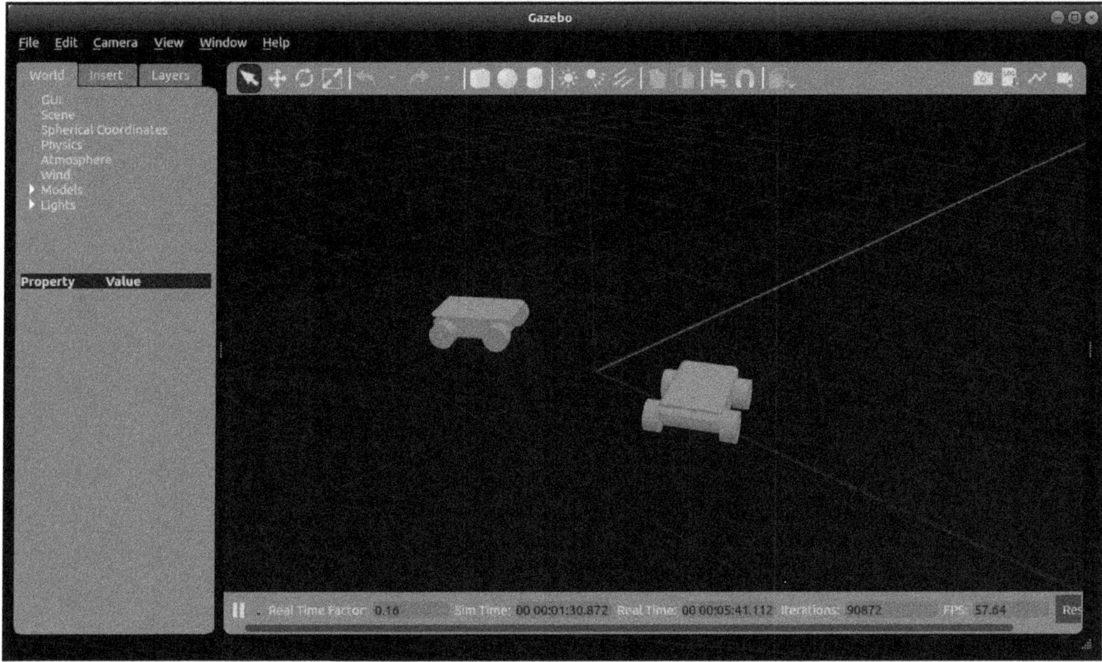

Robots moving in the gazebo

In case you need to select specific topics for synchronizing, check the `master_sync.launch` file for the necessary parameter list. You can do that by specifying the required topics with the rosparam `sync_topics`.

Summary

In this chapter, you understood how multiple robots communicate with each other in the best possible ways in ROS. We saw how to communicate between nodes on the same network, as well as the limitations of this, and we saw how to communicate between robots on the same network using namespaces and the limitations of this. Finally, we saw how to communicate between robots with multiple roscores using the `multimaster_fkie` package. This helps increase robotics applications. This chapter helped us understand how to communicate between robots of similar and dissimilar kinds.

In the next chapter, we will learn how to handle the latest embedded hardware in ROS.

7
ROS on Embedded Platforms and Their Control

We now know how to build our own robots in simulation and how to control them using the ROS framework. You have also learned how to handle complex robot tasks and how to communicate between multiple robots. All of these concepts were tested out virtually and you should be comfortable with them. If you're wondering how to get them working with real robots or to build your own robots and control them via the ROS framework, this chapter will help you achieve that.

Anyone interested in robotics would definitely be familiar with names such as Arduino and Raspberry Pi. Such boards can be used to control the robot actuators individually or in a control loop by reading the sensors connected to them. But what exactly are those boards? How are they used? How different are they from each other and why does it matter to choose such specific boards? The answers to these questions are quite rational and you should be able to answer them by the end of this chapter.

In this chapter, we shall see how such embedded control boards and computes can be used in ROS. You will begin by understanding how different microcontrollers and processors work, followed by an introduction to a series of such boards that are commonly used by the Robotics community with practical examples. Later, you will learn how to set up ROS on such boards and use them along with your custom projects.

The following topics will be covered in this chapter:

- Understanding embedded boards
- Introduction to microcontroller boards
- Introduction to the **Single Board Computer** (**SBC**)
- Debian versus Ubuntu
- Setting up ROS on SBC
- Controlling GPIOs from ROS
- Benchmarking of SBC
- Getting started with Alexa and connecting with ROS

Technical requirements

Let's look into the technical requirements for this chapter:

- ROS Melodic Morenia on Ubuntu 18.04 (Bionic)
- Timelines and test platform:
    - **Estimated learning time**: On average, 150 minutes
    - **Project build time (inclusive of compile and run time)**: On average, 90-120 minutes (depending on setting up the hardware boards with the indicated requirements)
    - **Project test platform**: HP Pavilion laptop (Intel® Core™ i7-4510U CPU @ 2.00 GHz × 4 with 8 GB Memory and 64-bit OS, GNOME-3.28.2)

The code for this chapter is available at `https://github.com/PacktPublishing/ROS-Robotics-Projects-SecondEdition/tree/master/chapter_7_ws`.

Let's begin this chapter by understanding the different types of embedded boards.

Understanding embedded boards

In an application, if the software is embedded into hardware of a specific design, the application is called an **embedded systems application**. Embedded systems are found in most of our daily routine gadgets and electronics such as mobile phones, kitchen appliances, and consumer electronics. They are usually designed for specific purposes. One such application where embedded systems are quite famous is robotics.

The hardware boards that carry the software (say, firmware) and that are intended for such specific purposes are what we call embedded boards. They come in two flavors:

- **Microcontroller-based**: In microcontroller-based boards, the hardware constitutes a CPU, memory units, peripheral device connectivity through IOs, and communication interface, all in a single chip.
- **Microprocessor-based**: In microprocessor-based boards, the hardware majorly constitutes the CPU. The other components such as communication interface, peripheral device connectivity, and timers are all available but as separate modules.

There is another category that resembles a combination of these two and is called **System on Chip** (**SoC**). They are quite compact in size and usually target products that are small in size. Modern microprocessor boards, also called SBCs, consist of an SoC embedded on them, along with other components. A simple comparison of the these can be seen in the following table:

	Microcontroller (MCU)	Microprocessor(MPU)	SoC
OS	No	Yes	It may be MCU- or MPU-based. If MPU, then the OS would be compact and light.
Data/computing width	4, 8, 16, 32-bit	16, 32, 64-bit	16, 32, 64-bit.
Clock speed	≤ MHz	GHz	MHz - GHz.
Memory (RAM)	Often in KB, rarely in MB	512 MB - several GB	MB - GB.
Memory (ROM)	KB to MB (FLASH, EEPROM)	MB to TB (FLASH, SSD, HDD)	MB to TB (FLASH, SSD, HDD).
Cost	Low	High	High.
Example	Atmel 8051 microcontrollers, PIC, ATMEGA series microcontrollers	x86, Raspberry Pi, BeagleBone black	Cypress PSoc, Qualcomm Snapdragon.

Comparison of embedded boards

Now, let's look at the basic concepts we need to know about for embedded boards.

Important concepts

A general embedded system may constitute a lot of components in its architecture. Since our scope is robotics, let's try to understand embedded systems with respect to robotics:

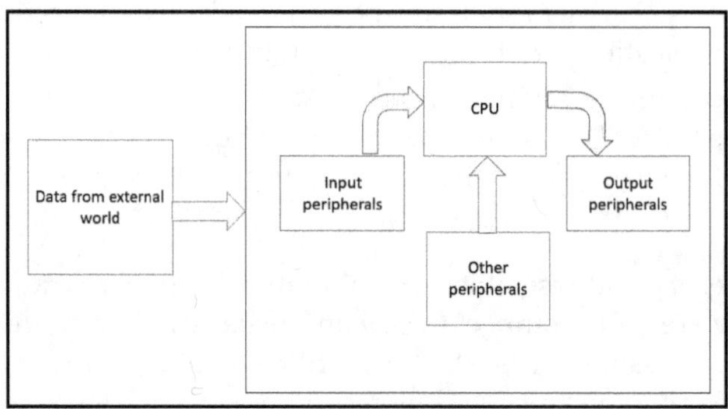

A simple embedded system representation

As you can see from the preceding diagram, the major components are as follows:

- **Input peripherals**: These could be sensors such as lidars, cameras, ultrasound, or infrared sensors that provide information about the environment. User interaction through a UI, joystick, or keypad could also be a part of input peripherals.
- **Output peripherals**: These could be actuator controls such as rotating wheels or a link's motion through mechanisms or LCD screens or displays.
- **CPU**: This is necessary for computing models or running algorithms, and memory is needed to save this information either temporarily or permanently for operation.
- **Other peripherals**: These could be communication interfaces, such as SPI, I2C, or RS-485, which take place either between individual components of the system or with other such systems in the network; or USB and network interfaces such as Ethernet or Wi-Fi.

The overall embedded block that constitutes these hardware components, along with the software layers on top, is the one we will use for our robotics application. As we mentioned in the *Understanding embedded boards* section, they're available as a microcontroller, processor, or SoC. This block supports techniques so that we can understand the information that's perceived by the environment, perform computations on such data, and convert them into necessary control actions. Now, let's differentiate microcontrollers from microprocessors in robotics.

How different are microcontrollers and microprocessors in robotics?

A microcontroller is usually targeted toward a single or specific process, while a microprocessor is targeted toward running multiple processes at a time. This is possible because a microprocessor runs an OS that allows multiple processes to run at the same time, whereas microcontrollers run bare-metal (without an OS). In robotics, microcontrollers are usually used to control the actuators via a motor driver circuit or to communicate sensor information back into the system.

A microprocessor could typically do both at the same time, making the application look realtime. Microcontrollers are less expensive than microprocessors and, at times, are much faster than processors. However, microprocessors can run heavy computations that are required in applications such as image processing or model-training, unlike microcontrollers. Robots usually constitute a combination of both microcontroller and microprocessor systems because they are comprised of multiple blocks that are responsible for specific functionalities. This combination would lead to a simple robot-agnostic, modular, efficient, and fail-safe solution.

What matters while choosing such boards

There are plenty of microcontroller- and processor-based boards available in the market today that are feature-rich and advantageous in their own ways. They usually depend on what the application demands. Some of the specifications to look out for in microcontroller boards are as follows:

- Microcontroller type
- Number of digital and analog I/Os
- Memory
- Clock speed
- Operating voltage
- Current per volt

Some of the specifications to look out for in microprocessor boards are as follows:

- SoC
- CPU and GPU
- Memory (RAM and ROM)
- GPIO

- Networking support
- USB ports

What matters the most while choosing such boards is as follows:

- Understanding the application and end user requirements
- Examining the hardware and software components in that requirement
- Designing and developing the application requirement with available features
- Evaluating the application virtually and finalizing on the requirements
- Choosing the board with the right features that meet our requirements, with long term support and upgradable features
- Deploying our application with that board

Now, let's look into some interesting microcontroller and microcomputer boards that are available in the market.

Introduction to microcontroller boards

In this section, we will look at some of the popular microcontroller boards and microcomputers that can be used in robots:

- Arduino Mega
- STM32
- ESP8266
- ROS-supported embedded boards

Arduino Mega

Arduino is one of the most popular embedded controller boards that can be used in robots. It is mainly used for prototyping electronics projects and robots. The boards mainly contain AVR series controllers, in which its pins are mapped as Arduino board pins. The main reason for the Arduino board's popularity is its programming and easiness in prototyping. The Arduino APIs and packages are very easy to use. So, we can prototype our application without much difficulty. The Arduino programming IDE is based on a software framework called Wiring (`http://wiring.org.co/`), where we can code using C/ C++ in a simplified way. The code is compiled using C/C++ compilers. Here is an image of a popular Arduino board, the Arduino Mega:

Arduino Mega (source: https://www.flickr.com/photos/arakus/8114424657. Image by Arkadiusz Sikorski. Licensed under Creative Commons CC-BY-SA 2.0: https://creativecommons.org/licenses/by-sa/2.0/legalcode)

There are multiple Arduino boards available. We'll learn how to choose the best one in the next section.

How to choose an Arduino board for your robot

The following are some of the important specifications of this board that may be useful when selecting an Arduino board for your robot:

- **Speed**: Almost all Arduino boards work under 100 MHz. Most of the controllers on boards are 8 MHz and 16 MHz. If you want to do some serious processing, such as implementing a PID on a single chip, then the Arduino may not be the best choice, especially if we want to run it at a higher rate. The Arduino is best suited for simple robot control. It is best for tasks such as controlling a motor driver and servo, reading from analog sensors, and interfacing serial devices using protocols such as **Universal Asynchronous Receiver/Transmitter** (**UART**), **Inter-Integrated Circuit** (**I2C**), and **Serial Peripheral Interface** (**SPI**).
- **GPIO pins**: Arduino boards provide different kinds of I/O pins to developers, such as **General Purpose Input/Output** (**GPIO**), **Analog-to-Digital Converter** (**ADC**), and **Pulse Width Modulation** (**PWM**), I2C, UART, and SPI pins. We can choose Arduino boards according to our pin requirements. There are boards with a pin count from 9 to 54. The more pins the board has, the larger the size of the board.
- **Working voltage levels**: There are Arduino boards that work on TTL (5V) and CMOS (3.3V) voltage levels. For example, if the robot sensors are working only in 3.3V mode and our board is 5V, then we have to either convert 3.3V into the 5V equivalent using a level shifter or use an Arduino working at 3.3V. Most Arduino boards can be powered from USB.

- **Flash memory**: Flash memory is an important aspect when selecting an Arduino board. The output hex file that's generated by the Arduino IDE may not be optimized compared to the hex of the embedded C and assembly code. If your code is too big, it is better to go for higher flash memory, such as 256 KB. Most basic Arduino boards have only 32 KB of flash memory, so you should be aware of this issue before selecting the board.
- **Cost**: One of the final criteria is, of course, the cost of the board. If your requirement is just for a prototype, you can be flexible; you can take any board. But if you are making a product using this, cost will be a constraint.

Now, let's look at another board called STM32.

STM32

What do we do if the Arduino is not enough for our robotic applications? No worries; there are advanced ARM-based controller boards available, for example, STM32 microcontroller-based development boards such as NUCLEO and **Texas Instrument** (**TI**) microcontroller-based boards such as Launchpads. The STM32 is a family of 32-bit microcontrollers from a company called STMicroelectronics (`http://www.st.com/content/st_com/en.html`).

They manufacture microcontrollers based on different ARM architectures, such as the Cortex-M series. The STM32 controllers offer a lot more clock speed than Arduino boards. The range of STM32 controllers are from 24 MHz to 216 MHz, and the flash memory sizes are from 16 KB to 2 MB. In short, STM32 controllers offer a stunning configuration with a wider range of features than the Arduino. Most boards work at 3.3V and have a wide range of functionalities on the GPIO pins. You may be thinking about the cost now, right? But the cost is also not high: the price range is from $2 to $20. There are evaluation boards available in the market to test these controllers. Some famous families of evaluation boards are as follows:

- **STM32 nucleo boards**: The nucleo boards are ideal for prototyping. They are compatible with Arduino connectors and can be programmed using an Arduino-like environment called Mbed (`https://www.mbed.com/en/`).
- **STM32 discovery kits**: These boards are very cheap and come built-in with components such as an accelerometer, microphone, and LCD. The Mbed environment is not supported on these boards, but we can program the board using IAR, Keil, and **Code Composer Studio** (**CCS**).
- **Full evaluation boards**: These kinds of boards are comparatively expensive and are used to evaluate all the features of the controller.

- **Arduino-compatible boards**: These are Arduino header-compatible boards that have STM32 controllers. Examples of these boards include Maple, OLIMEXINO-STM32, and Netduino. Some of these boards can be programmed using the Wiring language, which is used to program Arduino. One of the most commonly used STM32 boards that resembles the Arduino mini's look is the STM32F103C8/T6, shown here:

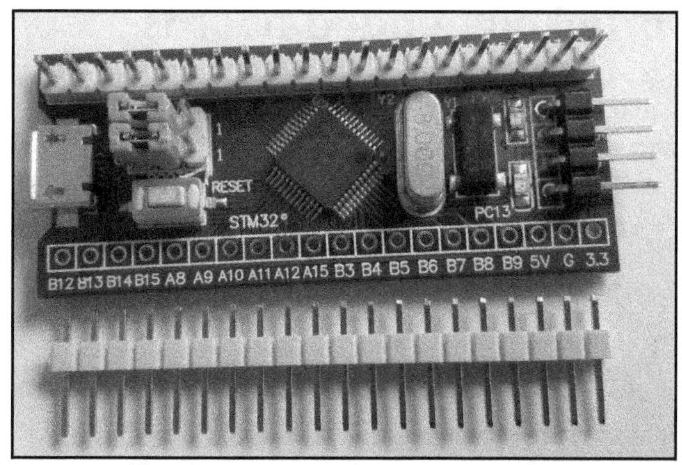

STM32F103C8/T6 image (source: https://fr.m.wikipedia.org/wiki/Fichier:Core_Learning_Board_module_Arduino_STM32_F103_C8T6.jpg. The image is taken by Popolon. Licensed under Creative Commons CC-BY-SA 4.0: https://creativecommons.org/licenses/by-sa/4.0/legalcode)

The next board we will be covering is ESP8266.

ESP8266

Another interesting microcontroller that is quite popular in the **Internet of Things** (**IoT**) sector is the ESP8266 chip. It is actually a Wi-Fi chip with microcontroller capabilities. It consists of an L-106 32-bit **RISC**-based (short for **Reduced Instruction Set Computing**) Tensilica microcontroller that runs at 80 MHz or overclocked to 160 MHz, with digital peripheral interfaces, antennas, and power modules embedded on a single chip.

Some of its features are that it has 2.4 GHz Wi-Fi (802.11 b/g/n, supporting WPA/WPA2), 16 GPIOs, a 10-bit ADC, and communication interfaces such as I2C, SPI, I2S, UART, and PWM. I2S and UART are available as shared pins, along with GPIOs. Memory size is a 64 KiB boot ROM, 32 KiB instruction RAM, and 80 KiB user data RAM. It also has 32 KiB instruction cache RAM and 16 KiB ETS system data RAM. External flash memory can be accessed using the SPI interface.

The ESP8266 comes in various form factors with different **SDKs** (short for **Software Development Kits**). For instance, there is a Lua scripting language-based firmware called NodeMCU that runs on ESP8266 SoC.

Now, let's look at a few ROS-supported embedded boards.

ROS-supported embedded boards

The ROS community has interesting embedded boards as well that can be directly coupled with sensors and actuators and provide ROS-based messages for interfacing with other ROS components with ease. A couple of such boards that we shall see here are the OpenCR board by Robotis and the Arbotix-Pro controller by Vanadium labs.

OpenCR

OpenCR, otherwise called the open source control module, is an STM32F7 series chip-based embedded board that is based on a very powerful ARM Cortex-M7 with a floating-point unit. It has an IMU (MPU9250) embedded on the board. It has 18 GPIO pins, 6 PWM IOs, and communication ports such as USB, a couple of UARTs, 3xRS485, and a CAN port. It has a 32-pin style berg connector like an Arduino board for connecting Arduino shields. The OpenCR board is shown here:

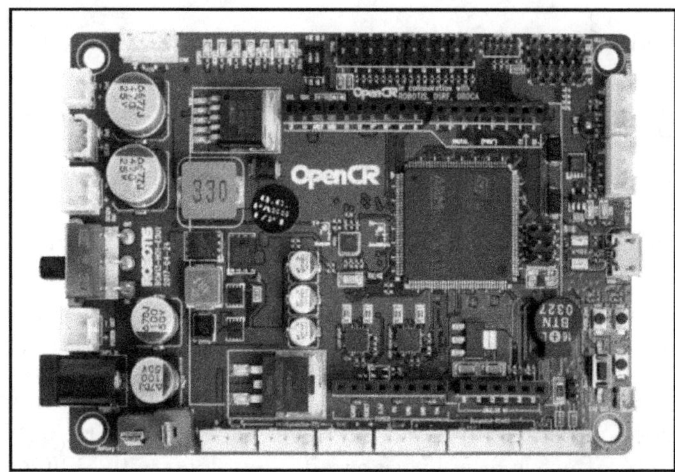

OpenCR board pinout (Source : http://wiki.ros.org/opencr. Image from ros.org. Licensed under Creative Commons CC-BY-3.0: https://creativecommons.org/licenses/by/3.0/us/legalcode)

You can find pinout details here: `http://emanual.robotis.com/docs/en/parts/controller/opencr10/#layoutpin-map`. This board is used in Turtlebot 3 and the ROS package integration is pretty similar to how Arduino connects with ROS. You can have a look at how Arduino-like microcontrollers can connect with ROS here: `http://wiki.ros.org/rosserial_arduino`.

More information on this board can be found at `http://emanual.robotis.com/docs/en/parts/controller/opencr10/`.

Arbotix-Pro

The Arbotix-Pro is another ROS supported controller based on the STM32F103 Cortex M3 32-bit ARM microprocessor. It has 16 GPIO pins for peripheral operations and communication buses such as TTL or I2C. It also has an Accelerometer/Gyroscope sensor embedded on the board, along with a single USB, UART, and interface support for XBee communication. This board is built toward using Dynamixel Robot actuators and supports both TTL or RS-485 communication, hence it's able to control AX, MX, and Dynamixel-PRO actuators. Power to the actuators could be provided independently through a dedicated 60 amp MOSFET, allowing power saving options while operational or in case of shutdown or error. You can find more information about the board at `https://www.trossenrobotics.com/arbotix-pro`.

The Arbotix ROS package can be found at `http://wiki.ros.org/arbotix`.

Now, let's look at a table to compare the configuration of the different boards.

Comparison table

Let's look at a simple comparison of the preceding boards for quicker reference:

Content	Arduino Mega	STM32F103C8/T6	ESP8266	OpenCR	Arbotix Pro
Microcontroller	ATMega 2560	ARM Cortex-M3	Tensilica L106 32-bit processor	STM32F746ZGT6 / 32-bit ARM Cortex®-M7 with FPU	STM32F103RE Cortex M3 32bit ARM

Operating Voltage	5V	2V ~ 3.6V	2.5V ~ 3.6V	5V	2V ~ 3.6V
Digital I/Os	54	37	16	8	16 ADC/GPIOs
PWM I/Os	15	12	-	6	
Analog I/Os	16	10	1	6	
Flash memory	256 KB	64 KB	External Flash (512 KB to 4 MB typically included)	2 MB	512 KB
Clock speed	16 MHz	72 MHz	24 MHz to 52 MHz	216 MHz	72 MHz

Comparison of preceding MCU boards

Now that we know the basics of microcontroller boards, let's look at single-board computers.

Introduction to single-board computers

In the previous sections, we looked at a series of microcontroller boards with varied features, which have all of the components embedded in a single chip. What if you want your robot or drone to perform modern-day computations for path planning or obstacle avoidance? Microcontroller boards help us read sensor signals and control the actuator signals, provided they follow simple math libraries for understanding and processing sensors.

What if you want to perform intense computation, store incoming sensor values over time, and then perform computations on such stored values? That is where SBC comes into play. SBCs can be loaded with an OS and could perform computations like a modern-day desktop, in a small form factor setup. They can't immediately replicate a modern-day computer as desktop computers nowadays are attached to additional components such as a GPU or powerful network cards that help solve our problems.

SBC, on the other hand, can be found in simple applications such as robotics, drones, ATM machines, virtual slot machines, or advertisement display televisions. Today, there is a wide number of such SBCs available and they are still being researched and developed with more features, such as GPU support or a volatile memory of more than 2 GB. In fact, some boards support modern computation techniques such as machine learning algorithms or are preloaded with deep learning neural networks.

Let's look into some interesting SBCs and their features. We have classified them into two categories:

- CPU boards
- GPU boards

Let's begin with CPU boards.

CPU boards

Let's look at the following commonly used compute boards with feature comparison:

- Tinkerboard S
- BeagleBone Black
- Raspberry Pi

Tinkerboard S

One of the modern-day SBCs is the Tinkerboard S from Asus, which is the second iterated board after Tinkerboard. This board is widely used by the community because it holds the exact form factor as Raspberry Pi (another SBC discussed here).

This is a powerful $90 board embedded with a modern quad-core ARM-based processor—the Rockchip RK3288, 2GB of LPDDR3 dual-channel memory, onboard 16 GB eMMC and SD card interface (3.0) for faster read and write speeds for the OS, applications, and file storage. The board is powered by an ARM-based Mali™-T760 MP4 GPU featuring a powerful and energy-efficient design that supports computer vision, gesture recognition, image stabilization, and processing.

Other features include H.264 and H.265 playback support, as well as HD and UHD video playback. Unlike other SBCs that support fast Ethernet, this board supports Gigabyte Ethernet for network connectivity and the internet. Tinkerboard S is shown here:

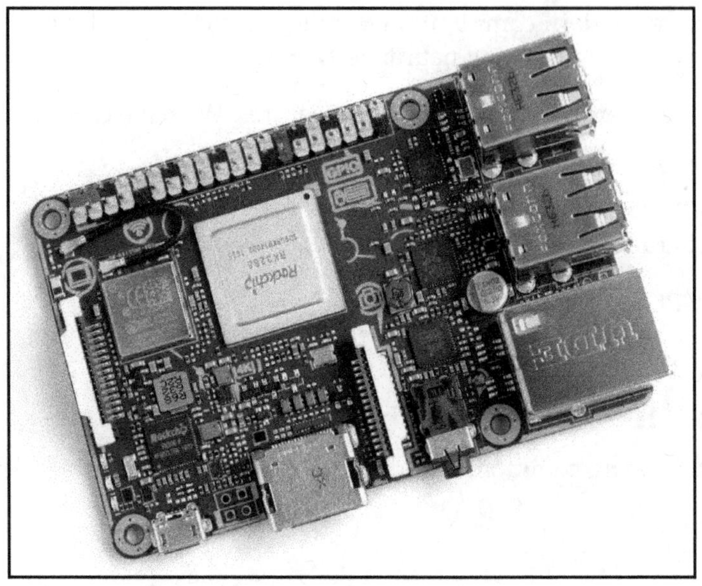

Tinkerboard S (source: https://www.flickr.com/photos/120586634@N05/32268232742. Image is taken by Gareth Halfacree. Licensed under Creative Commons CC-BY-SA 2.0: https://creativecommons.org/licenses/by-sa/2.0/legalcode)

It consists of a 40-pin color-coded GPIO pinout that helps beginners recognize the respective pin features. It also has an enhanced I2S interface with master and slave modes for improved compatibility and supports displays, touchscreens, Raspberry Pi cameras, and more. The OS that runs on this SBC is called TinkerOS, which is a Debian 9-based system. You will learn how to set up ROS on this OS in the upcoming sections.

BeagleBone Black

A credit card-sized SBC of low cost and that has good community support is the BeagleBone Black board from the Beagleboard organization. This board comes with an AM335x 1GHz ARM® Cortex-A8 processor with 512 MB DDR3 RAM support. It has an onboard 4 GB 8-bit eMMC flash storage that could hold an OS.

Other features include PowerVR SGX530, a GPU that supports a 3D graphics accelerator, a NEON float point accelerator, and two 32-bit PRU microcontrollers. This board has two 46 pin headers with USB client, host Ethernet, and HDMI connectivity options. BeagleBone Black is shown here:

BeagleBone Black (source: https://www.flickr.com/photos/120586634@N05/14491195107. Image by Gareth Halfacree. Licensed under Creative Commons CC-BY-SA 2.0: https://creativecommons.org/licenses/by-sa/2.0/legalcode)

Another robotics-specific board with loaded features is the Beablebone Blue. It comes with the same processor but with additional features such as Wi-Fi, Bluetooth, IMU with barometer support, power regulation and state-of-charge LEDs for a 2-cell LiPo, H-Bridge motor drivers that can connect 4 DC motors+encoders, 8 servos, and all of the commonly-needed buses for additional peripherals such as GPS, DSM2 radio, UARTs, SPI, I2C, 1.8V analog, and 3.3V GPIOs.

The board supports a Debian version of OS called Armbian. It also has Ubuntu OS support, which we will see in the upcoming sections when we set up ROS.

Raspberry Pi

You would have definitely heard about Raspberry Pi and we have seen this board in the previous edition of this book as well. This is the most commonly used SBC and has a huge fan base (community support). It is quite easy to use and set up. The latest version of this SBC is the Raspberry pi 4 (shown in the following photograph).

It is a $40 board that comes with Broadcom BCM2711, Quad-core Cortex-A72 (ARM v8) 64-bit SoC @ 1.5GHz with 1 GB, 2 GB, or 4 GB LPDDR4-3200 SDRAM support with 40 pin GPIO pinout. Unlike the old boards, this board supports Gigabit Ethernet, along with 2.4 GHz and 5.0 GHz IEEE 802.11ac wireless and Bluetooth 5.0. It also has a dual-display port that supports resolutions up to 4K via a pair of micro-HDMI ports and can decode hardware video at up to 4Kp60:

Raspberry Pi 4 (source: https://commons.wikimedia.org/wiki/File:Raspberry_Pi_4_Model_B_-_Side.jpg. Image by Laserlicht. Licensed under Creative Commons CC-BY-SA 4.0: https://creativecommons.org/licenses/by-sa/4.0/legalcode)

The board comes with standard Debian OS support called Raspbian and can support Ubuntu OS as well.

Now, let's look at a table to compare the configuration of the different boards.

Comparison table

For quick and simple understanding, let's compare these boards' features::

Content	Tinkerboard S	BeagleBone Black	Raspberry pi 4
Processor	Rockchip Quad-Core RK3288 processor	Sitara AM3358BZCZ100 1 GHz	Broadcom BCM2711, Quad-core Cortex-A72 (ARM v8) 64-bit SoC @ 1.5GHz

Memory	2 GB Dual Channel DDR3	SD RAM of 512 MB DDR3L 800MHZ and 4 GB FLASH—8-bit Embedded MMC	1 GB, 2 GB or 4 GB LPDDR4-3200 SDRAM
Connectivity	4 x USB 2.0, RTL GB LAN, 802.11 b/g/n, Bluetooth V4.0 + EDR	1 x USB 2.0, 10/100 Mbit-RJ45	2 x USB 3.0 ports, 2 x USB 2.0 ports, Gigabit Ethernet, 2.4 GHz and 5.0 GHz IEEE 802.11ac wireless, Bluetooth 5.0, BLE
GPIO	Up to 28 GPIOs	Up to 69 GPIOs	Up to 28 GPIOs
SD card support	Yes	Yes	Yes

Comparison of preceding SBC boards

Now, let's look at a few popular GPU boards.

GPU boards

Unlike the SBCs with low key GPU support, there are SBCs with good GPU capabilities that support online image processing and computer vision algorithms, modern-day model training, and feature extractions. Let's look at the following boards from the NVIDIA Corporation:

- Jetson TX2
- Jetson Nano

Jetson TX2

Jetson TX2 is one of the most power-efficient and fastest compute boards available from the NVIDIA Corporation. This board is majorly targeted at AT computation. It hardly consumes 7.5 watts compared to a regular desktop GPU and goes up to 15 watts maximum. It has a 256-core NVIDIA Pascal™ GPU architecture-based GPU processor with 256 NVIDIA CUDA cores and has two Dual-Core NVIDIA Denver 64-Bit CPUs, along with Quad-Core ARM® Cortex®-A57 MPCore CPU processors. It comes with 8 GB 128-bit LPDDR4 memory and a 1866 MHx - 59.7 GB/s data transfer rate. It also comes with onboard storage of 32 GB eMMC 5.1.

This board is quite costly compared to the previous boards, because of which it might look like an overkill. However, no other boards can compute AI algorithms like this board can.

Jetson Nano

NVIDIA came up with a cheap alternative to that of the other SBCs we saw previously called Jetson Nano. This is a $99 board that is small and a powerful computer that lets us run multiple neural networks in parallel for applications such as image classification, object detection, segmentation, and speech processing. It comes with a 128-core Maxwell GPU and a Quad-Core ARM A57 @ 1.43 GHz CPU, with 4 GB 64-bit LPDDR4 25.6 GB/s memory. There is no eMMC but it comes with an SD card slot for storage or to run an OS. It has Gigabit Ethernet and HDMI 2.0 and eDP 1.4 display ports, and it supports I2C, I2S, SPI, and UART. It runs at as low as 5 watt power. Jetson Nano is shown here:

Jetson Nano

Now, let's look at a comparison between Jetson Nano and Jetson TX2.

Comparison table

The following table provides comparison between the available GPU SBCs:

Content	Jetson Nano	Jetson TX2
GPU	128-core NVIDIA Maxwell™ GPU	256-core NVIDIA Pascal™ GPU
CPU	Quad-Core ARM®Cortex®-A57 MPCore	Dual-Core NVIDIA Denver 2 64-Bit CPU and Quad-Core ARM®Cortex®-A57 MPCore
Memory	4GB 64-bit LPDDR4 Memory 1600MHz - 25.6 GB/s	8 GB 128-bit LPDDR4 Memory 1866MHz - 59.7 GB/s
AI Performance	472 GFLOPs	1.3 TFLOPs
Storage	16GB eMMC	32GB eMMC
Power	5W / 10W	7.5W / 15W

Comparison of the preceding GPU boards

The next section will help us differentiate Debian from Ubuntu.

Debian versus Ubuntu

Before we get into setting up ROS on these compute boards, let's try to understand the two most commonly used Linux distributions by the community—Debian and Ubuntu.

As you know, the projects in this book target ROS installed on top of Ubuntu. So, you should be familiar with Ubuntu by now. But how different is Debian? Not much—you simply need to know that Ubuntu is actually derived from Debian. Debian is one of the oldest OSes based on Linux kernels and acts as a base for most of the newer Linux distributions. Ubuntu was released by a private firm called Canonical who intended to produce an easy to use Linux distribution for daily use. The following diagram explains how Ubuntu is derived from Debian:

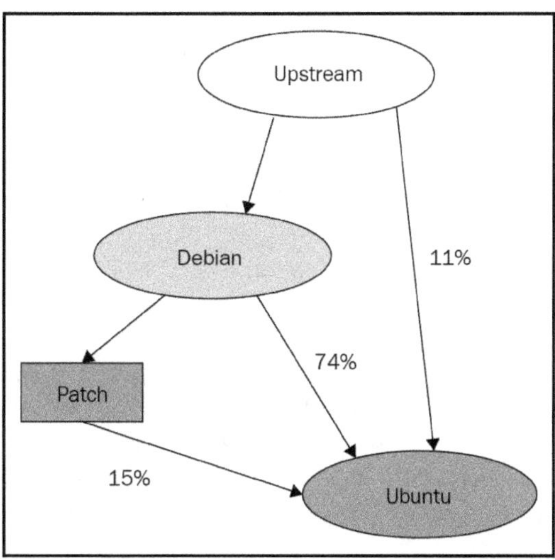

Ubuntu derivation from Debian

Some of the notable differences between both is in terms of software packages, ease of use or flexibility, stability, and support. While Ubuntu is easy to use with its GUI approach toward setting up packages, packages have to be manually set up in Debian. But Debian is extremely reliable while installing or upgrading packages compared to Ubuntu as it doesn't need to deal with new addons that act as bugs and lead to either screen blackouts or a popout error message with a sound. Notably, both have their own advantages and disadvantages.

When it comes to SBCs, Debian has been supported widely by the open source community compared to Ubuntu, which began releasing arm-based OSes recently. Notably, for ROS, Ubuntu is considered a better option because ROS was majorly developed on top of Ubuntu. Though ROS is supported on Debian today, the Debian kernels are one of the oldest and sometimes do not support new hardware. Hence, Ubuntu may have a slight edge over Debian. However, we shall see how to set up ROS on both Debian and Ubuntu in the upcoming sections.

Let's start with setting up ROS on Tinkerboard S.

Setting up ROS on Tinkerboard S

Let's learn how to set up ROS on Tinkerboard S step by step. We shall learn how to set up the Debian OS called Armbian on the SBC and then set up Ubuntu. Later, we learn how to install ROS on both. First, let's look at a few prerequisites.

Prerequisites

To set up any OS on this SBC, you will need the following hardware components:

- An SD card
- Rated power supply
- Optional case (to avoid electrostatic contact)

In terms of software requirements, assuming you're on your Ubuntu laptop, you would need an open source OS image writer called Etcher. You can download it from this website: `https://www.balena.io/etcher/`. There is no installation for this software, so once downloaded, extract the file to the necessary location and simply execute the file. Let's learn how to set up ROS on Tinkerboard Debian OS and Armbian.

Installing the Tinkerboard Debian OS

Interestingly, ROS didn't work on this OS through a regular installation setup. The following are suggestions for installing ROS on Armbian and for using a readily available image with Ubuntu 16.04 and ROS Kinetic.

Let's follow these steps to set up ROS on the Tinkerboard Debian OS:

1. To install the OS, follow these steps:
    1. Download the Tinkerboard Debian OS version image file from this website: `https://dlcdnets.asus.com/pub/ASUS/mb/Linux/Tinker_Board_2GB/20190821-tinker-board-linaro-stretch-alip-v2.0.11.img.zip`.
    2. Once the image is downloaded, insert the SD card into your system and open Etcher.
    3. Select the downloaded image and the specific SD card location and click **Flash!**. You should see the output shown here:

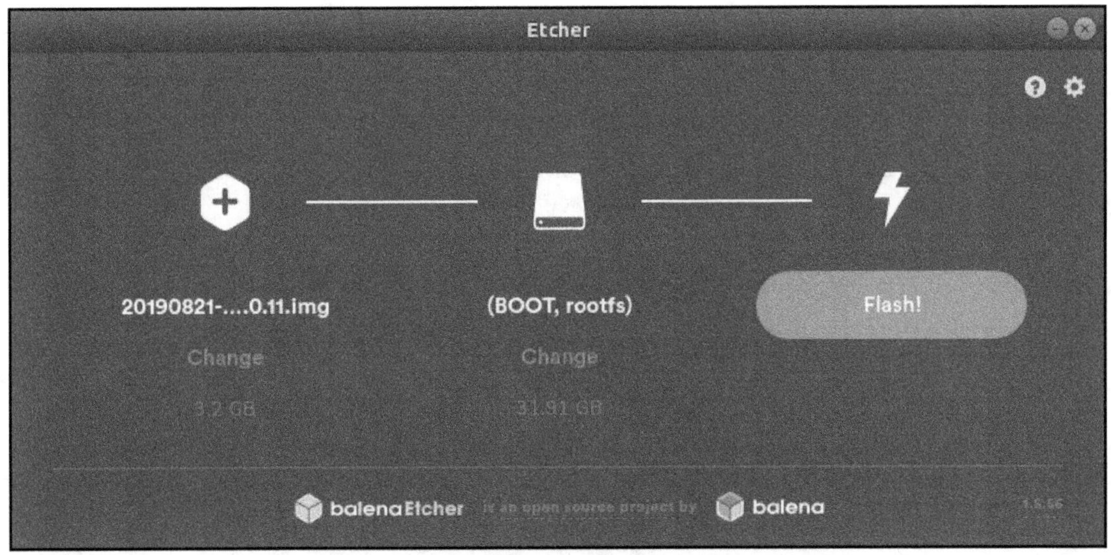

Etcher OS screenshot

    4. Once completed, load the SD card into the SBC.
2. Follow these steps to load the OS:
    1. Now that the SD card is loaded, power the SBC using an appropriate power supply adapter.
    2. You should see a series of loading commands and the OS should be booted up.

You could also try working with GPIOs in this OS, as indicated in the upcoming section. To use ROS, we need to set up Armbian and ROS on it.

Installing Armbian and ROS

Let's observe the following steps to set up ROS on the Armbian OS:

1. Installing the OS requires you to do the following:
    1. Download the Armbian Bionic version image file from this website: `https://dl.armbian.com/tinkerboard/Ubuntu_bionic_default_desktop.7z`.
    2. Once the image is downloaded, insert the SD card into your system and open Etcher.
    3. Select the downloaded image and the specific SD card location and click **Flash!**.
    4. Once completed, load the SD card into the SBC.
2. Loading the OS requires that you do the following:
    1. Now that the SD card is loaded, power the SBC using an appropriate power supply adapter.
    2. You should see a series of loading commands. Once the OS has been booted up, you should see the following screenshot:

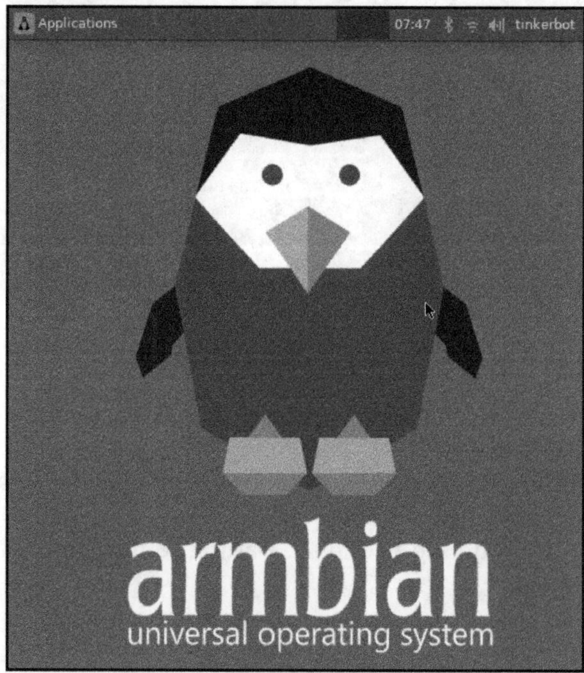

Tinkerboard with Armbian loaded

3. To install ROS on Armbian, do the following:
    1. Ensure that Tinkerboard accepts the software from `packages.ros.org` by setting up `sources.list` using the following command:

    ```
    $ sudo sh -c 'echo "deb http://packages.ros.org/ros/ubuntu
    (lsb_release -sc) main" > /etc/apt/sources.list.d/ros-
    latest.list'
    ```

    2. Set up keys using the following command:

    ```
    $ sudo apt-key adv --keyserver
    'hkp://keyserver.ubuntu.com:80' --recv-key
    C1CF6E31E6BADE8868B172B4F42ED6FBAB17C654
    ```

    3. Now, ensure the Debian package is up to date using the following command:

    ```
    $ sudo apt-get update
    ```

    4. Install ROS-desktop using the following command:

    ```
    $ sudo apt-get install ros-melodic-desktop
    ```

    5. Now, initialize `rosdep` using the following commands:

    ```
    $ sudo rosdep init

    $ rosdep update
    ```

    6. Once initialized, set up the environment using the following commands:

    ```
    $ echo "source /opt/ros/melodic/setup.bash" >> ~/.bashrc

    $ source ~/.bashrc
    ```

7. Now, you're done installing ROS on Armbian OS. Run `roscore` to see that the ROS version is installed. You should see the following Terminal output:

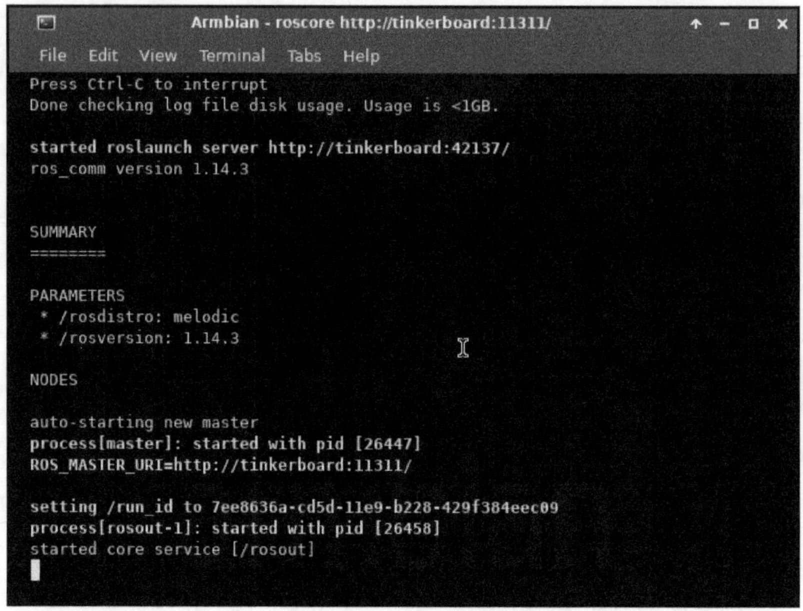

Armbian OS with ROS running (roscore)

Now that you have seen how to set up ROS on Armbian, let's look at a preloaded and available ROS image.

Installing using an available ROS image

There is an image available from Husarion, a robotic company that creates a rapid development platform for robots, called CORE2-ROS. The image is a pre-built image with ROS Kinetic on Ubuntu 16.04. You can download the image and burn the it using Etcher, which is available at `https://husarion.com/downloads/`.

For more information, look at `https://husarion.com/index.html`.

Now, let's learn how to set up ROS on BeagleBone Black.

Setting up ROS on BeagleBone Black

Let's learn how to set up ROS on BeagleBone Black step by step. We shall learn how to set up the Debian OS on the SBC and then set up Ubuntu. Later, we shall learn how to install ROS on both. First, let's look at a few prerequisites.

Prerequisites

To set up any OS on this SBC, you would need the following hardware components:

- An SD card
- Rated power supply
- Optional case (to avoid electrostatic contact)

In terms of software requirements, assuming you're on your Ubuntu laptop, you would need an open source OS image writer called Etcher. You could download it from this website: `https://www.balena.io/etcher/`. There is no installation for this software, so once downloaded, extract the file to the necessary location and simply execute the file.

Installing the Debian OS

Installing ROS on the Debian OS is quite a task and is not straightforward. If you're planning to use ROS on BeagleBone Black, I would suggest installing ROS via the Ubuntu setup in the upcoming section.

If you're interested in trying to install ROS, after installing the OS, you can try the steps at `http://wiki.ros.org/melodic/Installation/Debian`.

Let's follow these steps to set up the Debian OS on the BeagleBone Black Debian OS:

1. Follow these steps to install the OS:
    1. Download the BeagleBone Black Debian OS version image file from this website: `http://beagleboard.org/latest-images/`. We have downloaded Debian 9.5 2018-10-07 4GB SD LXQT (`http://debian.beagleboard.org/images/bone-debian-9.5-lxqt-armhf-2018-10-07-4gb.img.xz`) for this book.
    2. Once the image is downloaded, insert the SD card into your system and open Etcher.

   3. Select the downloaded image and the specific SD card location and click **Flash!**.
   4. Once completed, load the SD card into the SBC.

2. Follow these steps to load the OS:
   1. Now that the SD card is loaded, power the SBC using an appropriate power supply adapter.
   2. Since we're using BeagleBone Black for this example, you need to write the image to the on-board eMMC. Log in to the board using the following user credentials:

      ```
      username : debian
      password : temppwd
      ```

   3. Now, open the `/boot/uEnv.txt` file using the Nano editor for editing:

      ```
      $ sudo nano /boot/uEnv.txt
      ```

   4. Uncomment the following line by removing the # symbol:

      ```
      $ cmdline=init=/opt/scripts/tools/eMMC/init-eMMC-flasher-v3.sh
      ```

   5. Now, reboot the system. You should see a series of flashing lights that show that the OS is being flashed into the eMMC. Once the light stops blinking, remove the SD card and boot the board.

Installing Ubuntu and ROS

Previously, we saw how to burn the OS onto the eMMC from an SD card. For setting up Ubuntu on BeagleBone Black, you do not need to write the OS onto the eMMC. Instead, you can boot from the SD card itself every time you power the board:

1. Installing the OS requires these steps:
   1. Download the Ubuntu arm-based Bionic version image file from this website: `https://rcn-ee.com/rootfs/2019-04-10/microsd/bbxm-ubuntu-18.04.2-console-armhf-2019-04-10-2gb.img.xz`.
   2. Once the image is downloaded, insert the SD card into your system and open Etcher.

3. Select the downloaded image and the specific SD card location and click **Flash!**. You should see the output shown here:

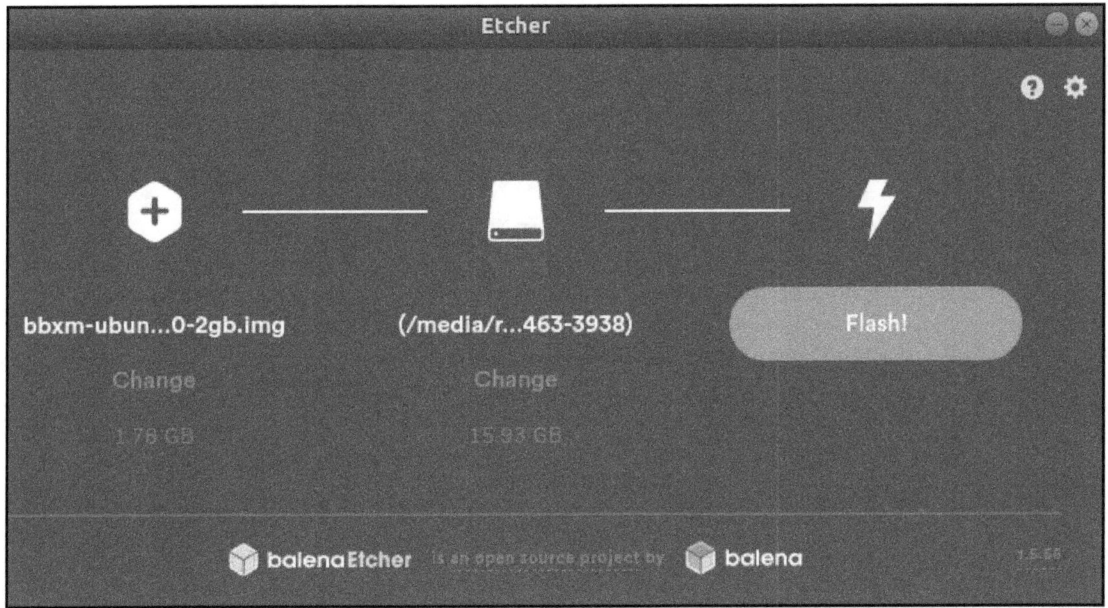

Etcher OS screenshot

Once completed, load the SD card into the SBC.

2. Follow these steps to load the OS:
    1. Now that the SD card is loaded, holding the boot button and power the SBC using an appropriate power supply adapter. Ensure that you hold the boot button while booting until the user LEDs begin to flash.
    2. You should see a series of loading commands. Once the OS is booted up, you should see a Terminal screen with temporary login details.
    3. Log in with the following user credentials:

    ```
    Username: ubuntu
    Password: ubuntu
    ```

    4. Note that there won't be any GUI, so you will need to install lightweight GUIs for ease of use. You can install the LDXE desktop using the following command:

    ```
    $ sudo apt-get install lxde
    ```

5. Remember that if you want to boot from micro SD and not the eMMC, you should make sure that you hold down the boot button and wait for the LED to flash.

3. Installing ROS is pretty straightforward, as we saw in the *Installing Armbian and ROS* section. Follow the same steps mentioned in that section.

Now, let's learn how set up ROS on Raspberry Pi 3/4.

Setting up ROS on Raspberry Pi 3/4

Setting up the OS is similar and pretty straightforward in Raspberry Pi 3/4. Raspberry Pi supports a list of OSes due to its popularity and application intention. In this book, we'll learn how to set up the commonly used OSes for robotics—Raspbian and Ubuntu MATE. First, let's look at the prerequisites.

Prerequisites

To set up any OS on this SBC, you will need the following hardware components:

- An SD card
- Rated power supply
- Optional case (to avoid electrostatic contact)

In terms of software requirements, assuming you're on your Ubuntu laptop, you would need an open source OS image writer called Etcher. You can download it from this website: `https://www.balena.io/etcher/`. There is no installation for this software, so once downloaded, extract the file to the necessary location and simply execute the file. Now, let's learn to install Raspbian and ROS.

Installing Raspbian and ROS

Follow the steps to set up ROS on the Raspbian OS:

1. Follow these steps, to install the OS:
    1. Download the Raspbian OS version image file from this website: `https://downloads.raspberrypi.org/raspbian_full_latest`.
    2. Once the image is downloaded, insert the SD card into your system and open Etcher.
    3. Select the downloaded image and the specific SD card location and click **Flash!**.
2. To load the OS, make sure that the SD card is loaded. Power the SBC using an appropriate power supply adapter. If everything is successful, you should see the following output:

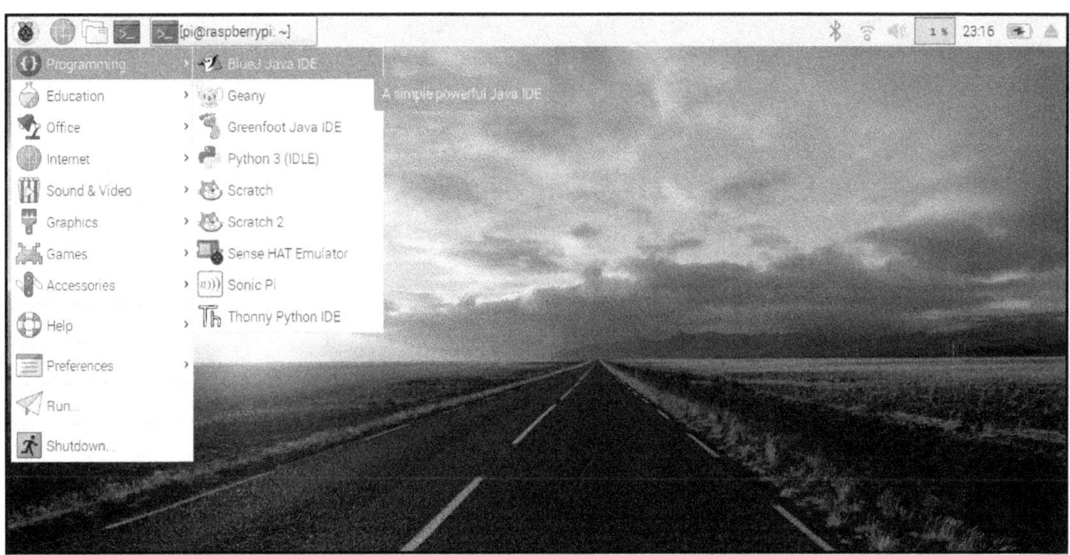

Raspbian booting

3. Installing ROS is pretty straightforward. Follow the same steps mentioned on this web page: `http://wiki.ros.org/melodic/Installation/Debian`.

Installing Ubuntu and ROS

Follow these steps to set up ROS on Ubuntu MATE:

1. To install the OS, follow these steps:
    1. Download the Ubuntu MATE bionic arm64-based version image file from this link: `https://ubuntu-mate.org/raspberry-pi/ubuntu-mate-18.04.2-beta1-desktop-arm64+raspi3-ext4.img.xz`.
    2. Once the image is downloaded, insert the SD card into your system and open Etcher.
    3. Select the downloaded image and the specific SD card location and click **Flash!**.
2. Loading the OS requires that the SD card is loaded. Power the SBC using an appropriate power supply adapter. If everything is successful, you should see the following output (the text and numbers in this image are intentionally illegible):

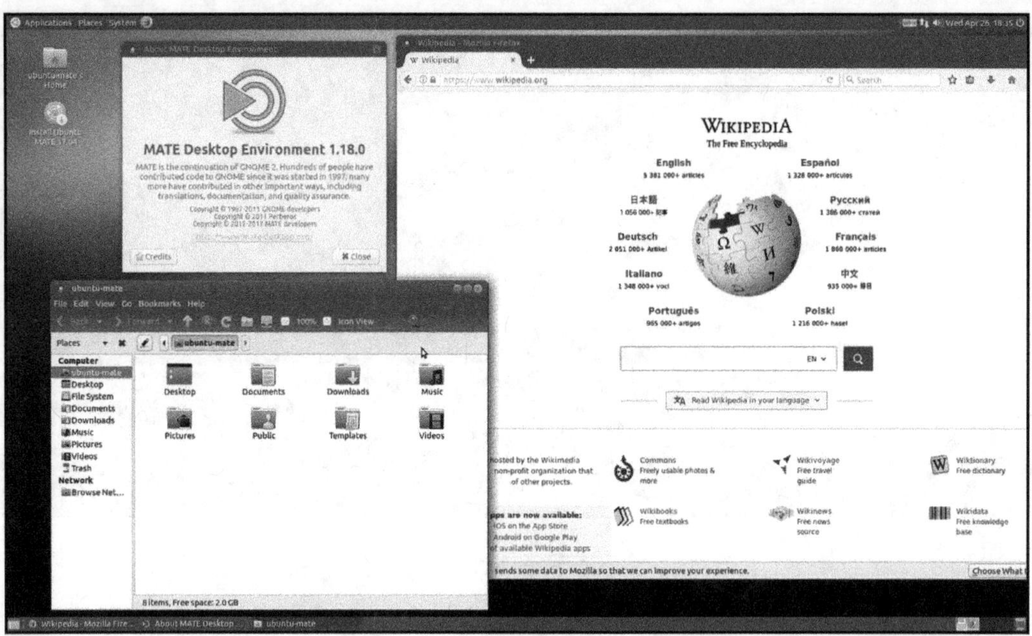

Ubuntu MATE booting

3. Installing ROS is pretty straightforward, as we saw in the *Installing Armbian and ROS* section. Follow the same steps mentioned in that section.

Now, let's learn how to set up ROS on Jetson Nano.

Setting up ROS on Jetson Nano

Let's learn how to set up ROS on a GPU-based SBC—Jetson Nano. This is going to be straightforward since we're going to set up Nano with Ubuntu. Follow these steps:

1. To install the OS, follow these steps:
    1. Download the OS image file from this link: `https://developer.nvidia.com/jetson-nano-sd-card-image-r322`.
    2. Once the image is downloaded, insert the SD card into your system and open Etcher.
    3. Select the downloaded image and the specific SD card location and click **Flash!**.
2. To load the OS, make sure that the SD card is loaded. Power the SBC using an appropriate power supply adapter. If everything is successful, you should see the following output:

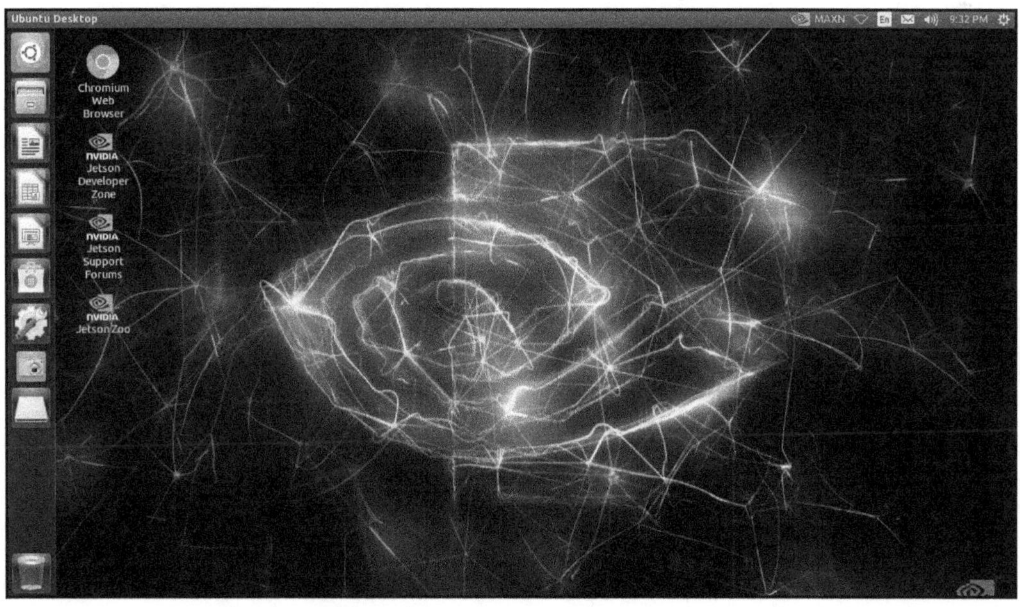

Jetson Nano screen

3. Installing ROS is pretty straightforward, as we saw in the *Installing ROS on the Tinkerbord Armbian OS* section. Follow the same steps mentioned in that section.

Let's now learn to control GPIOS from ROS.

Controlling GPIOS from ROS

Now that you have seen how different OSes are set up on the SBCs along with ROS, let's see how to take control of the GPIOs of the boards individually so that you can control input-output peripherals directly with them. Also, we shall write them as ROS nodes so that we can integrate them with other ROS applications. So, let's begin by looking at each board individually, starting with Tinkerboard S.

Tinkerboard S

Tinkerboard comes with GPIO API support in shell, Python, and C programming. Since we've working with ROS, let's learn how to control the GPIOs using Python. For more information on the GPIO pinout, please refer to the following link: `https://tinkerboarding.co.uk/wiki/index.php/GPIO#Python`.

GPIOs in Tinkerboard can be controlled via the Python GPIO library at `http://github.com/TinkerBoard/gpio_lib_python`. To install this library, use the following commands:

```
$ sudo apt-get install python-dev
$ git clone http://github.com/TinkerBoard/gpio_lib_python --depth 1
GPIO_API_for_Python
$ cd GPIO_API_for_Python/
$ sudo python setup.py install
```

Now that the Python library is installed, let's learn how to control an LED blink from a ROS topic. You can find the whole code here: `https://github.com/PacktPublishing/ROS-Robotics-Projects-SecondEdition/blob/master/chapter_7_ws/tinkerboard_gpio.py`. Check out the important details of the following code:

```
...

 import ASUS.GPIO as GPIO

 GPIO.setwarnings(False)
 GPIO.setmode(GPIO.ASUS)
 GPIO.setup(LED,GPIO.OUT)

...

 def ledBlink(data):
  if data.data = true:
     GPIO.output(LED, GPIO.HIGH)
     time.sleep(0.5)
     GPIO.output(LED, GPIO.LOW)
```

```
        time.sleep(0.5)
    else:
        GPIO.cleanup()

...
```

In the preceding code, we make use of the `ASUS.GPIO` library we had installed. After the necessary import statements, we set the mode to `GPIO.ASUS`. We then define the GPIO that we would use as OUT, meaning output (as it is LED in our case, whereas it would be IN if it was a switch). Then, we have the function to make the LED blink, in case the `/led_status` topic receives a Boolean `true` or `false`.

BeagleBone Black

For BeagleBone Black, there is a Python library from Adafruit that is similar to the one we saw with Tinkerboard's library.

For pinout details, please refer to `https://learn.adafruit.com/setting-up-io-python-library-on-beaglebone-black/pin-details`.

In Ubuntu, you can install the library using `pip` by following these steps:

1. Update and install dependencies:

```
$ sudo apt-get update
$ sudo apt-get install build-essential python-dev python-setuptools
python-pip python-smbus -y
```

2. Install the library using the `pip` command:

```
$ sudo pip install Adafruit_BBIO
```

In Debian, you can clone the repository and install it manually by following these steps:

1. Update and install the dependencies:

```
$ sudo apt-get update
$ sudo apt-get install build-essential python-dev python-setuptools
python-pip python-smbus -y
```

2. Clone the workspace and install the library manually:

```
$ git clone git://github.com/adafruit/adafruit-beaglebone-io-python.git
$ cd adafruit-beaglebone-io-python
$ sudo python setup.py install
```

Now that the Python library is installed, let's see how to control an LED blink from a ROS topic. You can find the whole code here: `https://github.com/PacktPublishing/ROS-Robotics-Projects-SecondEdition/blob/master/chapter_7_ws/beagleboneblack_gpio.py`. Check out the important details of the code:

```
...

import Adafruit_BBIO.GPIO as GPIO

LED  = "P8_10"
GPIO.setup(LED,GPIO.OUT)

def ledBlink(data):
 if data.data = true:
    GPIO.output(LED, GPIO.HIGH)
    time.sleep(0.5)
    GPIO.output(LED, GPIO.LOW)
    time.sleep(0.5)
 else:
    GPIO.cleanup()

...
```

In the preceding code, we make use of the `Adafruit_BBIO.GPIO` library we installed. After the necessary import statements, we define the GPIO that we would use as OUT, meaning output (as it is LED in our case; it would be IN if it was a switch). Then, we have the function to make the LED blink in case the `/led_status` topic receives a Boolean that's `true` or `false`.

Raspberry Pi 3/4

Controlling GPIOs is straightforward in Raspberry Pi 3/4. There are libraries such as GPIO zero (`https://gpiozero.readthedocs.io/`), pigpio (`http://abyz.me.uk/rpi/pigpio/`), and WiringPi (`http://wiringpi.com/`) that support GPIO access. We shall install GPIO zero for our example interaction with ROS. The pinout of Raspberry Pi GPIO is shown here:

Raspberry Pi2 GPIO Header

	Pin#	NAME	NAME	Pin#
Early Models	01	3.3v DC Power	DC Power 5v	02
	03	GPIO02 (SDA1 , I²C)	DC Power 5v	04
	05	GPIO03 (SCL1 , I²C)	Ground	06
	07	GPIO04 (GPIO_GCLK)	(TXD0) GPIO14	08
	09	Ground	(RXD0) GPIO15	10
	11	GPIO17 (GPIO_GEN0)	(GPIO_GEN1) GPIO18	12
	13	GPIO27 (GPIO_GEN2)	Ground	14
	15	GPIO22 (GPIO_GEN3)	(GPIO_GEN4) GPIO23	16
	17	3.3v DC Power	(GPIO_GEN5) GPIO24	18
	19	GPIO10 (SPI_MOSI)	Ground	20
	21	GPIO09 (SPI_MISO)	(GPIO_GEN6) GPIO25	22
	23	GPIO11 (SPI_CLK)	(SPI_CE0_N) GPIO08	24
	25	Ground	(SPI_CE1_N) GPIO07	26
Late Models	27	ID_SD (I²C ID EEPROM)	(I²C ID EEPROM) ID_SC	28
	29	GPIO05	Ground	30
	31	GPIO06	GPIO12	32
	33	GPIO13	Ground	34
	35	GPIO19	GPIO16	36
	37	GPIO26	GPIO20	38
	39	Ground	GPIO21	40

Rev. 1
26/01/2014

Raspberry Pi GPIO pinout (source: https://learn.sparkfun.com/tutorials/raspberry-gpio/gpio-pinout. Licensed under Creative Commons CC-BY-SA 4.0: https://creativecommons.org/licenses/by-sa/4.0/)

For more information, please refer to `https://learn.sparkfun.com/tutorials/raspberry-gpio/gpio-pinout`.

To install the GPIO library, use the following commands:

```
$ sudo apt update
$ sudo apt install python-gpiozero
```

In case you have Python 3 installed, use the following command:

```
$ sudo apt install python3-gpiozero
```

If you're using any other OS, you can install via `pip`:

```
$ sudo pip install gpiozero #for python 2
```

Or, you could use the following:

```
$ sudo pip3 install gpiozero #for python 3
```

Now that the Python library is installed, let's see how to control an LED blink from a ROS topic. You could find the whole code here: `https://github.com/PacktPublishing/ROS-Robotics-Projects-SecondEdition/blob/master/chapter_7_ws/raspberrypi_gpio.py`. Check out the important details of the code here:

```
...
 from gpiozero import LED

 LED  = LED(17)
 GPIO.setup(LED,GPIO.OUT)

 def ledBlink(data):
  if data.data = true:
     LED.blink()
  else:
     print("Waiting for blink command...")
...
```

In the preceding code, we make use of the `gpiozero` library we installed. After the necessary import statements, we define the GPIO that we would use as OUT, meaning output (as it is LED in our case, but would be IN if it was a switch). Then, we have the function to make the LED blink, in case the `/led_status` topic receives a Boolean that's `true` or `false`.

Jetson Nano

For controlling GPIO pins in Jetson Nano, we can make use of the Jetson GPIO Python library (`https://github.com/NVIDIA/jetson-gpio`). You can install the library using a simple `pip` command shown here:

```
$ sudo pip install Jetson.GPIO
```

Now that the Python library is installed, let's learn how to control an LED blink from a ROS topic. You can find the whole code here: `https://github.com/PacktPublishing/ROS-Robotics-Projects-SecondEdition/blob/master/chapter_7_ws/jetsonnano_gpio.py`. Check out the important details of the code here:

```
...

 import RPi.GPIO as GPIO

 LED  = 12
 GPIO.setup(LED,GPIO.OUT)

 def ledBlink(data):
  if data.data = true:
     GPIO.output(LED, GPIO.HIGH)
     time.sleep(0.5)
     GPIO.output(LED, GPIO.LOW)
     time.sleep(0.5)
  else:
     GPIO.cleanup()

...
```

In the preceding code, we make use of the `RPi.GPIO` library we installed. After the necessary import statements, we define the GPIO that we would use as OUT, which means output (since it is LED in our case, but would be IN if it was a switch). Then, we have the function to make the LED blink in case the `/led_status` topic receives a Boolean that's `true` or `false`. In the next section, we will cover benchmarking embedded boards.

Benchmarking embedded boards

Now that you have seen some embedded boards, have you ever wondered what board to choose for your robotics project? Which board would have the perfect amount of power and capacity to run your application? All these questions can be answered using a simple test called a **benchmark**. Benchmarking or a benchmark is a test that is used to examine and judge the capabilities of such embedded boards. They're a group of tests that screen specific capabilities such as CPU clock speed, graphics, and memory.

Note that benchmarks do not give accurate results but give precise results that help us to understand specific aspects of SBC, as we mentioned earlier. There are a bunch of such benchmarks available online. You can see a few tests related to our projects here. We shall make use of the benchmark suites from PTS (`http://www.phoronix-test-suite.com/`). The results are automatically published to an online platform called open benchmarking (`openbenchmarking.org`), where users can store their results publicly or privately.

Let's look at what they're used for.

The machine learning test suite (`https://openbenchmarking.org/suite/pts/machine-learning`) is mainly used for popular pattern recognition and computational learning algorithms and is a suite that consists of a group of tests such as the following:

- Caffe (`https://openbenchmarking.org/test/pts/caffe`) is a benchmark of the Caffe deep learning framework that currently supports the AlexNet and GoogleNet model.
- Shoc (`https://openbenchmarking.org/test/pts/shoc`) is the CUDA and OpenCL version of Vetter's Scalable HeterOgeneous computing benchmark suite.
- R benchmark (`https://openbenchmarking.org/test/pts/rbenchmark`) is a fast-running survey of general R performance.
- NumPy (`https://openbenchmarking.org/test/pts/numpy`) is used to obtain general NumPy performance.
- scikit-learn (`https://openbenchmarking.org/test/pts/scikit-learn`) is used to obtain scikit-learn performance.
- PlaidML (`https://openbenchmarking.org/test/pts/plaidml`) uses the PlaidML deep learning framework for benchmarking.
- Leela chess zero (`https://openbenchmarking.org/test/pts/lczero`) is a chess engine automated via neural networks. It is specifically used for OpenCL, CUDA + cuDNN, and BLAS (CPU-based) benchmarking.

Other individual test cases are as follows:

- SciMark (`https://openbenchmarking.org/test/pts/scimark2-1.3.2`) is a test that is used for scientific and numerical computing. The test is made up of fast Fourier transform, Jacobi successive over-relaxation, Monte Carlo, sparse matrix multiply, and dense LU matrix factorization benchmarks.
- PyBench (`https://openbenchmarking.org/test/pts/pybench-1.1.3`) is a test that reports average test times for different functions such as `BuiltinFunctionCalls` and `NestedForLoops`. The test runs PyBench each time for 20 rounds.
- OpenSSL (`https://openbenchmarking.org/test/pts/openssl-1.9.0`) is an open source toolkit that implements **Secure Sockets Layer** (**SSL**) and **Transport Layer Security** (**TLS**) protocols. The test measures RSA 4096-bit performance.
- Himeno Benchmark (`https://openbenchmarking.org/test/pts/himeno-1.1.0`) is a linear solver of pressure Poisson using a point-Jacobi method.

You can have a look at a variety of other such tests that would suit your requirements at `https://openbenchmarking.org/`. These tests are simple to run, and you can install them using the following steps:

1. Update the Debian system and install the necessary dependencies using the following commands:

```
$ sudo apt-get update
$ sudo apt-get install php5-cli
```

2. Download and install PTS:

```
$ git clone
https://github.com/phoronix-test-suite/phoronix-test-suite/
$ cd phoronix-test-suite
$ git checkout v5.8.1
$ sudo ./install-sh
```

3. To run any of the tests, install the test using the following command. To run the benchmark, use the `benchmark` keyword instead of `install`:

```
phoronix-test-suite install pts/<TEST>
phoronix-test-suite benchmark pts/<TEST>
```

4. To run the PyBench test, use the following commands:

```
phoronix-test-suite install pts/pybench
phoronix-test-suite benchmark pts/pybench
```

Unless it is a graphics test, the GUI (for example, X server: `http://askubuntu.com/questions/66058/how-to-shut-down-x`) should be shut down. Press *Ctrl* + *Alt* + *F1* to get to a different console and enter `$ sudo service lightdm stop`.

In the next section, we will cover Alexa and connecting it to ROS.

Getting started with Alexa and connecting with ROS

Before I tell you about Alexa, go to `https://echosim.io/welcome`, log in with your Amazon account if you have one, or create an account. Once you're done setting up the account, ask the following questions and find out the following about the replies you get:

- What is Alexa?
- How is it useful?

Fascinating? I'm sure it might be for those who are new to Alexa but, on the other hand, it might seem pretty common as you would've seen something similar with other voice services such as `OK Google` or `Hey Siri`.

Alexa is an intelligent cloud-based voice service from Amazon (`https://www.amazon.com/`). It is available as a voice service API as well with Amazon certified devices such as those found `https://www.amazon.com/Amazon-Echo-And-Alexa-Devices/b?ie=UTF8node=9818047011`, which help people interact with the technology they use every day. We can build simple voice-based experiences (called skills) with the tools available from Amazon here: `https://developer.amazon.com/alexa`.

To interact with Alexa, let's look at a simple way of creating an Alexa skillset (`https://developer.amazon.com/public/solutions/alexa/alexa-skills-kit`).

Alexa skill-building requirements

Let's follow these steps for skill-building:

1. We shall make use of a Python micro-framework called Flask-Ask (`https://flask-ask.readthedocs.io/en/latest/`), which simplifies our lives when it comes to developing Alexa skills.

It is a Flask extension that helps build Alexa skills quickly and with ease. Having prior knowledge of Flask (`https://www.fullstackpython.com/flask.html`) is not mandatory. You can install `flask-ask` using the `pip` command:

```
$ pip install flask-ask
```

2. Create a developer account with Alexa (`https://developer.amazon.com/`). It is free to create and try out Alexa skills. Alexa skills should run either behind a public HTTPS server or a Lambda function (`https://aws.amazon.com/lambda/`).
3. We shall make use of an open source command-line program called ngrok (`http://ngrok.com/`) for this purpose as it helps in opening a secure tunnel to the localhost and exposes it behind the HTTPS endpoint.
4. You need to create an account and follow the setup and installation steps in the getting started page (`https://dashboard.ngrok.com/get-started`), as shown here:

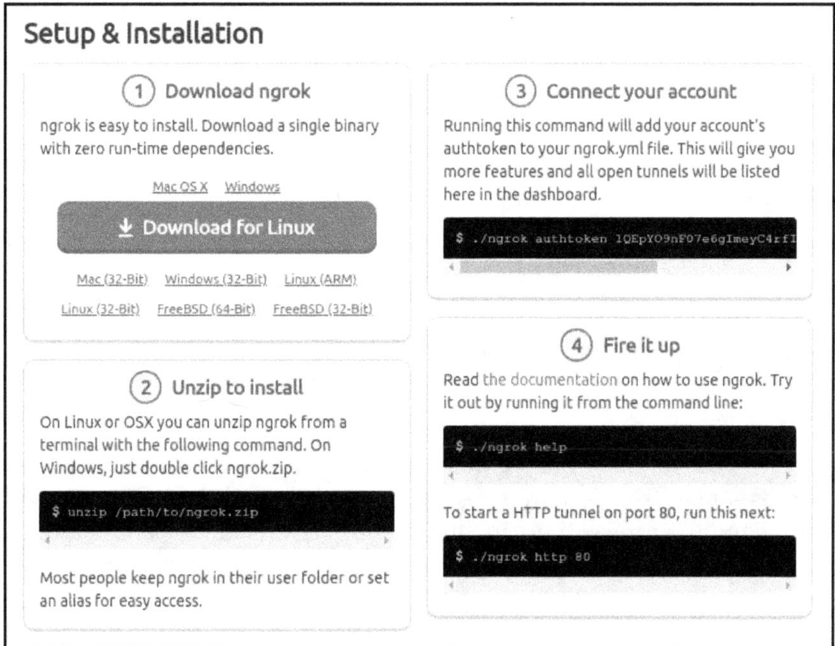

Setting and installing ngrok

Now, let's learn how to create an Alexa skill.

Creating a skill

Now that you have completed the preceding steps, let's follow these steps to create an Alexa skill.

Let's create a file called alexa_ros_test.py and copy the code into our repository at https://github.com/PacktPublishing/ROS-Robotics-Projects-SecondEdition/blob/master/chapter_7_ws/alexa_ros_test.py into it. Let's look at the important details of the code here:

```
...

threading.Thread(target=lambda: rospy.init_node('test_code',
disable_signals=True)).start()
pub = rospy.Publisher('test_pub', String, queue_size=1)

app = Flask(__name__)
ask = Ask(app, "/")
logging.getLogger("flask_ask").setLevel(logging.DEBUG)

@ask.launch
def launch():
    welcome_msg = render_template('welcome')
    return question(welcome_msg)

@ask.intent('HelloIntent')
def hello(firstname):
    text = render_template('hello', firstname=firstname)
    pub.publish(firstname)
    return statement(text).simple_card('Hello', text)

...
```

In the preceding code walkthrough, we made use of render_template from Flask to import custom responses from a stored YAML file. We started the ROS node as a parallel thread and initialized the topics, the app, and the loggers. Then, we created decorators to run a launch (def launch()) and response message (def hello()). After that, we started the app.

Let's create another file named template.yaml and copy the following contents into it:

```
welcome: Welcome, This is a simple test code to evaluate one of my
abilities. How do I call you?
hello: Hello, {{ firstname }}
```

Now, run `roscore` in one Terminal and the Python program in another Terminal using the following command:

```
$ python code.py
```

While running the Python code in another Terminal, ensure you have `initros1` invoked so that you don't see any ROS-related errors.

In a new Terminal, open ngrok to generate a public HTTPS endpoint to the localhost at port `5000`:

```
$ ./ngrok http 5000
```

You should see the following screen:

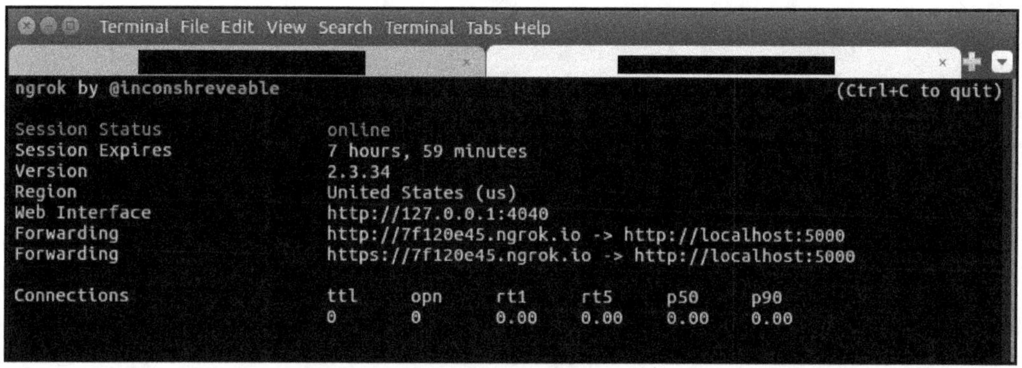

ngrok status message

Note down the HTTPS endpoint link at the end (in this case, it is `https://7f120e45.ngrok.io`). Once done, go to the Alexa developer account page (`https://developer.amazon.com/alexa/console/ask?`) and execute the following steps:

Please note that every time you close the ngrok Terminal, a new endpoint is created.

1. In the Alexa developer account page, select **Create Skill**. Enter the **Skill Name** as `Alexa ros` and select a custom model and **Provision your own** as the method to host the skill's backend resources.
2. Choose **Start from scratch**. You should see the following screen (the text and numbers in this image are intentionally illegible):

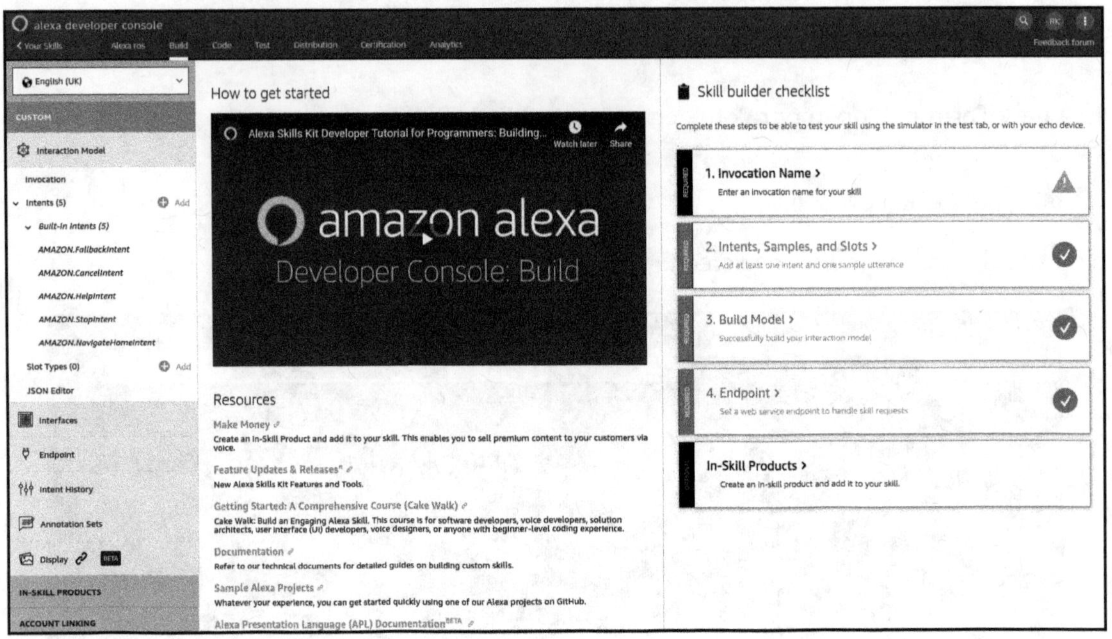

Alexa developer account page

3. Select **Invocation Name** on the right-hand side and set the name as `test code`.

Once done, on the left-hand side, select **Intents** to add a new intent. Create a custom intent with the name `HelloIntent` and add the utterances as shown in the following screenshot. Also, ensure that the slot type is `AMAZON.firstname`, as shown here:

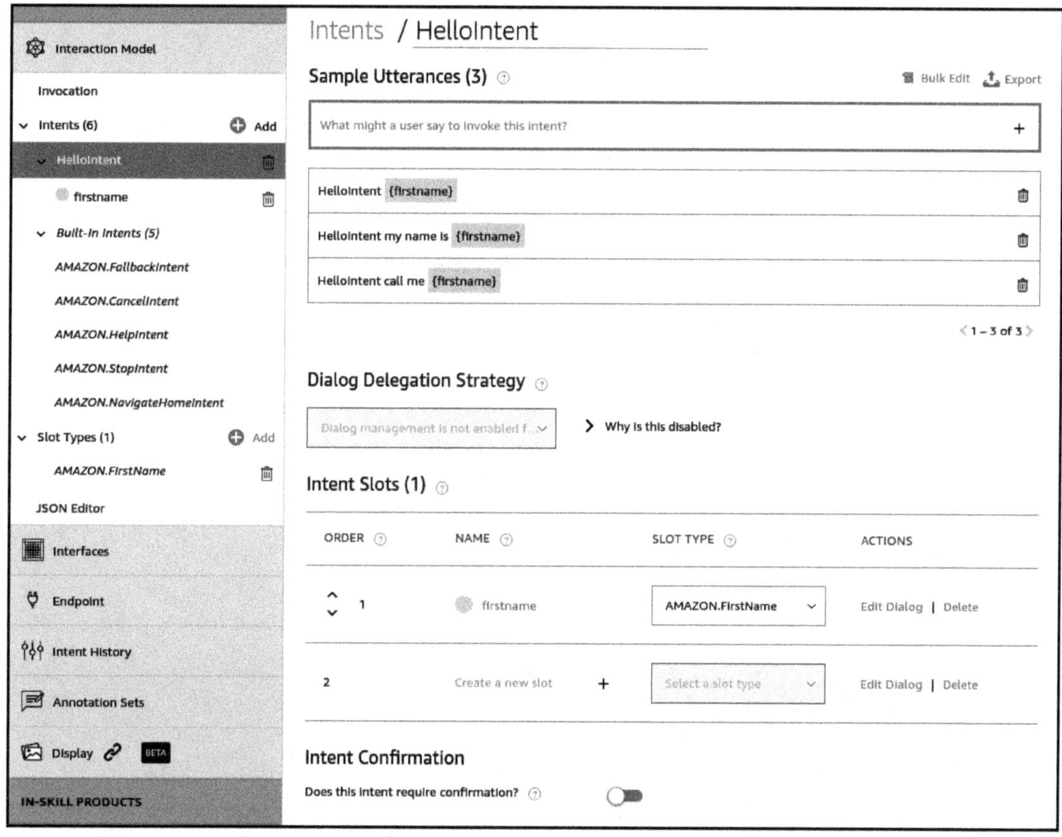

Utterances reference

4. You're done. Now, save the model and build it using the **Build Model** button. Once you're done, you should see the following screen:

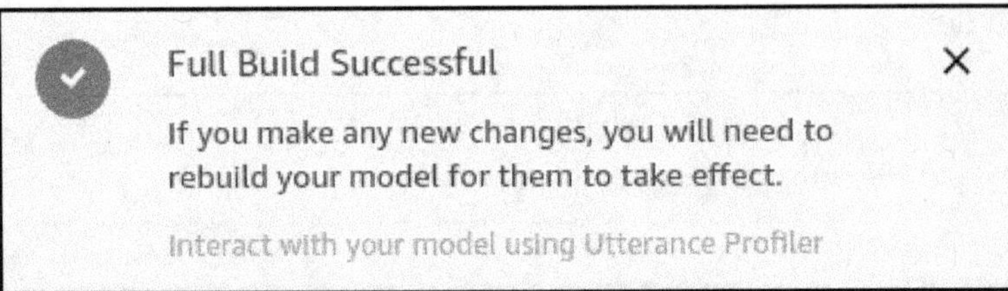

Full Build Successful
If you make any new changes, you will need to rebuild your model for them to take effect.
Interact with your model using Utterance Profiler

Build successful

5. Once the build is successful, go to **Endpoint** on the left-hand side and select the **HTTPS** radio button. Enter the `https` endpoint link you noted in the ngrok Terminal and paste that into the `Default Region` textbox.
6. Select the **My development endpoint is a sub-domain of a domain that has a wild card certificate...** option. Once done, save the endpoints using the **Save Endpoints** button.
7. You're all set now! Assuming you've started `roscore`, the Python code—`alexa_ros_test.py`—and ngrok can begin the test on the developer page if we browse to the **Test** tab.
8. In the **Test** tab, select **Development** from the drop-down list box. Now, you're ready to use the Alexa simulator. You can start by typing the following command:

```
Alexa, launch test code
```

You should hear Alexa reply with *Welcome, this is a simple test code to evaluate one of my abilities. What do I call you?*, since we have provided this as the template in the YAML file.

Type: `Alexa, Call me ROS.`

You should hear Alexa say *Hello ROS*.

You should also be able to see the data being published as `rostopic`, as shown here (the text and numbers in this image are intentionally illegible):

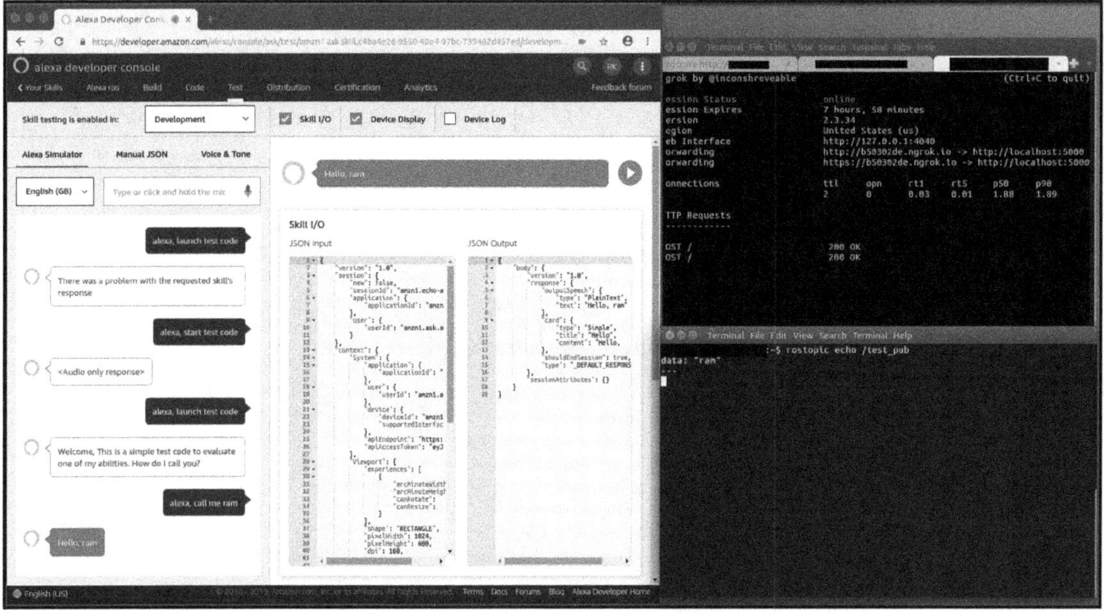

Overall testing

That's it! You have now created an Alexa skill successfully and learned how to publish simple data with ROS. You could make use of the GPIO API usage that we saw for the embedded boards and connect Alexa with them to take control of the devices that are connected to them.

Summary

In this chapter, we saw the differences between a microcontroller, a microprocessor, and a single-board computer. After understanding their differences, we looked at the boards that are commonly used by the robotics community. Also, we covered how to use such boards with ROS and take control of their GPIOs to control external peripherals such as motors or sensors. We looked at different types of GPU boards that have a good advantage over normal SBCs. To select a suitable embedded board for any application, it's suggested that you try out benchmarking, where we can choose boards based on test results. We also learned how to create a skill with Alexa and connect it with ROS. This chapter helped us to install ROS and exploit its hardware capabilities for robot applications.

In the next chapter, we will learn how to provide intelligence to our robot through a well-known machine learning technique called reinforcement learning.

8
Reinforcement Learning and Robotics

So far, we have been looking at how a robot or a group of robots can handle a certain application. We began by creating robots, defining links, and programming robots to handle tasks in an appropriate manner for a certain application. We also learned how to handle multiple such robots for that application. What if we gave our robots intelligence like a human being so that they can sense, think, and act by understanding the actions they carry out in an environment? This chapter deals with one of the most important topics of machine learning called **reinforcement learning**, which might pave the way for artificial intelligence-based robot solutions.

In this chapter, you will be introduced to reinforcement learning, its usage with robotics, the algorithms used (such as Q-learning and SARSA), ROS-based reinforcement learning packages, and examples with robots (such as TurtleBot) in simulation.

In this chapter, we will cover the following topics:

- Machine learning
- Understanding reinforcement learning
- **Markov Decision Process** (**MDP**) and Bellman equation
- Reinforcement learning algorithms
- Reinforcement learning in ROS

Technical requirements

Let's look into the technical requirements for this chapter:

- ROS Melodic Morenia on Ubuntu 18.04, with Gazebo 9 (preinstalled), and ROS-2 Dashing Diademata
- ROS packages: OpenAI Gym, `gym-gazebo`, and `gym-gazebo2`
- Timelines and test platform:
    - **Estimated learning time**: On average, 120 minutes
    - **Project build time (inclusive of compile and run time)**: On average, 60-90 minutes
    - **Project test platform**: HP Pavilion laptop (Intel® Core™ i7-4510U CPU @ 2.00 GHz × 4 with 8 GB Memory and 64-bit OS, GNOME-3.28.2)

The code for this chapter is available at `https://github.com/PacktPublishing/ROS-Robotics-Projects-SecondEdition/tree/master/chapter_8_ws/taxi_problem`.

Let's begin this chapter with a simple introduction to machine learning.

Introduction to machine learning

Before we dive into the topic, let's understand what machine learning is. Machine learning is a subset of **Artificial Intelligence** (**AI**) and, as the name suggests, is the ability of a machine to learn by itself, provided it has a preexisting dataset or has processed information in the past. The very next question you may be thinking of is, how does a machine learn or how can we train a machine to learn? The machine learning paradigm constitutes three different types of learning, as follows:

- Supervised learning
- Unsupervised learning
- Reinforcement learning

Let's cover them in detail, starting with supervised learning.

Supervised learning

In supervised learning, a user provides the machine with the desired input-output pairs as training data, that is, the data fed is into the machine as training data and is labeled. Whenever the machine encounters inputs similar to what it was trained with, it can define or classify the output according to the labeling it learned from the training data. Based on the labeled dataset, the machine would be able to find patterns in a new set of data.

One of the best examples of supervised learning is **classification**, where we can label a series of datasets as certain objects. Based on the labeled features and accuracy of the data being fed to the machine (training data), the machine would be able to classify newer datasets as those objects. Another example of supervised learning is **regression**, where, based on a training dataset and correlation between some parameters, the system can predict the output. Weather forecasting or detecting medical anomalies in human beings are scenarios where supervised learning is used.

Unsupervised learning

Unsupervised learning is one of the most intriguing methods of learning that is widely being researched. Using unsupervised learning, the machine would be able to group or cluster a given dataset, but without any labels. This is helpful, especially when the machine is handling large volumes of data. The machine won't know what the data is (as there are no labels) but would be able to cluster the data based on similar or dissimilar features.

Unsupervised learning is more useful in real life as it requires less or no human intervention compared to supervised learning. Unsupervised learning can sometimes act as an initial step for supervised learning problems. Unsupervised learning is useful in applications such as recommendation systems on the web, which provide recommendations based on user interest or analysis in a retail scenario where retail giants can forecast a product's need based on sales numbers.

Reinforcement learning

Usually, people often confuse reinforcement learning with supervised learning. Reinforcement learning is a learning technique where the machine learns on the go and doesn't need training data, unlike supervised learning. The machine tries to understand the application by interacting with the application directly and stores a history of information, along with a reward system that indicates whether the machine did the right thing or not.

Reinforcement learning can be found in games, where the computer tries to understand the user's action and reacts accordingly to ensure that it wins. Reinforcement learning is also being tested out in self-driving cars and robotics. As this chapter specifically deals with robotics, we shall look into this learning technique more deeply. Reinforcement learning can become supervised learning if the machine has understood enough from the environment it is interacting with. You shall understand this more clearly in the upcoming sections when we talk about *explore versus exploit*.

Now, let's learn more about reinforcement learning, which will be the main concept of this chapter.

Understanding reinforcement learning

As we have already seen, reinforcement learning is an on-the-go learning technique. Let's consider a simple analogy to understand reinforcement learning. Think about a nine-month-old baby trying to get up and walk.

The following diagram represents our analogy:

Baby walking analogy

The first step the baby does is try to get up by pressing their legs toward the ground. Then, they try to balance themselves and try to hold still. If this is successful, you would see a smile on the baby's face. Now, the baby takes one step forward and tries to balance itself again. If, while trying, the baby lost balance and fell down, then there is a chance that the baby might frown or cry. The baby may either give up walking if it doesn't have the motivation to walk or may try once again to get up and walk. If the baby was successful taking two steps forward, you might see a bright smile, along with a happy sound from the baby. The more steps the baby takes, the more confident they will become and they will eventually continue walking or, at times, even running:

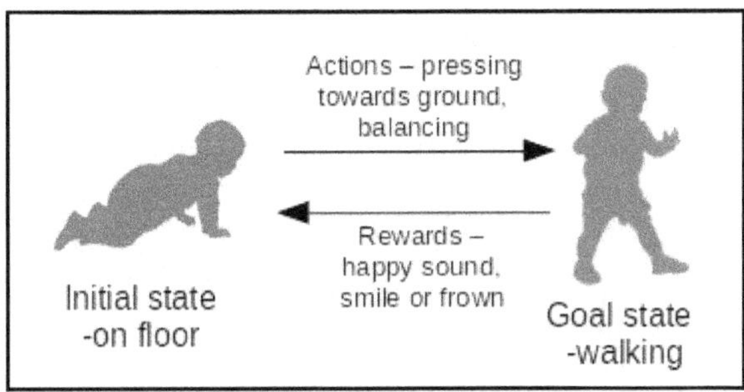

Transition representation

To start with, the baby is called an agent. The two states that the agent assumes are represented in the preceding diagram. The goal of the agent is to get up and walk in an environment such as a living room or bedroom. The steps that the agent takes or does such as getting up, balancing, and walking are called actions. The agent's smiles or frowns are treated as rewards (positive and negative rewards) and the happy sound that came from the baby while walking is a reward that probably indicates the agent is doing the right thing. The motivation that the agent receives to get up and walk once again or to stay put on the ground explains the explore-exploit concept in reinforcement learning. What do we mean by this? Let's look at this in detail now.

Explore versus exploit

Let's imagine that the baby is actually walking to find their way toward a toy and has ended up in the bedroom. There are a couple of toys in that room the baby sees, which might make them happy. What if the baby tried to walk outside the bedroom and moved toward the storeroom where there is a basket full of toys? The baby would be much happier than with the few toys in the bedroom.

The former scenario, where the baby walked into the bedroom in a random manner and settled having found the toys, is called exploitation. Here, the baby has seen the toys and hence logically feels comfortable. But if the baby tried to experiment by searching outside that room, they may have ended up in the storeroom, where there are more toys than the bedroom, and this would've made the baby much happier and satisfied. This is called exploration. Here, as indicated earlier, there are chances that the baby may either find a basket full of toys that would overjoy them (abruptly increase the reward) or may also end up not finding even a single toy, which might upset the baby (reduce the reward).

This is what makes reinforcement learning more interesting. Based on the agent's interaction with the environment, the agent can either exploit and increase its rewards steadily or can explore and increase or reduce the rewards abruptly. Both come with advantages and limitations. If the agent is supposed to train in such a way that it needs less time (or episodes) to increase its rewards, then it is suggested to exploit. If the agent doesn't worry about the time taken to increase its rewards but this is necessary to experiment all around the environment, then it is suggested to explore. There are certain algorithms in reinforcement learning that help to decide whether to explore or exploit. This can be solved with the famous multi-armed bandits problem.

Reinforcement learning formula

Let's look at the important components of a reinforcement learning problem, one by one:

- **Agent**: The agent is the one that interacts with the environment by performing necessary steps or actions.
- **Environment**: The environment is the one in which the agent interacts. Some of the available environments are deterministic, stochastic, fully observable, continuous, discrete, and many more.
- **State**: State is the condition or position in which the agent is currently exhibiting or residing. The state is usually a result of the agent's action.
- **Action**: This is something the agent does while interacting with the environment.
- **Reward**: This is the outcome of an action the agent performs to transition to a state. The reward may be positive or negative.

A reinforcement learning problem can be understood with the following diagram:

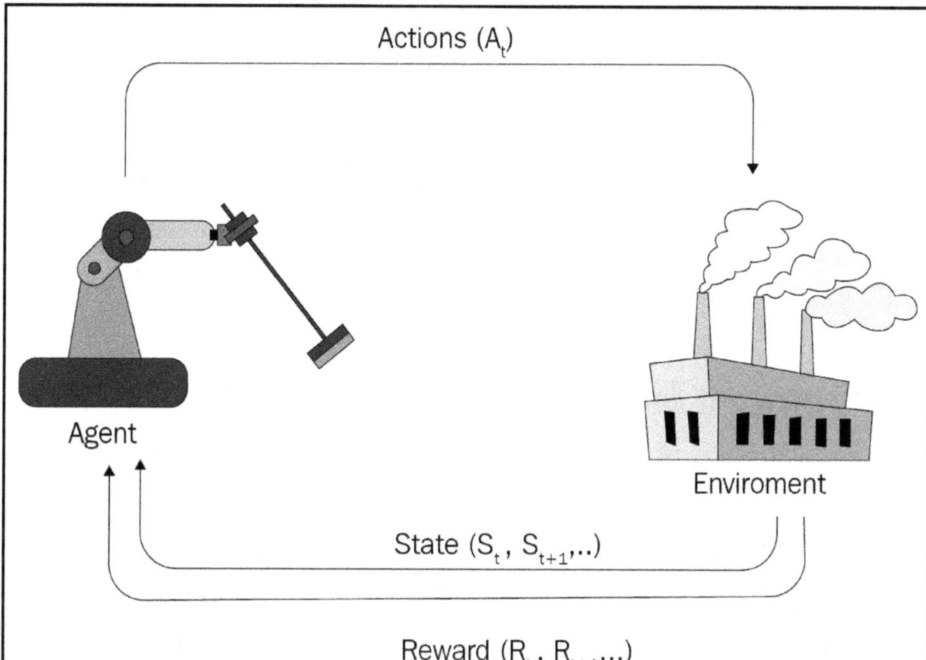

Agent and environment—general diagram of reinforcement learning

As shown in the preceding diagram, the overall reinforcement learning algorithm works as follows:

1. Let's assume that our agent is a mobile manipulator moving in an environment from one place to another.
2. Considering the agent's state as the initial state (S_t), the agent interacts with the environment by performing an action (A_t). This can be visualized as the robot making a decision to move forward.
3. As a result of performing the action, the agent transitions from the initial state to another (S_{t+1}) and receives a reward (R_t). This may mean that the robot has moved from one point to another without any obstruction, and so treats the trajectory path as a valid path.
4. Based on the reward received, the agent understands whether the action that was performed was the right choice or not.

5. If the action that was performed was right, the agent may continue to do the same action (A_t) to increase its reward (R_{t+1}). If the action that was performed was wrong, it keeps the action-state result in its history and continues trying a different action.
6. Now, the cycle continues (back to *step 1*, where the agent once again interacts with the environment by performing an action). Hence, in this way, if the robot's aim was to reach a certain destination, for every action the robot takes, it reads the state and reward and evaluates whether it was a success or not. This way, the robot goes through a series of such cyclic checks and eventually reaches the goal.

We will now look at reinforcement learning platforms.

Reinforcement learning platforms

There are many reinforcement learning platforms available for experimenting with reinforcement learning problems in simulation. We can build and define the environment and the agent and emulate certain behaviors for the agent to see how the agent learns to solve the problem. Some of the available learning platforms are as follows:

- **OpenAI Gym**: This is one of the most commonly used toolkits for experimenting with reinforcement learning algorithms. It is a simple and easy to use tool with prebuilt environments and agents. It is open source and supports frameworks such as TensorFlow, Theano, and Keras. There is support for ROS using the OpenAI Gym package, which we will see in the upcoming chapters. More information can be found here: `https://github.com/openai/gym`.
- **OpenAI Universe**: This is an extension of OpenAI Gym. This tool provides more realistic and complex environments, like those of PC games. More information can be found here: `https://github.com/openai/universe`.
- **DeepMind Lab**: This is another cool tool that provides realistic and science fiction style environments that are customizable. More information can be found here: `https://github.com/deepmind/lab`.
- **ViZDoom**: This is a *Doom*—a PC game-based tool that supports multi-agent support, but limited to only one environment (the *Doom* game environment). More information can be found here: `https://github.com/mwydmuch/ViZDoom`.
- **Project Malmo**: This is a sophisticated experimentation platform built on top of the Minecraft game by Microsoft that supports research in reinforcement learning and AI. More information can be found here: `https://github.com/Microsoft/malmo`.

Now, let's look at reinforcement learning in robotics.

Reinforcement learning in robotics

Suppose we consider industrial robots: the most complex task for a robot arm is to plan kinematics in a constrained space. Tasks can be like picking an object from a basket or bin, assembling asymmetric components one over the other, trying to enter a confined space, and carrying out sensitive tasks such as welding. Wait, what are we talking about? Aren't all of these already in place and the robots are actually doing a great job in these applications? That's totally right, but what happens is the robot is guided by a human worker to move to a certain position in such confined spaces. This is simply because it is, at times, intricate to mathematically model such environments and hence they are solved using robot kinematics. As a result, the robot arm would have a hard time trying to solve inverse kinematics and may have a chance of colliding with the environment. This may cause damage to both the environment and the workpiece.

This is where reinforcement learning can be helpful. The robot can initially take the help of human workers to understand the application and environment and learn its trajectory. Once the robot has learned, it can begin operating without any human intervention.

To look at research about using robot arms to open doors, look at the following link, which talks about research that was conducted in this area by Google: `https://spectrum.ieee.org/automaton/robotics/artificial-intelligence/google-wants-robots-to-acquire-new-skills-by-learning-from-each-other`.

It's not just industrial robot arms: reinforcement learning can be used in different applications, such as in mobile robots to plan their navigation or in aerial robotics to help to stabilize drones with heavy payloads.

The next section will give us an introduction to MDP and the Bellman equation.

MDP and the Bellman equation

In order to solve any reinforcement learning problem, the problem should be defined or modeled as a MDP. A Markov property is termed by the following condition: the future is independent of the past, given the present. This means that the system doesn't depend on any past history of data and the future depends only on the present data. The best example to explain this with is rain prediction. Here, we're considering an analogy and not an actual rain estimation model.

There are various methods in which rain estimation work that may or may not need historical data for estimating "rain measure." We're not going to measure anything here but are instead going to predict whether it is going to rain or not. Hence, considering the MDP equation in terms of this analogy, the equation needs the current state to understand the future and doesn't depend on the past—you can see the current state and predict the next state. If the current state is cloudy, then there is a chance or probability that it may rain in the next state, which is irrespective of the previous state, which may have been sunny or cloudy itself.

A MDP model is nothing but the following tuple, ($S_t, A_t, P^a_{ss'}, R_t, \gamma^k$), where S_t is our state space, A_t is the action list, $P^a_{ss'}$ is the state probability transition, R_t is the reward, and γ^k is the discount factor. Let's look at these new terminologies.

The state probability transition function is given with this equation:

$$P^a_{ss'} = \mathbb{P}[S_{t+1} = s' | S_t = s, A_t = a]$$

The reward for s, given the state and action, is given with this equation:

$$R^a_s = \Sigma[R_{t+1} | S_t = s, A_t = a]$$, or

$$R^a_s = \sum_{k=0}^{\infty} \gamma^k R_{t+k+1}$$

γ^k is introduced because there are chances that the rewards that the agent receives can be never-ending. Hence, it acts as a means to control immediate and future rewards. A γ^k value closer to 0 means that the agent needs to give importance to immediate rewards, and a value closer to 1 means the agent needs to give importance to future rewards. Also, note that the former would require fewer episodes or time compared to the latter.

The aim of reinforcement learning is to find the optimal policy and value function for the agent to maximize the rewards and move closer to the goal:

- **Policy function (π)**: The policy is the way in which the agent decides what action to perform to transition from the current state. The function that maps this transition is called the policy function.
- **Value function (V^π)**: The value function helps the agent determine whether it needs to stay in that particular state or transition to another state. It is usually the total reward received by the agent, starting from the initial state.

The preceding two definitions play the most crucial role in solving a problem using reinforcement learning. They help set the trade-off for the explore or exploit nature of the agent. Some of the different policies that help the agent select the appropriate actions are ϵ-greedy, ϵ-soft, and softmax. Discussion on different policies is out of the scope of this chapter; however, we will be looking at how value functions are solved. Value functions (V^{π}) are represented as follows:

$$V^{\pi} = E_{\pi}[R_t | S_t = s]\text{, or}$$

$$V^{\pi}(s) = E_{\pi}[\sum_{k=0}^{\infty} \gamma^k R_{t+k+1} | S_t = s]$$

From the preceding equation, it is evident that the value function depends on the policy and varies based on the policy we choose. These value functions are usually a collection of rewards or a value denoted in a table with states. The state with the highest value is usually chosen. If the table constitutes actions along with states and values in the table, then the table is called a Q-table and the state-action value function, also called a Q-function, is shown as follows:

$$Q^{\pi}(s,a) = E_{\pi}[\sum_{k=0}^{\infty} \gamma^k R_{t+k+1} | S_t = s, A_t = a]$$

The Bellman equation is used to find the optimal value function. The final equation is given like this:

$$V^{\pi}(s) = \sum_{a} \pi(s,a) \sum_{s'} \mathbb{P}^a_{ss'} [R^a_{ss'} + \gamma V^{\pi}(s')]$$

Here, $\pi(s,a)$ is the policy for the given state and action.

For Q-function, the equation is given here:

$$Q^{\pi}(s,a) = \sum_{s'} \mathbb{P}^a_{ss'} [R^a_{ss'} + \gamma \sum_{a'} Q^{\pi}(s',a')]$$

The preceding equation is also called a Bellman optimal equation.

Now that we know the basics of reinforcement learning, let's look at a few reinforcement learning algorithms.

Reinforcement learning algorithms

MDP models can be solved in many ways. One of the methods to solve MDP is using Monte Carlo prediction, which helps in predicting value functions and control methods for further optimization of those value functions. This method is only for time-bound tasks or episodic tasks. The problem with this method is that if the environment is large or the episodes are long, the time taken to optimize the value functions takes a while as well. However, we won't be discussing Monte Carlo methods in this section. Instead, we shall look at a more interesting model-free learning technique that is actually a combination of Monte Carlo methods and dynamic programming. This technique is called temporal difference learning. This learning can be applied to non-episodic tasks as well and doesn't need any model information to be known in advance. Let's discuss this technique in detail with the help of an example.

Taxi problem analogy

Let's consider the famous taxi problem, which is available in OpenAI Gym as a test environment (`https://gym.openai.com/envs/Taxi-v2/`). The taxi environment is as follows:

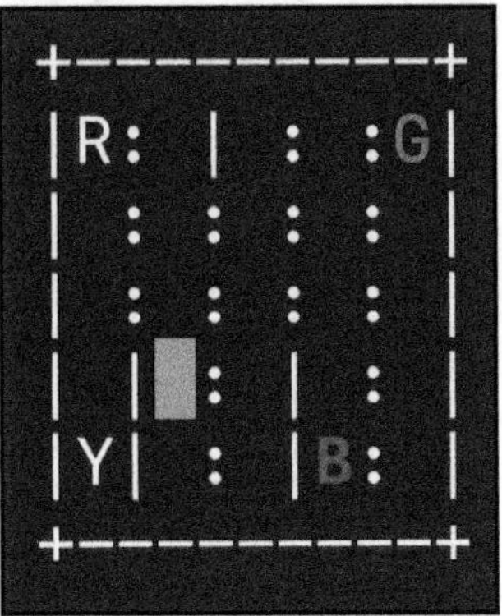

Taxi environment

This is a 5 x 5 grid-based environment where there are certain locations labeled **R**, **G**, **Y**, and **B**. The green cursor is the taxi. The taxi has six possible actions:

- Move south (intuitively, move down)
- Move north (intuitively, move up)
- Move east (intuitively, move right)

- Move west (intuitively, move left)
- Pick up passenger
- Drop off passenger

The pickup and dropoff locations are chosen randomly and are shown in blue and violet, respectively. The aim of the taxi is to go to the pickup location wherever the taxi is located in the grid-space and then to the dropoff location.

In the next section, we will cover the **Temporal Difference** (**TD**) prediction.

TD prediction

Now that we have an example, let's look at the mathematical expression of the TD prediction equation:

$$V(s) \leftarrow V(s) + \alpha(r + \gamma V(s') - V(s))$$

The preceding equation states that the value of the current state, $V(s)$, is given by the sum of the value of the current state, $V(s)$, and the learning rate α times the TD error. Wait, what is the TD error? Let's take a look:

$$r + \gamma V(s') - V(s)$$

It is the difference between the current state value and the predicted state reward, where the predicted state reward is represented as the sum of the reward of the predicted state and discount times' value of the predicted state. The predicted state reward is also known as the TD target. The TD algorithm for evaluating the value function is given as follows:

```
Input: the policy π to be evaluated

Initialize V(s) arbitrarily
 Repeat (for each episode):
  Initialize S
  Repeat (for each step of episode):
    A ← action given by π for S
    Take action A, observe R, S'
    V (S) ← V (S) + α[R + γV (S) - V (S')]
    S ← S'
  until S is Terminal
```

Algorithm explanation

The algorithm begins with initializing an empty value (mostly, `0`). Then, for every episode, a state (`S`) is initialized, followed by a loop for every time step, where an action (`A`) is chosen based on policy gradient techniques. Then, the updated state (`S'`) and reward (`R`) are observed and we begin solving the value function using the preceding mathematical equation (TD equation). Finally, the updated state (`S'`) is equated to the current state (`S`) and the loop continues until the current state (`S`) is terminated.

Now, let's try to implement this in our analogy. Let's initialize all the values of our states as zeros (as shown in the following table):

State	Value
(1, 1)	0
(1, 2)	0
(1, 3)	0
...	...
(4, 4)	0

Table with zero values

Let's assume that the taxi is in the position (2, 2). The rewards for each action are given as follows:

```
0- Move south : Reward= -1
1- Move north : Reward= -1
2- Move east : Reward= -1
3- Move west : Reward= -1
4- Pickup passenger : Reward= -10
5- Dropoff passenger : Reward= -10
```

The learning rate is usually within 0 and 1 and cannot be equal to 0, and we have already seen that the discount factor is within 0 and 1 as well, but can be zero. Let's consider that the learning rate is 0.5 and the discount is 0.8 for this example. Let's assume an action is chosen at random (usually, the e-greedy algorithm is used to define the policy, π), say, north. For the action north, the reward is `-1` and the next state or predicted state (`S'`) is (1, 2). Hence, the value function of (2, 2) is as follows:

```
V(S) ← 0+0.5*[(-1)+0.8*(0)-0]
 => V(S) ← -0.5
```

Now, the updated table is as follows:

State	Value
(1, 1)	0
(1, 2)	0
(1, 3)	0
...	0
(2, 2)	-0.5
...	...
(4, 4)	0

Updated table

Likewise, the algorithm tries to calculate until the episode terminates.

Hence, TD prediction is advantageous compared to Monte Carlo methods and dynamic programming, but there is the possibility that the random policy generation can sometimes make TD prediction take more time in terms of solving the problem. In the preceding example, the taxi may result in many negative rewards and may take time as well to arrive at a perfect solution. There are other ways in which the predicted values can be optimized. In the next section, we will cover TD control, which can achieve this.

TD control

As we mentioned previously, TD control is used to optimize the value function. There are two types of TD control techniques:

- Off-policy learning: Q-learning
- On-policy learning: SARSA

Let's look at them in detail.

Off-policy learning – the Q-learning algorithm

One of the most common and well-known off-policy learning techniques is the Q-learning algorithm. Remember the Q-function we saw earlier? Instead of storing values alone for each state, the Q-learning algorithm stores the action and state pairs with a value. This value is called a Q-value and the table in which it is updated is called a Q-table.

The Q-learning equation is given as follows:

$$Q(s,a) \leftarrow Q(s,a) + \alpha(r + \gamma_{max} Q(s', a') - Q(s,a))$$

The preceding equation is similar to the TD prediction equation we saw earlier. The difference is it has a state-action pair, $Q(s,a)$, instead of a value, $V(s)$. Also, as you may have noticed, there is a *max* function that assumes the maximum state-action pair value and the pair itself. The Q-learning algorithm is given as follows:

```
Initialize Q(s, a), for all s ∈ S, a ∈ A(s), arbitrarily, and Q(Terminal-
state, ·) = 0
 Repeat (for each episode):
   Initialize S
   Repeat (for each step of episode):
     Choose A from S using policy derived from Q (e.g., e-greedy)
     Take action A, observe R, S'
     Q(S, A) ← Q(S, A) + α[R + γ max Q(S', a') - Q(S, A)]
     S ← S'
   until S is Terminal
```

Algorithm explanation

The algorithm begins with initializing initial values (mostly `0`) for the Q-function, `Q(S, A)`. Then, for every episode, a state (`S`) is initialized, followed by a loop for every time step, where an action (`A`) is chosen based on policy gradient techniques. Then, the updated state (`S'`) and reward (`R`) are observed and we begin filling the Q table with values by solving the Q-learning equation. Finally, the updated state (`S'`) is equated as the current state (`S`) and the loop continues until the current state (`S`) is terminated.

Now, let's try to implement this in our analogy. Let's start the table so that all the values are zeros:

State	Value
(0, 0)	[0., 0., 0., 0., 0., 0.],
(0, 1)	[0., 0., 0., 0., 0., 0.],
(0, 2)	[0., 0., 0., 0., 0., 0.],
...,	
(4, 2)	[0., 0., 0., 0., 0., 0.],
(4, 3)	[0., 0., 0., 0., 0., 0.],
(4, 4)	[0., 0., 0., 0., 0., 0.]]

Q-table with initial values

Note that the actions are mapped with respect to each state. There are 500 possible states and six actions have been defined; hence, you can see a matrix of 500 x 6 of all zeros. The action is chosen using the e-greedy technique (which we shall see in code soon). Let's consider that the pickup location is *Y* and that the drop location is *B* in the taxi environment. Here, the learning rate is 0.4 and the discount is 0.9, and the taxi is at (2, 2). This is represented by a numeral in code (say, 241, which we shall see in the code soon). Let's assume that the random action that was chosen was east, that is, 2. Here, the next state would be (2, 3), and our equation would be as follows:

```
Q(2,2;2) ← 0 + 0.4*[(-1)+0.999*max(0,0,0,0,0)-0]
 => Q(2,2;2) ← -0.4
```

Our table becomes the following:

State	Value
(0,0)	[0., 0., 0., 0., 0., 0.],
(0,1)	[0., 0., 0., 0., 0., 0.],
(0,2)	[0., 0., 0., 0., 0., 0.],
...,	
(2,2)	[0., 0., -0.4, 0., 0., 0.]
...,	
(4,2)	[0., 0., 0., 0., 0., 0.],
(4,3)	[0., 0., 0., 0., 0., 0.],
(4,4)	[0., 0., 0., 0., 0., 0.]]

Updated Q-table

Likewise, the table gets updated for certain time intervals. To be more specific about time intervals, each move the taxi takes is considered a step. We allow a maximum of 2,500 steps for our taxi to try out possibilities to reach the goal. These 2,500 steps constitute one episode. Hence, the taxi actually tries 2,500 actions to reach the goal. If successful, then the reward is noted. If the taxi isn't able to reach the goal in this many steps, then the episode simply doesn't take the reward into consideration.

Our final Q-table may look like this:

State	Value
(0,0)	[1. 1. 1. 1. 1. 1.],
(0,1)	[-3.04742139 -3.5606193 -3.30365009 -2.97499059 12.88138434 -5.56063984],
(0,2)	[-2.56061413 -2.2901822 -2.30513928 -2.3851121 14.91018981 -5.56063984],
...,	
(4,2)	[-0.94623903 15.92611592 -1.16857666 -1.25517116 -5.40064 -5.56063984],

(4,3)	[-2.25446383 -2.28775779 -2.35535871 -2.49162523 -5.40064 -5.56063984],
(4,4)	[-0.28086999 0.19936016 -0.34455025 0.1196 -5.40064 -5.56063984]]

Q-Table with values after 3,000 episodes

The preceding explanation is shown in the following code:

```
alpha = 0.4
gamma = 0.999
epsilon = 0.9
episodes = 3000
max_steps = 2500

def qlearning(alpha, gamma, epsilon, episodes, max_steps):
    env = gym.make('Taxi-v2')
    n_states, n_actions = env.observation_space.n, env.action_space.n
    Q = numpy.zeros((n_states, n_actions))
    timestep_reward = []
    for episode in range(episodes):
        print "Episode: ", episode
        s = env.reset()
        a = epsilon_greedy(n_actions)
        t = 0
        total_reward = 0
        done = False
        while t < max_steps:
            t += 1
            s_, reward, done, info = env.step(a)
            total_reward += reward
            a_ = np.argmax(Q[s_, :])
            if done:
                Q[s, a] += alpha * ( reward - Q[s, a] )
            else:
                Q[s, a] += alpha * ( reward + (gamma * Q[s_, a_]) - Q[s, a]
)
            s, a = s_, a_
    return timestep_reward
```

The `epsilon_greedy` function is given as follows:

```
def epsilon_greedy(n_actions):
 action = np.random.randint(0, n_actions)
 return action
```

The code should be pretty straightforward to understand. There are a few Gym APIs that might be new. `gym.make()` allows us to select the available environment in OpenAI Gym. We choose `Taxi-v2` in this case. The environment usually has an initial position for the agent. You can read that position using the `reset()` instance call. In order to view the environment, you can use the `render()` instance call. One more interesting instance call is `'step()'`, which returns the current state, the reward at that state if the state is complete, and probability information. Hence, OpenAI Gym makes our life as easy as possible with these APIs.

Let's see how the same example works with the on-policy learning technique.

On-policy learning – the SARSA algorithm

SARSA (short for **State, Action, Reward, State, Action**) technique is an on-policy learning technique. There is one marginal difference between SARSA and Q-learning. In Q-learning, we chose the epsilon greedy technique to choose a policy (say, an action), and while computing the Q-value for that state, we choose the next state action based on the maximum value of available Q-values for every action in that state. In SARSA, we use epsilon greedy once again to choose a state action instead of using the *max* function. The SARSA-learning equation is given as follows:

$$Q(s,a) \leftarrow Q(s,a) + \alpha(r + \gamma Q(s',a') - Q(s,a))$$

The preceding equation is similar to the TD prediction equation we saw earlier as well. As indicated, the only difference is not using the *max* function and randomly choosing a state-action pair. The SARSA algorithm is given as follows:

```
Initialize Q(s, a), for all s ∈ S, a ∈ A(s), arbitrarily, and Q(Terminal-
state, ·) = 0
 Repeat (for each episode):
   Initialize S
   Choose A from S using policy derived from Q (e.g., e-greedy)
   Repeat (for each step of episode):
     Take action A, observe R, S'
     Choose A' from S'using policy derived from Q (e.g., e-greedy)
     Q(S, A) ← Q(S, A) + α[R + γQ(S', A) - Q(S, A)]
     S ← S'; A ← A';
   until S is Terminal
```

Algorithm explanation

The algorithm begins with initializing initial values (mostly `0`) for the Q-function, `Q(S, A)`. Then, for every episode, a state (`S`) is initialized, an action is chosen from the given list based on policy gradient techniques (such as the e-greedy method), and is followed by a loop for every time step, where the updated state (`S'`) and reward (`R`) are observed for the chosen action. Then, a new action (`A'`) is chosen again from the given list using policy gradient techniques and we begin filling the Q-table with values by solving the equation. Finally, the updated state (`S'`) is equated as the current state (`S`), the new action (`A'`), is equated as the current action (`A`) and the loop continues until the current state (`S`) is terminated.

Now, let's try to implement this in our analogy. Let's start the table with all the values as zeros:

State	Value
(0,0)	[0., 0., 0., 0., 0., 0.],
(0,1)	[0., 0., 0., 0., 0., 0.],
(0,2)	[0., 0., 0., 0., 0., 0.],
...,	
(4,2)	[0., 0., 0., 0., 0., 0.],
(4,3)	[0., 0., 0., 0., 0., 0.],
(4,4)	[0., 0., 0., 0., 0., 0.]]

Table with initial values

Considering the same conditions we chose for Q-learning, our equation is given as follows:

```
Q(2,2;2) ← 0 + 0.4*[(-1)+0.999*(0)-0]
 => Q(2,2;2) ← -0.4
```

Our table becomes the following:

StateValue

(0,0)	[0., 0., 0., 0., 0., 0.],
(0,1)	[0., 0., 0., 0., 0., 0.],
(0,2)	[0., 0., 0., 0., 0., 0.],
...,	
(2,2)	[0., 0., -0.4, 0., 0., 0.]
...,	
(4,2)	[0., 0., 0., 0., 0., 0.],
(4,3)	[0., 0., 0., 0., 0., 0.],
(4,4)	[0., 0., 0., 0., 0., 0.]]

Updated table

You might not have seen much difference in this example, but you can logically make out that, instead of using the *max* function, we're using the e-greedy technique to rechoose the action at random, because of which the algorithm can be time-consuming or can solve it immediately, provided it chooses the right random value. Likewise, the table gets updated for certain time intervals (already explained in *Off-policy learning – the Q-learning algorithm* section). Our final table may look like this:

State	Value
(0,0)	[1. 1. 1. 1. 1. 1.],
(0,1)	[1.21294807 2.30485594 1.73831 2.84424473 9.01048181 -5.74954889],
(0,2)	[3.32374208 -2.67730041 2.0805796 1.83409763 8.14755201 -7.0017296],
...,	
(4,2)	[-0.94623903 10.93045652 -1.11443659 -1.1139482 -5.40064 -3.16039984],
(4,3)	[-6.75173275 2.75158375 -7.07323206 -7.49864668 -8.74536711 -11.97352065],
(4,4)	[-0.42404557 -0.35805959 -0.28086999 18.86740811 -5.40064 -5.56063984]]

Table with values after 3,000 episodes

The preceding explanation is shown in the following code:

```
alpha = 0.4
gamma = 0.999
epsilon = 0.9
episodes = 3000
max_steps = 2500

def sarsa(alpha, gamma, epsilon, episodes, max_steps):
```

```
env = gym.make('Taxi-v2')
n_states, n_actions = env.observation_space.n, env.action_space.n
Q = numpy.zeros((n_states, n_actions))
timestep_reward = []
for episode in range(episodes):
    print "Episode: ", episode
    s = env.reset()
    a = epsilon_greedy(n_actions)
    t = 0
    total_reward = 0
    done = False
    while t < max_steps:
        t += 1
        s_, reward, done, info = env.step(a)
        total_reward += reward
        a_ = epsilon_greedy(n_actions)
        if done:
            Q[s, a] += alpha * ( reward - Q[s, a] )
        else:
            Q[s, a] += alpha * ( reward + (gamma * Q[s_, a_]) - Q[s, a]
)
        s, a = s_, a_
return timestep_reward
```

You can change the a_ variable, which also uses epsilon greedy to chose the next policy. Now that we have seen how both algorithms are represented, let's look at the preceding analogy in simulation. For this, we need to install the OpenAI Gym, NumPy, and pandas libraries.

Installing OpenAI Gym, NumPy, and pandas

Follow these steps to install OpenAI Gym, NumPy, and pandas:

1. If you have Python 3.5 installed, then you can install gym using the pip command:

   ```
   $ pip install gym
   ```

   In case of a Permission denied error, try to give user permissions:

   ```
   $ pip install gym -U
   ```

2. For this example, we need NumPy and pandas, and you can install both using the following command:

```
$ python -m pip install --user numpy scipy matplotlib ipython
jupyter pandas sympy nose
```

Alternatively, you can use the following command:

```
$ sudo apt-get install python-numpy python-scipy python-matplotlib
ipython ipython-notebook python-pandas python-sympy python-nose
```

Now that your setup is complete, let's look at both algorithms in action.

Q-learning and SARSA in action

Now that we have seen both of the algorithms mathematically, let's see how they work in simulation. We shall make use of the `Taxi-v2` environment that's available in the OpenAI Gym library.

The whole code is available in our repository: https://github.com/PacktPublishing/ROS-Robotics-Projects-SecondEdition/tree/master/chapter_8_ws/taxi_problem.

You can test the code by simply using Python:

```
$ python taxi_qlearn.py
```

Or, you can use this:

```
$ python taxi_sarsa.py
```

You should see a series of training phases for those episodes. Once this has been done, the rendering begins, where you can actually see the taxi simulation, as shown here:

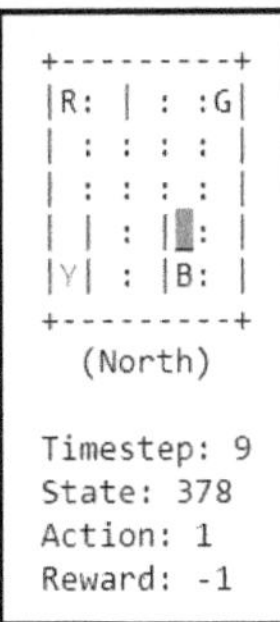

Taxi simulation

As you can see, the taxi goes to the pickup zone and dropoff zone in the quickest possible manner. Also, you can see that there is not much difference between both of them. Over a period of time, you can see that both have similar scores and don't show a radical difference in terms of the rewards that are acquired.

Now that we have covered reinforcement learning algorithms, let's learn more about reinforcement learning in ROS.

Reinforcement learning in ROS

So far, we have seen how to implement reinforcement learning algorithms such as Q-learning and SARSA in OpenAI Gym. Now, we shall look into the following examples and implementations of reinforcement learning in ROS:

- `gym-gazebo` by Erlerobot
- `gym-gazebo2` by Acutronic robotics

Let's look at them in detail.

gym-gazebo

`gym-gazebo` is an OpenAI Gym extension for Gazebo. This extension makes use of an ROS-Gazebo combination to train robots with reinforcement learning algorithms. In the previous chapters, we saw how ROS and Gazebo can be used to solve robotics applications or proof of concepts in simulations that, to an extent, emulate reality. This extension helps us to use this combination in controlling such robots in simulation. A simplified block diagram representation of `gym-gazebo` is shown as follows:

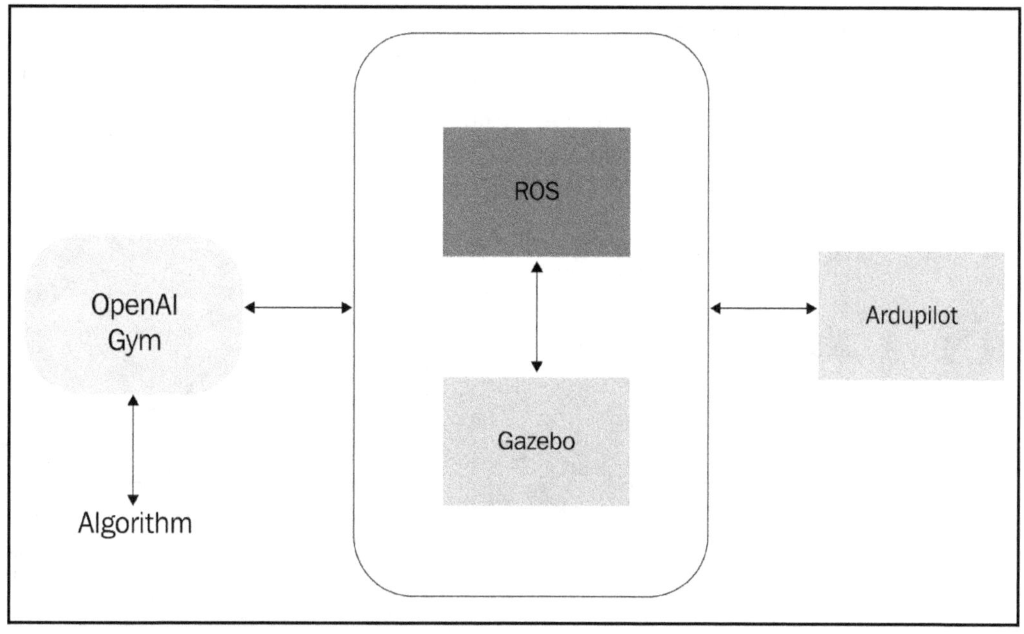

gym-gazebo architecture

This is the underlying architecture of `gym-gazebo`. It consists of three main blocks—out of these, we already know Gazebo and ROS and their advantages. The third block makes use of the OpenAI Gym libraries and helps define environments and robots that are understood by ROS and Gazebo. The creators have created a set of custom environments in this block that can be understood by Gazebo. Alternatively, ROS can be used to control the robots in Gazebo through the plugins that we saw in `Chapter 3`, *Building an Industrial Mobile Manipulator*. Robots are defined as catkin workspaces that establish the Gazebo-ROS connection. Many worlds and environments have been created by this team. However, we're going to see only one robot with one environment in this book.

You can find more such information in this whitepaper: `https://arxiv.org/pdf/1608.05742.pdf`.

We will now look at TurtleBot and its environment.

TurtleBot and its environment

As mentioned earlier, there are many different robots and environments available in the `gym-gazebo` extension. Here, we're going to discuss one of the most used mobile robotics research platform called TurtleBot 2. `gym-gazebo` has TurtleBot 2 packages implemented in it. We're going to use a TurtleBot 2 that has a laser sensor, as shown here:

TurtleBot with a laser sensor

There are four environments available for TurtleBot 2. We will consider the `GazeboCircuit2TurtlebotLIDAR-v0` environment for our example, which has straight tracks and 90 degree turns. The environment is as follows:

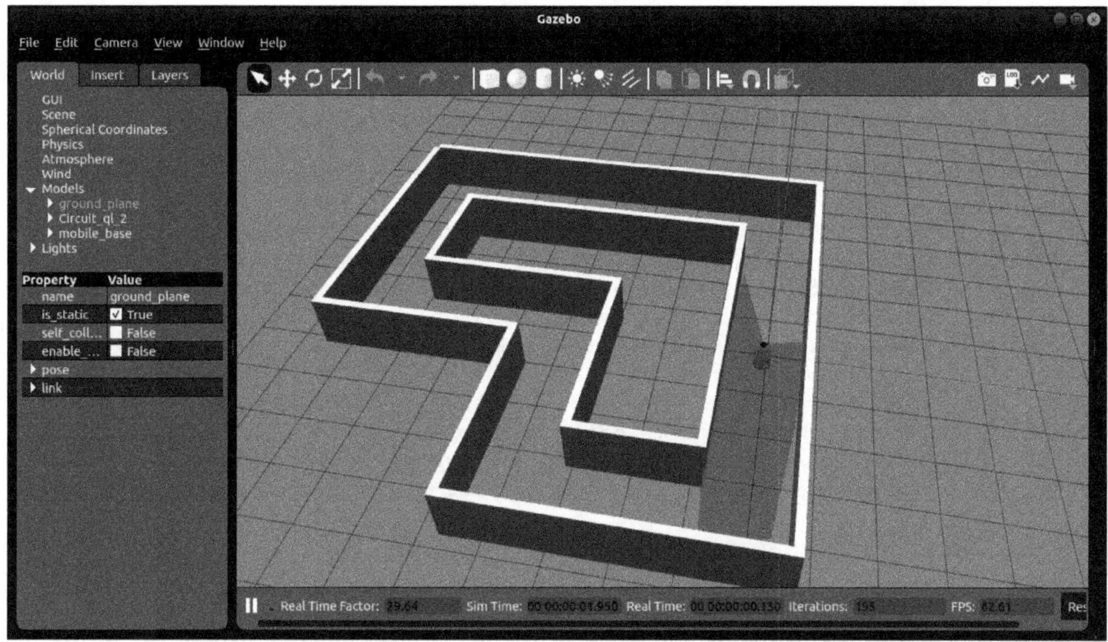

TurtleBot-2 environment—a simple maze with 90 degree turns and straight tracks

The aim of the TurtleBot is to move inside the environment freely without hitting the walls. The TurtleBot tries to discretize and indicate its state using the laser sensor. The actions of the TurtleBot are to move forward, left, or right, where linear velocities are 0.3 m/s and angular velocities are 0.05 m/s. The smaller velocity values are to accelerate the learning process. If the TurtleBot moves forward successfully, then it receives a reward of +5; if it turns left or right, it receives a reward of +1; and if it collides, it receives a reward of -200. Let's learn how to make use of this `gym-gazebo` TurtleBot example.

Installing gym-gazebo and its dependencies

Let's learn how to install `gym-gazebo` and its dependencies. We shall try to install these packages as source installations. Follow these steps:

1. Let's install the necessary dependencies using the following command:

```
$ sudo apt-get install python-pip python3-vcstool python3-pyqt4
pyqt5-dev-tools libbluetooth-dev libspnav-dev pyqt4-dev-tools
libcwiid-dev cmake gcc g++ qt4-qmake libqt4-dev libusb-dev libftdi-
dev python3-defusedxml python3-vcstool ros-melodic-octomap-msgs
ros-melodic-joy ros-melodic-geodesy ros-melodic-octomap-ros ros-
melodic-control-toolbox ros-melodic-pluginlib ros-melodic-
trajectory-msgs ros-melodic-control-msgs ros-melodic-std-srvs ros-
melodic-nodelet ros-melodic-urdf ros-melodic-rviz ros-melodic-kdl-
conversions ros-melodic-eigen-conversions ros-melodic-tf2-sensor-
msgs ros-melodic-pcl-ros ros-melodic-navigation ros-melodic-sophu
```

2. Now, let's install some Python dependencies:

```
$ sudo apt-get install python-skimage
$ sudo pip install h5py
$ sudo pip install kera
```

3. In case you have a GPU, then install `$ pip install tensorflow-gpu`; otherwise, install `$ pip install tensorflow`.
4. Now, let's clone the `gym-gazebo` packages for our workspace and compile them:

```
$ mkdir ~/chapter_8_ws/
$ cd ~/chapter_8_ws/
$ git clone https://github.com/erlerobot/gym-gazebo
$ cd gym-gazebo
$ sudo pip install -e
```

5. Once the dependencies are installed, compile the workspace using the bash command:

```
$ cd gym-gazebo/gym_gazebo/envs/installation
$ bash setup_melodic.bash
```

Once the compilation is successful, you should see the following command window:

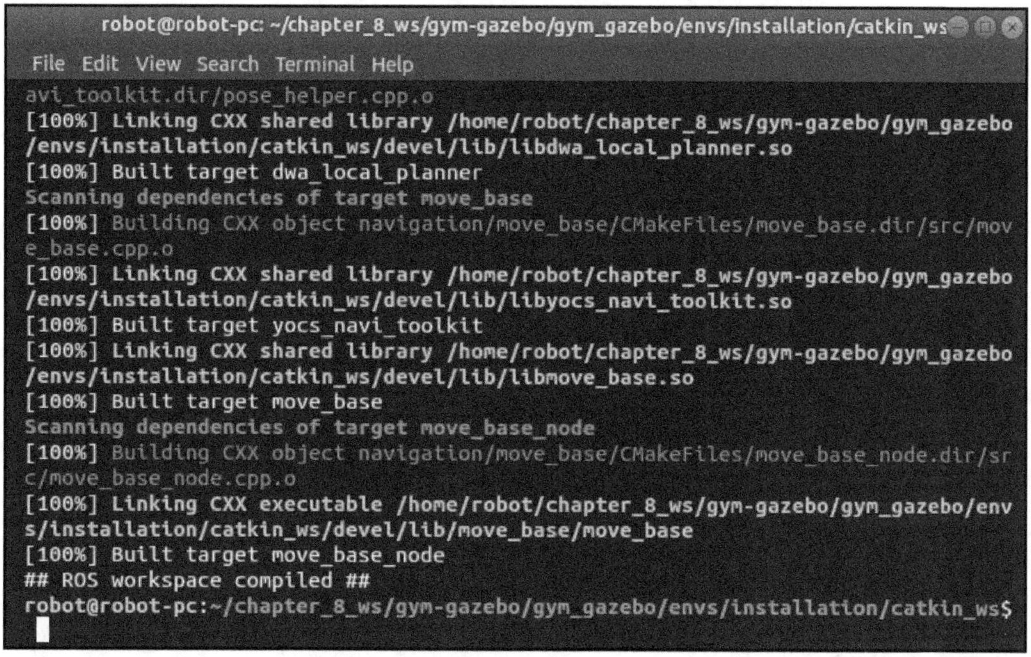

```
robot@robot-pc: ~/chapter_8_ws/gym-gazebo/gym_gazebo/envs/installation/catkin_ws
File Edit View Search Terminal Help
avi_toolkit.dir/pose_helper.cpp.o
[100%] Linking CXX shared library /home/robot/chapter_8_ws/gym-gazebo/gym_gazebo
/envs/installation/catkin_ws/devel/lib/libdwa_local_planner.so
[100%] Built target dwa_local_planner
Scanning dependencies of target move_base
[100%] Building CXX object navigation/move_base/CMakeFiles/move_base.dir/src/mov
e_base.cpp.o
[100%] Linking CXX shared library /home/robot/chapter_8_ws/gym-gazebo/gym_gazebo
/envs/installation/catkin_ws/devel/lib/libyocs_navi_toolkit.so
[100%] Built target yocs_navi_toolkit
[100%] Linking CXX shared library /home/robot/chapter_8_ws/gym-gazebo/gym_gazebo
/envs/installation/catkin_ws/devel/lib/libmove_base.so
[100%] Built target move_base
Scanning dependencies of target move_base_node
[100%] Building CXX object navigation/move_base/CMakeFiles/move_base_node.dir/sr
c/move_base_node.cpp.o
[100%] Linking CXX executable /home/robot/chapter_8_ws/gym-gazebo/gym_gazebo/env
s/installation/catkin_ws/devel/lib/move_base/move_base
[100%] Built target move_base_node
## ROS workspace compiled ##
robot@robot-pc:~/chapter_8_ws/gym-gazebo/gym_gazebo/envs/installation/catkin_ws$
```

Successful gym-gazebo compilation

The preceding screenshot shows a successful `gym-gazebo` compilation. We will now test the TurtleBot-2 environment.

Testing the TurtleBot-2 environment

Now, we're all ready to see the training visually. We can start the example by running the following commands:

```
$ cd gym-gazebo/gym_gazebo/envs/installation/
$ bash turtlebot_setup.bash
$ cd gym-gazebo/examples/turtlebot
$ python circuit_turtlebot_lidar_qlearn.py
```

You should see a Terminal window as shown in the following screenshot, that starts the training. If you want to see the TurtleBot training itself visually, open `gzclient` in another Terminal:

```
$ cd gym-gazebo/gym_gazebo/envs/installation/
$ bash turtlebot_setup.bash
$ export GAZEBO_MASTER_URI=http://localhost:13853
$ gzclient
```

You should see the TurtleBot trying to learn how to navigate in that environment, as shown in the following screenshot, through a series of episodes and respective rewards:

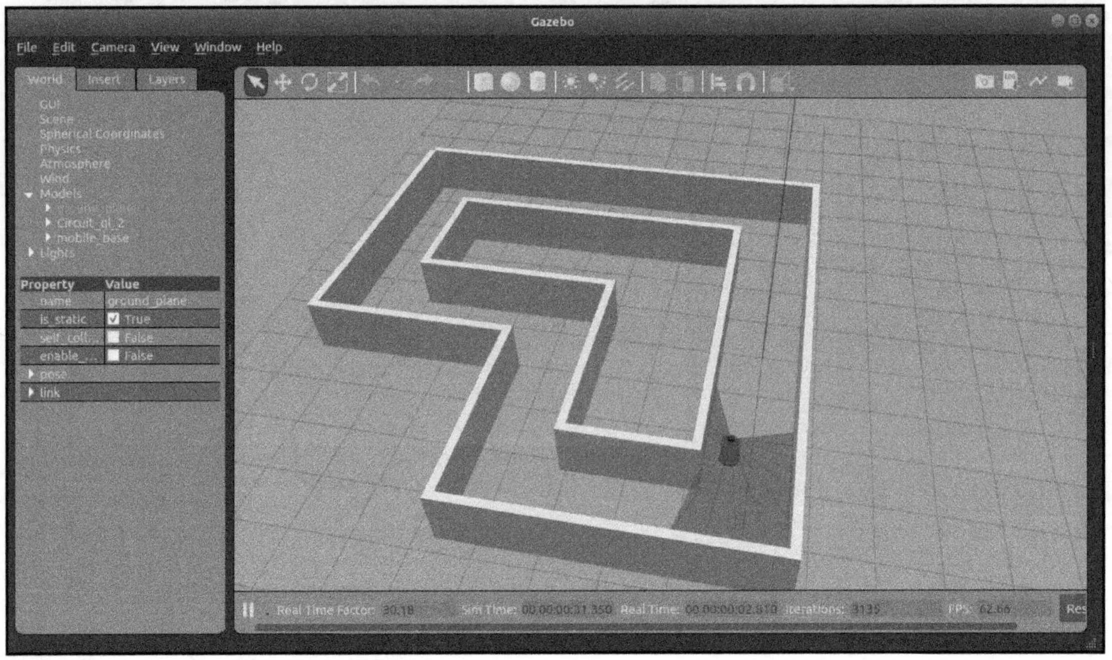

TurtleBot-2 training in the given environment

After, say, 500 episodes, you should see that the TurtleBot can move almost halfway through the environment without colliding. After, say, 2,000 episodes, you should see that the TurtleBot is able to complete one lap inside the environment without colliding with the environment.

A Terminal with rewards over specific episodes is as follows:

```
/home/robot/chapter_8_ws/gym-gazebo/gym_gazebo/envs/assets/launch/GazeboCircuit2TurtlebotLida...
File Edit View Search Terminal Tabs Help
/home/robot/chapter_8_ws/gym-gazebo/gym_g... ×    robot@robot-pc: ~/chapter_8_ws/gym-gazebo/g... ×
EP: 140 - [alpha: 0.2 - gamma: 0.8 - epsilon: 0.74] - Reward: -120     Time: 0:05:51
EP: 141 - [alpha: 0.2 - gamma: 0.8 - epsilon: 0.74] - Reward: -163     Time: 0:05:53
EP: 142 - [alpha: 0.2 - gamma: 0.8 - epsilon: 0.74] - Reward: -103     Time: 0:05:56
EP: 143 - [alpha: 0.2 - gamma: 0.8 - epsilon: 0.74] - Reward: -66     Time: 0:06:01
EP: 144 - [alpha: 0.2 - gamma: 0.8 - epsilon: 0.74] - Reward: -99     Time: 0:06:05
EP: 145 - [alpha: 0.2 - gamma: 0.8 - epsilon: 0.73] - Reward: -87     Time: 0:06:08
EP: 146 - [alpha: 0.2 - gamma: 0.8 - epsilon: 0.73] - Reward: -123     Time: 0:06:10
EP: 147 - [alpha: 0.2 - gamma: 0.8 - epsilon: 0.73] - Reward: -107     Time: 0:06:14
EP: 148 - [alpha: 0.2 - gamma: 0.8 - epsilon: 0.73] - Reward: -168     Time: 0:06:16
EP: 149 - [alpha: 0.2 - gamma: 0.8 - epsilon: 0.73] - Reward: -70     Time: 0:06:20
EP: 150 - [alpha: 0.2 - gamma: 0.8 - epsilon: 0.73] - Reward: -108     Time: 0:06:22
EP: 151 - [alpha: 0.2 - gamma: 0.8 - epsilon: 0.73] - Reward: -66     Time: 0:06:26
EP: 152 - [alpha: 0.2 - gamma: 0.8 - epsilon: 0.73] - Reward: -79     Time: 0:06:31
```

Terminal with rewards over specific episodes

A graph of rewards over specific episodes is as follows:

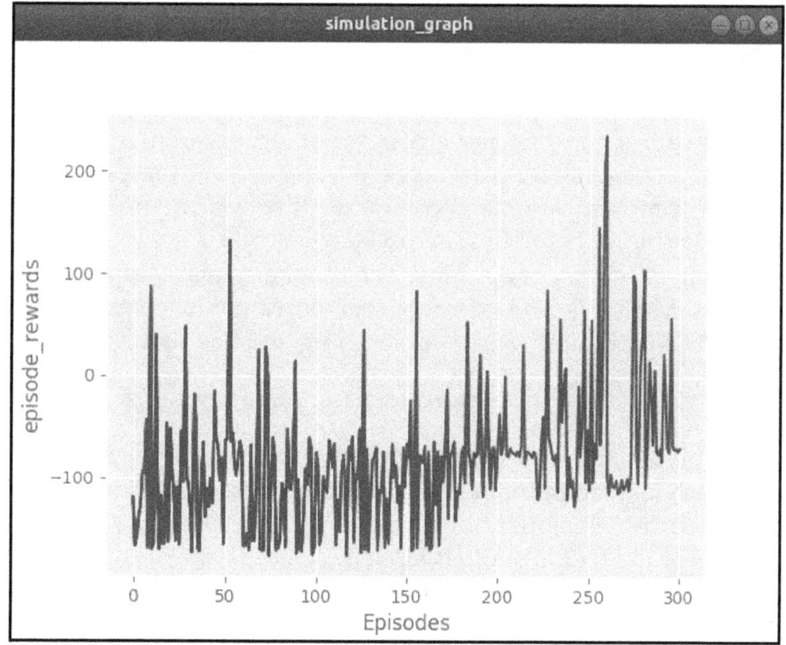

TurtleBot rewards graph

It is evident that the rewards are gradually increasing over time (episodes). Have a look at the whitepaper for more benchmark results and comparison with the SARSA algorithm.

gym-gazebo2

An important piece of information about `gym-gazebo` is that the repository has been archived and is not supported anymore. However, the same is tested in Ubuntu 18.04 and is working fine with ROS-Melodic. It should act as a good start to understand reinforcement learning implementation in ROS.

The redesigned and upgraded version of `gym-gazebo` is the `gym-gazebo2` extension by Acutronic robotics. This new version discusses the new ROS-2 based architecture and summarizes results using another reinforcement learning technique called **Proximal Policy Optimization** (**PPO**). The authors have tested this with a ROS-2 compliant robot arm called **MARA** (short for **Modular Articulated Robot ARM**) and have achieved accuracies of millimeter scale while achieving kinematics. More information on this extension can be found in this paper: `https://arxiv.org/pdf/1903.06278.pdf`. Currently, there is only one robot-environment pair has been implemented and tested, so let's see how this example works. The next section introduces MARA and its environment.

MARA and its environment

MARA is a cool collaborative robot arm that was built using H-ROS components, meaning it has ROS-2 in its actuators, sensors, and any of the other representative modules inside it. Hence, each module can provide industrial-grade features, for example, ROS 2 features such as deterministic communication and component life cycle management. Due to the nature of H-ROS integration, the robot arm can be connected with additional sensors and actuators, as well as an easy system upgrade. The robot arm also supports a good number of industrial grippers. MARA is a 6-DoF robot arm with a payload capability of 3 kg and a tool speed of 1 m/s. The robot arm weighs about 21 kg and has a reach of 0.65 meters.

There are four environments in this reinforcement learning implementation:

- **MARA**: This is the simplest environment and is where the robot's tool center tries to reach a given point in space. The environment is reset if a collision is detected but the collision is not modeled with the reward function. The orientation of the tool is omitted as well.
- **MARA Orient**: The environment considers the translation and rotation of the end-effector of the robot and resets if a collision is detected. These are also not modeled into the reward system.

- **MARA Collision**: This environment is like the MARA environment (only translation), but with collisions modeled in the reward function. The robot arm receives a punishment if the robot arm collides and gets reset to the initial pose.
- **MARA Collision Orient**: This environment is a combination of MARA Collision and Orient, where the robot arm's translation and orientation are considered and the collision is modeled in the reward function. This is one of the most complex environment implementations in this package.

Let's see how to use these packages in action.

Installing gym-gazebo2 and dependencies

In order to install `gym-gazebo2`, follow these steps:

1. Install the dependencies first:
    1. You need to install the latest Gazebo version, which is 9.9.0 at the time of writing. You can install this using the following command:

    ```
    $ curl -sSL http://get.gazebosim.org | sh
    ```

    2. Once the latest version has been installed, install the following ROS-2 dependencies:

    ```
    $ sudo apt install -y ros-dashing-action-msgs ros-dashing-message-filters ros-dashing-yaml-cpp-vendor ros-dashing-urdf ros-dashing-rttest ros-dashing-tf2 ros-dashing-tf2-geometry-msgs ros-dashing-rclcpp-action ros-dashing-cv-bridge ros-dashing-image-transport ros-dashing-camera-info-manager
    ```

    3. Since we're using the Dashing Diademata version, OpenSplice RMW is not available yet, so install it using the following command:

    ```
    $ sudo apt install ros-dashing-rmw-opensplice-cpp
    ```

    4. Now, install the Python dependencies using the following command:

    ```
    $ sudo apt install -y build-essential cmake git python3-colcon-common-extensions python3-pip python-rosdep python3-vcstool python3-sip-dev python3-numpy wget
    ```

5. Install TensorFlow using the following command:

```
pip3 install tensorflow
```

6. Install additional utilities such as `transform3d`, `billard`, and `psutil` using the following command:

```
$ pip3 install transforms3d billiard psutil
```

7. Finally, install FAST-RTPS dependencies using the following command:

```
$ sudo apt install --no-install-recommends -y libasio-dev
libtinyxml2-dev
```

2. Now that you've installed the dependencies, let's compile the MARA workspace. Follow the given set of commands:

```
mkdir -p ~/ros2_mara_ws/src
cd ~/ros2_mara_ws
wget
https://raw.githubusercontent.com/AcutronicRobotics/MARA/dashing/ma
ra-ros2.repos
vcs import src < mara-ros2.repos
wget
https://raw.githubusercontent.com/AcutronicRobotics/gym-gazebo2/das
hing/provision/additional-repos.repos
vcs import src < additional-repos.repos
# Avoid compiling erroneus package
touch
~/ros2_mara_ws/src/orocos_kinematics_dynamics/orocos_kinematics_dyn
amics/COLCON_IGNORE
# Generate HRIM dependencies
cd ~/ros2_mara_ws/src/HRIM
sudo pip3 install hrim
hrim generate models/actuator/servo/servo.xml
hrim generate models/actuator/gripper/gripper.xml
```

3. Now, compile the workspace using the following commands:

```
source /opt/ros/dashing/setup.bash
cd ~/ros2_mara_ws
colcon build --merge-install --packages-skip
individual_trajectories_bridge
# Remove warnings
touch ~/ros2_mara_ws/install/share/orocos_kdl/local_setup.sh
~/ros2_mara_ws/install/share/orocos_kdl/local_setup.bash
```

4. Assuming you've already installed Open AI Gym already, install the `gym-gazebo2` packages using the following commands:

```
cd ~ && git clone -b dashing
https://github.com/AcutronicRobotics/gym-gazebo2
cd gym-gazebo2
pip3 install -e .
```

Have a look at this site for updated installations in case you face any issues: `https://github.com/AcutronicRobotics/gym-gazebo2/blob/dashing/INSTALL.md`.

Now, let's test the MARA environment.

Testing the MARA environment

You can begin by initially training the agent. You can train MARA using the `ppo2_mlp.py` script, as follows:

```
$ cd ~/ros2learn/experiments/examples/MARA
$ python3 train_ppo2_mlp.py
```

You can try out the `-h` argument to get all the available commands for this script.

Now that the robot arm has been trained, you can test the trained policy using the following commands:

```
$ cd ~/ros2learn/experiments/examples/MARA
$ python3 run_ppo2_mlp.py -g -r -v 0.3
```

Summary

In this chapter, we understood what reinforcement learning is and how it stands out from other machine learning algorithms. The components of reinforcement learning were individually explained and we enhanced our understanding through examples. Then, reinforcement learning algorithms were introduced, both practically and mathematically, using suitable examples. We also saw reinforcement learning implementations in ROS, where robots such as TurtleBot 2 and the MARA robot arm were used in application-specific environments, and we understood how they're implemented and their usage. This chapter acted as a simple and gentle introduction to machine learning and its usage in ROS.

In the next chapter, we will see how deep we can dive into machine learning methods to make the agent more effective in terms of learning and achieving its goal.

9
Deep Learning Using ROS and TensorFlow

You may have come across deep learning many times on the web. Most of us are not fully aware of this technology, and many people are trying to learn it too. So, in this chapter, we are going to see the importance of deep learning in robotics and how we can implement robotics applications using deep learning and ROS. We begin with understanding how deep learning works and is implemented, followed by a glimpse of commonly used tools and libraries for deep learning. We will learn to install TensorFlow for Python and embed TensorFlow APIs in ROS. We will learn to carry out image recognition using ROS and TensorFlow. Later, you will come across practical examples using these libraries and interfacing them with ROS.

Here are the main topics we are going to discuss in this chapter:

- Introducing deep learning and its applications
- Deep learning for robotics
- Software frameworks and programming languages for deep learning
- Getting started with Google TensorFlow
- Installing TensorFlow for Python
- Embedding TensorFlow APIs in ROS
- Image recognition using ROS and TensorFlow
- Introduction to scikit-learn
- Implementing SVM using scikit-learn
- Embedding SVM on a ROS node
- Implementing an SVM-ROS application

Technical requirements

Let's look into the technical requirements for this chapter:

- ROS Melodic Morenia on Ubuntu 18.04, with Gazebo 9 (preinstalled)
- ROS packages: `cv_bridge` and `cv_camera`
- Libraries: TensorFlow (0.12 and above, not tested with 2.0 yet), scikit-learn, and more
- Timelines and rest platform:
    - **Estimated learning time**: On average, 120 minutes
    - **Project build time (inclusive of compile and run time)**: On average, 60-90 minutes
    - **Project test platform**: HP Pavilion laptop (Intel® Core™ i7-4510U CPU @ 2.00GHz × 4 with 8 GB Memory and 64-bit OS, GNOME-3.28.2)

The code for this chapter is available at `https://github.com/PacktPublishing/ROS-Robotics-Projects-SecondEdition/tree/master/chapter_9_ws`.

Let's begin this chapter by understanding deep learning and its applications.

Introduction to deep learning and its applications

What actually is deep learning? It is a buzzword in neural network technology. What is a neural network then? An artificial neural network is a computer software model that replicates the behavior of neurons in the human brain. A neural network is one way to classify data. For example, if we want to classify an image based on whether it contains an object or not, we can use this method.

There are several other computer software models for classification such as logistic regression and **Support Vector Machine** (**SVM**); a neural network is one of them. So, why are we not calling it a neural network instead of deep learning? The reason is that, in deep learning, we use a large number of artificial neural networks. So, you may ask, *why was it not possible before?* The answer is: to create a large number of neural networks (multilayer perceptron), we may need a high amount of computational power. So, how has it become possible now? It's because of the availability of cheap computational hardware. Will computational power alone do the job? No, we also need a large dataset to train with.

When we train a large set of neurons, it can learn various features from the input data. After learning the features, it can predict the occurrence of an object or anything we have taught to it.

To teach a neural network, we can either use the supervised learning method or go unsupervised. In supervised learning, we have a training dataset with input and its expected output. These values will be fed to the neural network, and the weights of the neurons will be adjusted in such a way that it can predict which output it should generate whenever it gets particular input data. So, what about unsupervised learning? This type of algorithm learns from an input dataset without having corresponding outputs. The human brain can work in a supervised or unsupervised way, but unsupervised learning is more predominant in our case. The main applications of deep neural networks are in the classification and recognition of objects, such as image recognition and speech recognition.

In this book, we are mainly dealing with supervised learning for building deep learning applications for robots. The next section will give us an introduction to deep learning for robotics.

Deep learning for robotics

Here are the main areas in robotics where we can apply deep learning:

- **Deep-learning-based object detector**: Imagine a robot wants to pick a specific object from a group of objects. What can be the first step in solving this problem? It should identify the object first, right? We can use image processing algorithms such as segmentation and Haar training to detect an object, but the problem with those techniques is that they are not scalable and can't be used for many objects. Using deep learning algorithms, we can train a large neural network with a large dataset. It can have good accuracy and scalability compared to other methods. Datasets such as ImageNet (`http://image-net.org/`), which have a large collection of image datasets, can be used for training. We also get trained models that we can just use without training. We will look at an ImageNet-based image recognition ROS node in an upcoming section.
- **Speech recognition**: If we want to command a robot to perform some task using our voice, what will we do? Will the robot understand our language? Definitely not. But using deep learning techniques, we can build a more accurate speech recognition system compared to the existing **Hidden Markov Model** (**HMM**)-based recognizer. Companies such as Baidu (`http://research.baidu.com/`) and Google (`http://research.google.com/pubs/SpeechProcessing.html`) are trying hard to create a global speech recognition system using deep learning techniques.

- **SLAM and localization**: Deep learning can be used to perform SLAM and localization of mobile robots, which perform much better than conventional methods.
- **Autonomous vehicles**: The deep learning approach in self-driving cars is a new way of controlling the steering of vehicles using a trained network in which sensor data can be fed to the network and corresponding steering control can be obtained. This kind of network can learn by itself while driving.

One of the companies doing a lot in deep reinforcement learning is DeepMind owned by Google. They introduced a method to ace the Atari 2600 games to an extremely high level with only the raw pixels and score as inputs (`https://deepmind.com/research/dqn/`). AlphaGo is another computer program developed by DeepMind, which can even beat a professional human Go player (`https://deepmind.com/research/alphago/`).

Let's now look at a few deep learning libraries.

Deep learning libraries

There are some popular deep learning libraries used in research and commercial applications. Let's look at them individually here:

- **TensorFlow**: This is an open source software library for numerical computation using data flow graphs. The TensorFlow library (`https://www.tensorflow.org/`) is designed for machine intelligence and developed by the Google Brain team. The main aim of this library is to perform machine learning and deep neural network research. It can be used in a wide variety of other domains as well.
- **Theano**: Theano is an open source Python library (`http://deeplearning.net/software/theano/`) allowing us to streamline and assess mathematical expressions including multidimensional arrays proficiently. Theano was essentially created by the machine learning group at the Montreal University, Canada.

- **Torch**: Torch is a scientific computing framework with more support for machine learning algorithms with GPU as a priority. It's very efficient, being built on the scripting language LuaJIT and has an underlying C/CUDA implementation (`http://torch.ch/`).
- **Caffe**: Caffe (`http://caffe.berkeleyvision.org/`) is a deep learning library made with a focus on modularity, speed, and expression. It was developed by the **Berkeley Vision and Learning Centre** (**BVLC**).

The next section will help us to get started with TensorFlow.

Getting started with TensorFlow

As we discussed, TensorFlow is an open source library mainly designed for fast numerical computing. This library mainly works with Python and was released by Google. TensorFlow can be used as a foundation library to create deep learning models. We can use TensorFlow both for research and development and in production systems. The good thing about TensorFlow is it can run on a single CPU all the way up to a large-scale distributed system of hundreds of machines. It also works well on GPUs and mobile devices.

You can check out the TensorFlow library at the following link: `https://www.tensorflow.org/`.

Let's now learn to install TensorFlow on Ubuntu 18.04 LTS.

Installing TensorFlow on Ubuntu 18.04 LTS

Installing TensorFlow is not a tedious task if you have a fast internet connection. The main tool we need to have is `pip`, which is a package management tool used for managing and installing software packages in Python.

You can get the latest installation instructions for Linux from the following link: `https://www.tensorflow.org/install/install_linux`.

Let's begin installing TensorFlow through the following steps:

1. Here is the command to install `pip` on Ubuntu:

```
$ sudo apt-get install python-pip python-dev
```

2. After installing `pip`, you can install TensorFlow using the following command:

```
$ pip install tensorflow
```

   This would install the latest and stable TensorFlow library.

   If you would like a GPU version, then use this:

```
$ pip install tensorflow-gpu
```

3. You have to execute the following command to set a bash variable called `TF_BINARY_URL`. This is for installing the correct binaries for our configuration. The following variable is for the Ubuntu 64 bit, Python 2.7, CPU-only version:

```
$ export
TF_BINARY_URL=https://storage.googleapis.com/tensorflow/linux/cpu/t
ensorflow-0.11.0-cp27-none-linux_x86_64.whl
```

   If you have an NVIDIA GPU, you may need a different binary. You may also need to install CUDA toolkit 8.0 cuDNN v5 for installing this, as shown here:

```
$ export
TF_BINARY_URL=https://storage.googleapis.com/tensorflow/linux/gpu/t
ensorflow-0.11.0-cp27-none-linux_x86_64.whl
```

   Check out these references for installing TensorFlow with NVIDIA acceleration:

   - `http://www.nvidia.com/object/gpu-accelerated-applications-tensorflow-installation.html`
   - `https://alliseesolutions.wordpress.com/2016/09/08/install-gpu-tensorflow-from-sources-w-ubuntu-16-04-and-cuda-8-0-rc/`

4. After defining the bash variable, use the following command to install the binaries for Python 2:

```
$ sudo pip install --upgrade $TF_BINARY_URL
```

Check out this link for installing cuDNN: `https://developer.nvidia.com/cudnn`.

If everything works fine, you will get the following kind of output in the Terminal:

```
robot@robot-pc:~$ sudo pip install --upgrade $TF_BINARY_URL
The directory '/home/robot/.cache/pip/http' or its parent directory is not owned
 by the current user and the cache has been disabled. Please check the permissio
ns and owner of that directory. If executing pip with sudo, you may want sudo's
-H flag.
The directory '/home/robot/.cache/pip' or its parent directory is not owned by
he current user and caching wheels has been disabled. check the permissions and
owner of that directory. If executing pip with sudo, you may want sudo's -H flag
.
Collecting tensorflow==0.11.0rc1 from https://storage.googleapis.com/tensorflow/
linux/cpu/tensorflow-0.11.0rc1-cp27-none-linux_x86_64.whl
  Downloading https://storage.googleapis.com/tensorflow/linux/cpu/tensorflow-0.1
1.0rc1-cp27-none-linux_x86_64.whl (39.8MB)
    100% |████████████████████████████████| 39.8MB 42kB/s
Collecting mock>=2.0.0 (from tensorflow==0.11.0rc1)
  Downloading mock-2.0.0-py2.py3-none-any.whl (56kB)
    100% |████████████████████████████████| 61kB 122kB/s
Collecting protobuf==3.0.0 (from tensorflow==0.11.0rc1)
  Downloading protobuf-3.0.0-py2.py3-none-any.whl (342kB)
    100% |████████████████████████████████| 348kB 176kB/s
Collecting numpy>=1.11.0 (from tensorflow==0.11.0rc1)
  Downloading numpy-1.11.2-cp27-cp27mu-manylinux1_x86_64.whl (15.3MB)
    67% |█████████████████████           | 10.4MB 321kB/s eta 0:00:16
```

Installing TensorFlow

If everything has been properly installed on your system, you can check it using a simple test. Open a Python Terminal, execute the following lines, and check whether you are getting the results shown in the following screenshot:

```
robot@robot-pc: ~
robot@robot-pc:~$ python
Python 2.7.11+ (default, Apr 17 2016, 14:00:29)
[GCC 5.3.1 20160413] on linux2
Type "help", "copyright", "credits" or "license" for more informati
on.
>>> import tensorflow as tf
>>> hello = tf.constant('Hello, TensorFlow!')
>>> sess = tf.Session()
>>> print(sess.run(hello))
Hello, TensorFlow!
>>> a = tf.constant(12)
>>> b = tf.constant(34)
>>> print(sess.run(a+b))
46
>>>
```

Testing a TensorFlow installation

We will look at an explanation of the code in the next section.

Here is our hello world code in TensorFlow:

```
import tensorflow as tf
hello = tf.constant('Hello, TensorFlow!')
sess = tf.Session()
print (sess.run(hello))
a = tf.constant(12)
b = tf.constant(34)
print(sess.run(a+b))
```

The output of the preceding code is as shown in the preceding screenshot. In the next section, we will look at a few important concepts of TensorFlow.

TensorFlow concepts

Before you start programming using TensorFlow functions, you should understand its concepts. Here is a block diagram of TensorFlow concepts demonstrated using the addition operation in TensorFlow:

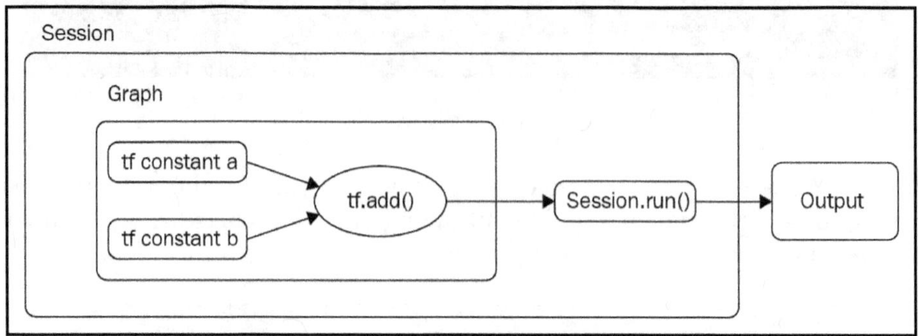

Block diagram of TensorFlow concepts

Let's look at each of the concepts shown in the preceding diagram.

Graph

In TensorFlow, all computations are represented as graphs. A graph consists of nodes. The nodes in a graph are called **operations (ops)**. An op or node can take tensors. Tensors are basically typed multi-dimensional arrays. For example, an image can be a tensor. So, in short, the TensorFlow graph has a description of all of the computation required.

In the preceding example, the ops of the graphs are as follows:

```
hello = tf.constant('Hello, TensorFlow!')
a = tf.constant(12)
b = tf.constant(34)
```

These `tf.constant()` methods create a constant op that will be added as a node in the graph. You can see how a string and integer are added to the graph.

Session

After building the graph, we have to execute it, right? For computing a graph, we should put it in a session. A `Session` class in TensorFlow places all ops or nodes onto computational devices such as CPU or GPU.

Here is how we create a `Session` object in TensorFlow:

```
sess = tf.Session()
```

For running the ops in a graph, the `Session` class provides methods to run the entire graph:

```
print(sess.run(hello))
```

It will execute the op called `hello` and print `Hello, TensorFlow` in the Terminal.

Variables

During execution, we may need to maintain the state of the ops. We can do so by using `tf.Variable()`. Let's check out an example declaration of `tf.Variable()`. This line will create a variable called `counter` and initialize it to scalar value `0`:

```
state = tf.Variable(0, name="counter")
```

Here are the ops to assign a value to the variable:

```
one = tf.constant(1)
update = tf.assign(state, one)
```

If you are working with variables, we have to initialize them all at once using the following function:

```
init_op = tf.initialize_all_variables()
```

After initialization, we have to run the graph for putting this into effect. We can run the previous ops using the following code:

```
sess = tf.Session()
sess.run(init_op)
print(sess.run(state))
sess.run(update)
```

Fetches

To fetch the outputs from the graph, we have to execute the `run()` method, which is inside the `Session` object. We can pass the ops to the `run()` method and retrieve the output as tensors:

```
a = tf.constant(12)
b = tf.constant(34)
add = tf.add(a,b)
sess = tf.Sessions()
result = sess.run(add)
print(result)
```

In the preceding code, the value of `result` will be 12+34 (46).

Feeds

Until now, we have been dealing with constants and variables. We can also feed tensors during the execution of a graph. Here we have an example of feeding tensors during execution. For feeding a tensor, first, we have to define the feed object using the `tf.placeholder()` function.

After defining two feed objects, we can see how to use it inside `sess.run()`:

```
x = tf.placeholder(tf.float32)
y = tf.placeholder(tf.float32)

output = tf.mul(input1, input2)

with tf.Session() as sess:
  print(sess.run([output], feed_dict={x:[8.], y:[2.]}))

# output:
# [array([ 16.], dtype=float32)]
```

Let's start coding using TensorFlow.

Writing our first code in TensorFlow

We are again going to write basic code that performs matrix operations such as matrix addition, multiplication, scalar multiplication, and multiplication with a scalar from 1 to 99. The code is written for demonstrating the basic capabilities of TensorFlow, which we have discussed previously.

Have a look at this code in our repository here: `https://github.com/PacktPublishing/ROS-Robotics-Projects-SecondEdition/blob/master/chapter_9_ws/basic_tf_code.py`.

As we know, we have to import the `tensorflow` module to access its APIs. We are also importing the `time` module to provide a delay in the loop:

```
import tensorflow as tf
import time
```

Here is how to define variables in TensorFlow. We are defining the `matrix_1` and `matrix_2` variables, two 3 x 3 matrices:

```
matrix_1 = tf.Variable([[1,2,3],[4,5,6],[7,8,9]],name="mat1")
matrix_2 = tf.Variable([[1,2,3],[4,5,6],[7,8,9]],name="mat2")
```

In addition to the preceding matrix variables, we are defining a constant and a scalar variable called `counter`. These values are used for scalar multiplication operations. We will change the value of `counter` from 1 to 99, and each value will be multiplied by a matrix:

```
scalar = tf.constant(5)
number = tf.Variable(1, name="counter")
```

The following is how we define strings in `tf`. Each string is defined as a constant:

```
add_msg = tf.constant("\nResult of matrix addition\n")
mul_msg = tf.constant("\nResult of matrix multiplication\n")
scalar_mul_msg = tf.constant("\nResult of scalar multiplication\n")
number_mul_msg = tf.constant("\nResult of Number multiplication\n")
```

The following are the main ops in the graph doing the computation. The first line will add two matrices, the second line will multiply those same two, the third will perform scalar multiplication with one value, and the fourth will perform scalar multiplication with a scalar variable.

Here is the code for the same:

```
mat_add = tf.add(matrix_1,matrix_2)
mat_mul = tf.matmul(matrix_1,matrix_2)
mat_scalar_mul = tf.mul(scalar,mat_mul)
mat_number_mul = tf.mul(number,mat_mul)
```

If we have `tf` variable declarations, we have to initialize them using the following line of code:

```
init_op = tf.initialize_all_variables()
```

Here, we are creating a `Session()` object:

```
sess = tf.Session()
```

This is one thing we hadn't discussed earlier. We can perform the computation on any device according to our priority. It can be a CPU or GPU. Here, you can see that the device is a CPU:

```
tf.device("/cpu:0")
```

This line of code will run the graph to initialize all variables:

```
sess.run(init_op)
```

In the following loop, we can see the running of a TensorFlow graph. This loop puts each op inside the `run()` method and fetches its results. To be able to see each output, we are putting a delay on the loop:

```
for i in range(1,100):

  print "\nFor i =",i

  print(sess.run(add_msg))
  print(sess.run(mat_add))

  print(sess.run(mul_msg))
  print(sess.run(mat_mul))

  print(sess.run(scalar_mul_msg))
  print(sess.run(mat_scalar_mul))
```

```
    update = tf.assign(number,tf.constant(i))
    sess.run(update)
    print(sess.run(number_mul_msg))
    print(sess.run(mat_number_mul))

    time.sleep(0.1)
```

After all of this computation, we have to release the `Session()` object to free up the resources:

```
sess.close()
```

The following is the output screenshot:

```
For i = 99

Result of matrix addition

[[ 2  4  6]
 [ 8 10 12]
 [14 16 18]]

Result of matrix multiplication

[[ 30  36  42]
 [ 66  81  96]
 [102 126 150]]

Result of scalar multiplication

[[150 180 210]
 [330 405 480]
 [510 630 750]]

Result of Number multiplication

[[ 2970  3564  4158]
 [ 6534  8019  9504]
 [10098 12474 14850]]
```

The output of basic TensorFlow code

Now that we have covered the basics of TensorFlow, let's now learn to carry out image recognition using ROS and TensorFlow.

Image recognition using ROS and TensorFlow

After discussing the basics of TensorFlow, let's start discussing how to interface ROS and TensorFlow to do some serious work. In this section, we are going to deal with image recognition using these two. There is a simple package to perform image recognition using TensorFlow and ROS. Here is the ROS package to do this: `https://github.com/qboticslabs/rostensorflow`.

This package was forked from `https://github.com/OTL/rostensorflow`. The package basically contains a ROS Python node that subscribes to images from the ROS webcam driver and performs image recognition using TensorFlow APIs. The node will print the detected object and its probability.

The image recognition is mainly done using a model called a deep convolution network. It can achieve high accuracy in the field of image recognition. An improved model we are going to use here is Inception v3 (`https://arxiv.org/abs/1512.00567`).

This model is trained for the **ImageNet Large Scale Visual Recognition Challenge (ILSVRC)** (`http://image-net.org/challenges/LSVRC/2016/index`) using data from 2012.

When we run the node, it will download a trained Inception v3 model to the computer and classify the object according to the webcam images. You can see the detected object's name and its probability in Terminal. There are a few prerequisites to run this node. Let's go through the dependencies.

Prerequisites

For running an ROS image recognition node, you should install the following dependencies. The first is `cv-bridge`, which helps us to convert an ROS image message into OpenCV image data type and vice versa. The second is `cv-camera`, which is one of the ROS camera drivers. Here's how to install them:

```
$ sudo apt-get install ros-melodic-cv-bridge ros-melodic-cv-camera
```

The next section will cover the ROS image recognition node.

The ROS image recognition node

You can download the ROS image recognition package from GitHub; it's also available in this book's code bundle. The `image_recognition.py` program can publish detected results in the `/result` topic, which is of the `std_msgs/String` type and is subscribed to image data from the ROS camera driver from the `/image` (`sensor_msgs/Image`) topic.

So, how does `image_recognition.py` work? First, take a look at the main modules imported to this node. As you know, `rospy` has ROS Python APIs. The ROS camera driver publishes ROS image messages, so here we have to import `Image` messages from `sensor_msgs` to handle those image messages.

To convert an ROS image to the OpenCV data type and vice versa, we need `cv_bridge` and, of course, the `numpy`, `tensorflow`, and `tensorflow imagenet` modules to classify images and download the Inception v3 model from `https://www.tensorflow.org/`.

Here are the imports:

```
import rospy
from sensor_msgs.msg import Image
from std_msgs.msg import String
from cv_bridge import CvBridge
import cv2
import numpy as np
import tensorflow as tf
from tensorflow.models.image.imagenet import classify_image
```

The following code snippet is the constructor for a class called `RosTensorFlow()`:

```
class RosTensorFlow():
    def __init__(self):
```

The constructor call has the API for downloading the trained Inception v3 model from `https://www.tensorflow.org/`:

```
        classify_image.maybe_download_and_extract()
```

Now, we are creating a TensorFlow `Session()` object, then creating a graph from a saved `GraphDef` file, and returning a handle for it. The `GraphDef` file is available in the code bundle:

```
        self._session = tf.Session()
        classify_image.create_graph()
```

This line creates a `cv_bridge` object for the ROS-OpenCV image conversion:

```
        self._cv_bridge = CvBridge()
```

Here are the subscriber and publisher handles of the node:

```
self._sub = rospy.Subscriber('image', Image, self.callback, queue_size=1)
        self._pub = rospy.Publisher('result', String, queue_size=1)
```

Here are some parameters used for recognition thresholding and the number of top predictions:

```
        self.score_threshold = rospy.get_param('~score_threshold', 0.1)
        self.use_top_k = rospy.get_param('~use_top_k', 5)
```

Here is the image call back in which a ROS image message is converted into an OpenCV data type:

```
    def callback(self, image_msg):
        cv_image = self._cv_bridge.imgmsg_to_cv2(image_msg, "bgr8")
        image_data = cv2.imencode('.jpg', cv_image)[1].tostring()
```

The following code runs the `softmax` tensor by feeding `image_data` as input to the graph. The `softmax:0` part is a tensor containing the normalized prediction across 1,000 labels:

```
softmax_tensor = self._session.graph.get_tensor_by_name('softmax:0')
```

The `DecodeJpeg/contents:0` line is a tensor containing a string providing JPEG encoding of the image:

```
        predictions = self._session.run(
            softmax_tensor, {'DecodeJpeg/contents:0': image_data})
        predictions = np.squeeze(predictions)
```

The following section of code will look for a matching object string and its probability and publish it through the topic called `/result`:

```
node_lookup = classify_image.NodeLookup()
top_k = predictions.argsort()[-self.use_top_k:][::-1]
for node_id in top_k:
    human_string = node_lookup.id_to_string(node_id)
    score = predictions[node_id]
    if score > self.score_threshold:
        rospy.loginfo('%s (score = %.5f)' % (human_string, score))
        self._pub.publish(human_string)
```

The following is the main code of this node. It simply initializes the class and calls the `main()` method inside the `RosTensorFlow()` object. The `main()` method will `spin()` the node and execute a callback whenever an image comes into the `/image` topic:

```
    def main(self):
        rospy.spin()
if __name__ == '__main__':
     rospy.init_node('rostensorflow')
     tensor = RosTensorFlow()
     tensor.main()
```

In the next section, we will learn to run the ROS image recognition node.

Running the ROS image recognition node

Let's go through how we can run the image recognition node:

1. First, you have to plug in a UVC webcam. Run `roscore`:

   ```
   $ roscore
   ```

2. Run the webcam driver:

   ```
   $ rosrun cv_camera cv_camera_node
   ```

3. Run the image recognition node, simply using the following command:

   ```
   $ python image_recognition.py image:=/cv_camera/image_raw
   ```

When we run the recognition node, it will download the inception model and extract it into the `/tmp/imagenet` folder. You can do it manually by downloading Inception v3 from the following link: `http://download.tensorflow.org/models/image/imagenet/inception-2015-12-05.tgz`. The source of the data is `https://www.tensorflow.org/datasets/catalog/overview#usage` and it is used under Apache License Version 2.0 (`https://www.apache.org/licenses/LICENSE-2.0`). You can copy this file into the `/tmp/imagenet` folder:

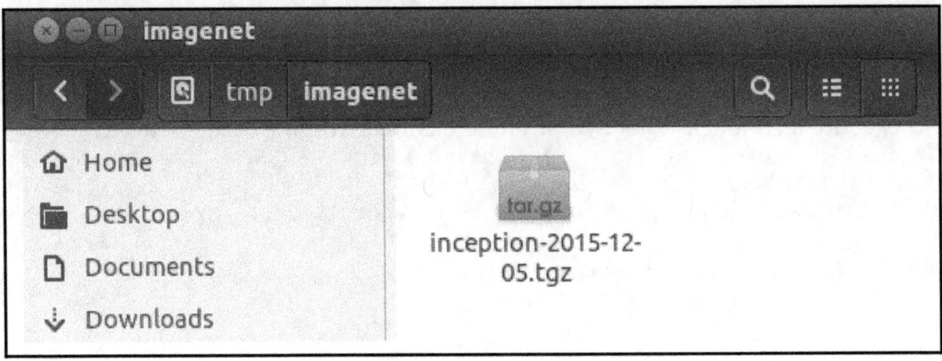

The inception model in the /tmp/imagenet folder

4. You can see the result by echoing the following topic:

```
$ rostopic echo /result
```

5. You can view the camera images using the following command:

```
$ rosrun image_view image_view image:= /cv_camera/image_raw
```

Here is the output from the recognizer. The recognizer detects the device as a cell phone:

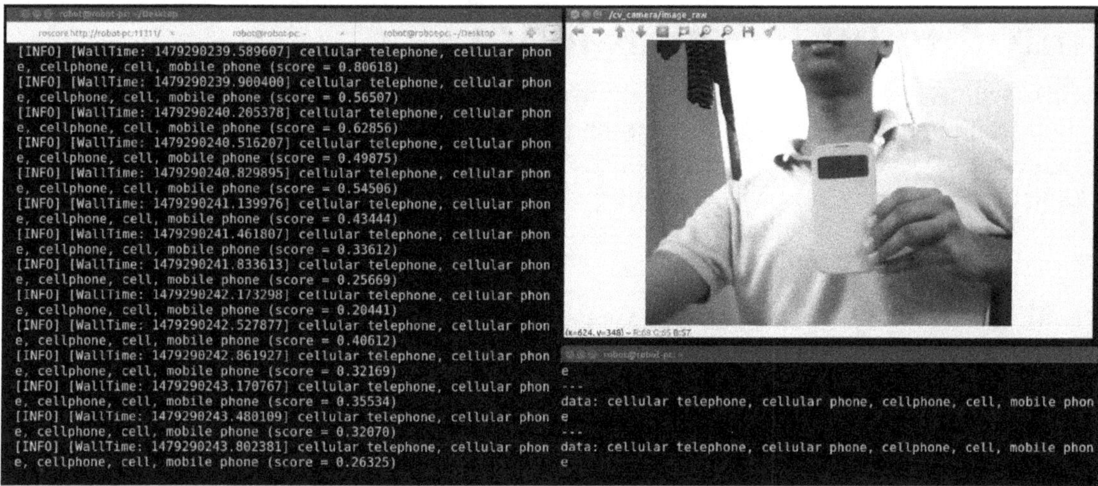

Output from the recognizer node

In the next detection, the object is detected as a water bottle:

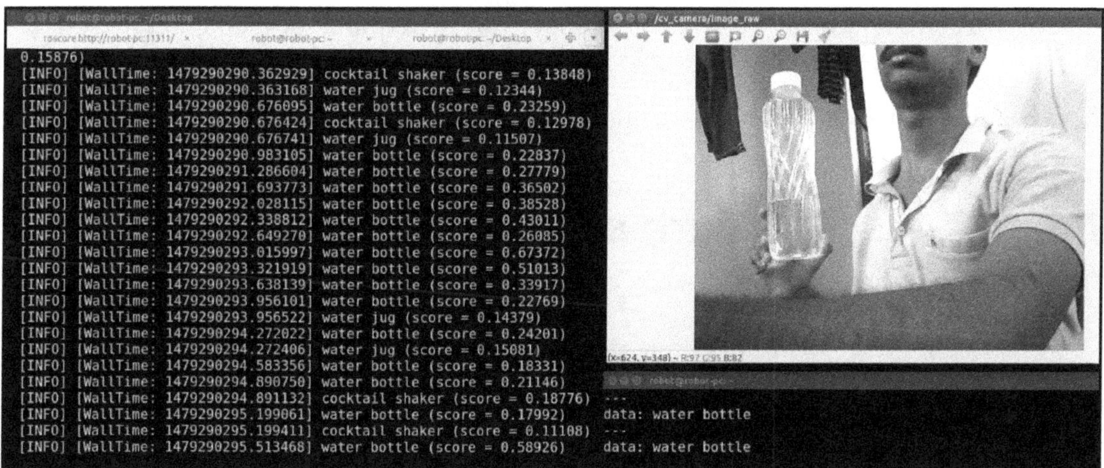

Output from the recognizer node detecting a water bottle

Let's now learn more about scikit-learn.

Introducing to scikit-learn

Until now, we have been discussing deep neural networks and some of their applications in robotics and image processing. Apart from neural networks, there are a lot of models available to classify data and make predictions. Generally, in machine learning, we can teach the model using supervised or unsupervised learning. In supervised learning, we train the model against a dataset, but in unsupervised, it discovers groups of related observations called clusters instead.

There are a lot of libraries available for working with other machine learning algorithms. We'll look at one such library called `scikit-learn`; we can use it to play with most of the standard machine learning algorithms and implement our own application. scikit-learn (`http://scikit-learn.org/`) is one of the most popular open source machine learning libraries for Python. It provides an implementation of algorithms for performing classification, regression, and clustering. It also provides functions to extract features from a dataset, train the model, and evaluate it. scikit-learn is an extension of a popular scientific python library called SciPy (`https://www.scipy.org/`). scikit-learn strongly binds with other popular Python libraries, such as NumPy and Matplotlib. Using NumPy, we can create efficient multidimensional arrays, and using Matplotlib, we can visualize the data. scikit-learn is well documented and has wrappers for performing SVM and natural language processing functions.

Let's begin with installing scikit-learn on Ubuntu 18.04 LTS.

Installing scikit-learn on Ubuntu 18.04 LTS

Installing scikit-learn on Ubuntu is easy and straightforward. You can install it either using `apt-get install` or `pip`.

Here is the command to install scikit-learn using `apt-get install`:

```
$ sudo apt-get install python-sklearn
```

We can install it using `pip` with the following command:

```
$ sudo pip install scikit-learn
```

After installing scikit-learn, we can test the installation with the following commands in Python Terminal:

```
>>> import sklearn
>>> sklearn.__version__
'0.19.1'
```

Congratulations, you have successfully set up scikit-learn! The next section will introduce you to SVM and its application in robotics.

Introduction to SVM and its application in robotics

We have set up scikit-learn, so what is next? Actually, we are going to discuss a popular machine learning technique called SVM and its applications in robotics. After discussing the basics, we can implement an ROS application using SVM.

So, what is SVM? It is a supervised machine learning algorithm that can be used for classification or regression. In SVM, we plot each data item in *n*-dimensional space along with its value. After plotting, it performs classification by finding a hyper-plane that separates those data points. This is how the basic classification is done! SVM can perform better for small datasets, but it does not do well if the dataset is very large. Also, it will not be suitable if the dataset has noisy data. SVM is widely used in robotics, especially in computer vision for classifying objects and various kinds of sensor data in robots.

In the next section, we will see how we can implement SVM and build an application using scikit-learn.

Implementing an SVM-ROS application

In this application, we are going to classify a sensor's data in three ways. The sensor values are assumed to be between 0 to 30,000 and we have a dataset that has the sensor value mapping. For example, for a sensor value, you can assign the value to 1, 2, or 3. To test the SVM, we are making another ROS node called a virtual sensor node, which can publish values between 0 to 30,000. The trained SVM model can classify the virtual sensor value. This method can be adopted for any kind of sensors for classifying its data. Before embedding SVM in ROS, here's some basic code in Python using `sklearn` to implement SVM.

The first thing is importing the `sklearn` and `numpy` modules. The `sklearn` module has the `svm` module, which is going to be used in this code, and `numpy` is used for making multi-dimensional arrays:

```
from sklearn import svm
import numpy as np
```

For training SVM, we need an input (predictor) and output (target); here, `X` is the input and `y` is the required output:

```
X = np.array([[-1, -1], [-2, -1], [1, 1], [2, 1]])
y = np.array([1, 1, 2, 2])
```

After defining `X` and `y`, just create an instance of an **SVM Classification** (**SVC**) object. Feed `X` and `y` to the SVC object for training the model. After feeding `X` and `y`, we can feed an input that may not be in `X`, and it can predict the `y` value corresponding to the given input:

```
model = svm.SVC(kernel='linear',C=1,gamma=1)
model.fit(X,y)
print(model.predict([[-0.8,-1]]))
```

The preceding code will give an output of `1`.

Now, we are going to implement an ROS application that does the same thing. Here, we are creating a virtual sensor node that can publish random values from 0 to 30,000. The ROS-SVM node will subscribe to those values and classify them using the preceding APIs. The learning in SVM is done from a CSV data file. You can view the complete application package in this book's code `https://github.com/PacktPublishing/ROS-Robotics-Projects-SecondEdition/tree/master/chapter_9_ws`; it's called `ros_ml`. Inside the `ros_ml/scripts` folder, you can see nodes such as `ros_svm.py` and `virtual_sensor.py`.

First, let's take a look at the virtual sensor node. The code is very simple and self-explanatory. It simply generates a random number between 0 and 30,000 and publishes it to the `/sensor_read` topic:

```
#!/usr/bin/env python
import rospy
from std_msgs.msg import Int32
import random

def send_data():
        rospy.init_node('virtual_sensor', anonymous=True)
    rospy.loginfo("Sending virtual sensor data")
    pub = rospy.Publisher('sensor_read', Int32, queue_size=1)
```

```
        rate = rospy.Rate(10) # 10hz

        while not rospy.is_shutdown():
          sensor_reading = random.randint(0,30000)
          pub.publish(sensor_reading)
          rate.sleep()

if __name__ == '__main__':
    try:
        send_data()
    except rospy.ROSInterruptException:
        pass
```

The next node is `ros_svm.py`. This node reads from a data file from a data folder inside the `ros_ml` package. The current data file is named `pos_readings.csv`, which contains the sensor and target values. Here is a snippet from that file:

```
5125,5125,1
6210,6210,1
...............

10125,10125,2
6410,6410,2
5845,5845,2
................

14325,14325,3
16304,16304,3
18232,18232,3
..................
```

The `ros_svm.py` node reads this file, trains the SVC, and predicts each value from the virtual sensor topic. The node has a class called `Classify_Data()`, which has methods to read the CSV file and train and predict it using scikit APIs.

We'll step through how these nodes are started:

1. Start `roscore`:

   ```
   $ roscore
   ```

2. Switch to the script folder of `ros_ml`:

   ```
   $ roscd ros_ml/scripts
   ```

3. Run the ROS SVM classifier node:

```
$ python ros_svm.py
```

4. Run the virtual sensor in another Terminal:

```
$ rosrun ros_ml virtual_sensor.py
```

Here is the output we get from the SVM node:

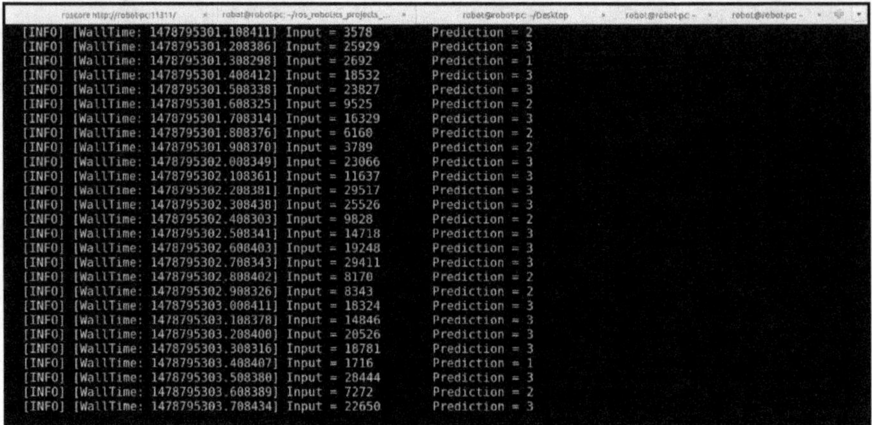

ROS-SVM node output

Summary

In this chapter, we mainly discussed the various machine learning techniques and libraries that can be interfaced with ROS. We started with the basics of machine learning and deep learning. Then, we started working with TensorFlow, which is an open source Python library mainly for performing deep learning. We discussed basic code using TensorFlow and later combined those capabilities with ROS for an image recognition application.

After discussing TensorFlow and deep learning, we discussed another Python library called scikit-learn, which is used for machine learning applications. We saw what SVM is and examined how to implement it using scikit-learn. Later, we implemented a sample application using ROS and scikit-learn for classifying sensor data. This chapter provided us with an overview of the integration of Tensor Flow in ROS for deep learning applications.

In the next chapter, we will look at how autonomous cars work and try simulating them in Gazebo.

10
Creating a Self-Driving Car Using ROS

In this chapter, we will discuss a technology that is trending in the robotics industry: driverless or self-driving cars. Many of you may have heard about this technology; those who haven't will get an introduction in the first section of this chapter.

In this chapter, we will first start with a software block diagram of a typical self-driving car. We will then learn how to simulate and interface self-driving car sensors in ROS. We will also cover the interfacing of a drive-by-wire car into ROS, visualize the car virtually, and read its sensor information. Creating a self-driving car from scratch is out of the scope of this book, but this chapter will give you an abstract idea of self-driving car components and tutorials to simulate it.

This chapter will cover the following important topics:

- Getting started with self-driving cars
- A software block diagram of a typical self-driving car
- Simulating and interfacing self-driving car sensors in ROS
- Simulating a self-driving car with sensors in Gazebo
- Interfacing a DBW car into ROS
- Introducing the Udacity open source self-driving car project
- The open source self-driving car simulator from Udacity

Technical requirements

Let's look into the technical requirements for this chapter:

- ROS Melodic Morenia on Ubuntu 18.04, with Gazebo 9 (preinstalled)
- ROS packages: `velodyne-simulator`, `sensor-sim-gazebo`, `hector-slam`, and others
- Timelines and test platforms:
    - **Estimated learning time**: On average, 90 to 120 minutes
    - **Project build time (inclusive of compile and runtime)**: On average, 90 minutes
    - **Project test platform**: HP Pavilion laptop (Intel® Core™ i7-4510U CPU @ 2.00 GHz × 4 with 8 GB memory and 64-bit OS, and GNOME-3.28.2)

The code for this chapter is available at `https://github.com/PacktPublishing/ROS-Robotics-Projects-SecondEdition/tree/master/chapter_10_ws`.

Let's now get started with understanding self-driving cars.

Getting started with self-driving cars

Just imagine a car driving itself without the help of anyone. Self-driving cars are like robot cars that can think and decide which path to navigate to reach their destination. You only need to specify the destination, and a robot car will take you there safely. To convert an ordinary car into a robotic car, we should add some robotic sensors to it. We know that, for a robot, there should be at least three important capabilities. It should be able to sense, plan, and act. Self-driving cars satisfy all of these requirements. We'll discuss all of the components we need to build a self-driving car. Before discussing the building of a self-driving car, let's go through some milestones in self-driving car development.

The history of autonomous vehicles

The concept of automating vehicles started long ago. Since 1930, people have been trying to automate cars and aircraft, but the hype around self-driving cars surfaced somewhere between 2004 and 2013. To encourage autonomous vehicle technology, the U.S. Department of Defense's research arm, DARPA, conducted a challenge called the DARPA Grand Challenge in 2004.

The aim of the challenge was to autonomously drive for 150 miles on a desert roadway.

In this challenge, no team was able to complete the goal, so they again challenged engineers in 2007, but this time, the aim was slightly different. Instead of a desert roadway, there was an urban environment spread across 60 miles. In this challenge, four teams were able to finish the goal.

The winner of the challenge was Team Tartan Racing from Carnegie Mellon University (`http://www.tartanracing.org/`). The second-place team was Stanford Racing from Stanford University (`http://cs.stanford.edu/group/roadrunner/`).

After the DARPA Challenge, car companies started working hard to implement autonomous driving capabilities in their cars. Now, almost all car companies have their own autonomous car prototype.

In 2009, Google started to develop its self-driving car project, now known as Waymo (`https://waymo.com/`). This project greatly influenced other car companies and was lead by Sebastian Thrun (`http://robots.stanford.edu/`), the former director of the Stanford Artificial Intelligence Laboratory (`http://ai.stanford.edu/`).

The car autonomously traveled around 2.7 million kilometers in 2016. Take a look at it:

The Google self-driving car (source: https://commons.wikimedia.org/wiki/File:Google_self_driving_car_at_the_Googleplex.jpg. Image by Michael Shick. Licensed under Creative Commons CC-BY-SA 4.0: https://creativecommons.org/licenses/by-sa/4.0/legalcode)

In 2015, Tesla Motors introduced a semi-autonomous autopilot feature in their electric cars. It enables hands-free driving mainly on highways and other roads. In 2016, NVIDIA introduced their own self-driving car (`http://www.nvidia.com/object/drive-px.html`), built using their AI car computer called NVIDIA-DGX-1 (`http://www.nvidia.com/object/deep-learning-system.html`). This computer was specially designed for the self-driving car and is the best for developing autonomous training driving models.

Other than self-driving cars, there are self-driving shuttles for campus mobility. A lot of start-ups are building self-driving shuttles now, and one of these start-ups is called Auro Robotics (`http://www.auro.ai/`).

One of the most common terms used when describing autonomous cars is the level of autonomy. Let's understand what this means in the next section

Levels of autonomy

Let's go through the different levels of autonomy used to describe an autonomous vehicle:

- **Level 0**: Vehicles having level 0 autonomy are completely controlled by a human driver. Most old cars belong in this category.
- **Level 1**: Vehicles with level 1 autonomy will have a human driver, but they will also have a driver assistance system that can either automatically control the steering system or acceleration/deceleration using information from the environment. All other functions have to be controlled by the driver.
- **Level 2**: In level 2 autonomy, the vehicle can perform both steering and acceleration/deceleration. All other tasks have to be controlled by the driver. We can say that the vehicle is partially automated at this level.
- **Level 3**: At this level, it is expected that all tasks will be performed autonomously, but at the same time, it is expected that a human will intervene whenever required. This level is called conditional automation.
- **Level 4**: At this level, there is no need for a driver; everything is handled by an automated system. This kind of autonomous system will work in a particular area under specified weather conditions. This level is called high automation.
- **Level 5**: This level is called full automation. At this level, everything is heavily automated and can work on any road and in any weather conditions. There is no need for a human driver.

Let's look at various components of a self-driving car in the next section.

Components of a typical self-driving car

The following diagram shows the important components of a self-driving vehicle. The list of parts and their functionalities will be discussed in this section. We'll also look at the exact sensor that was used in the autonomous car for the DARPA Challenge:

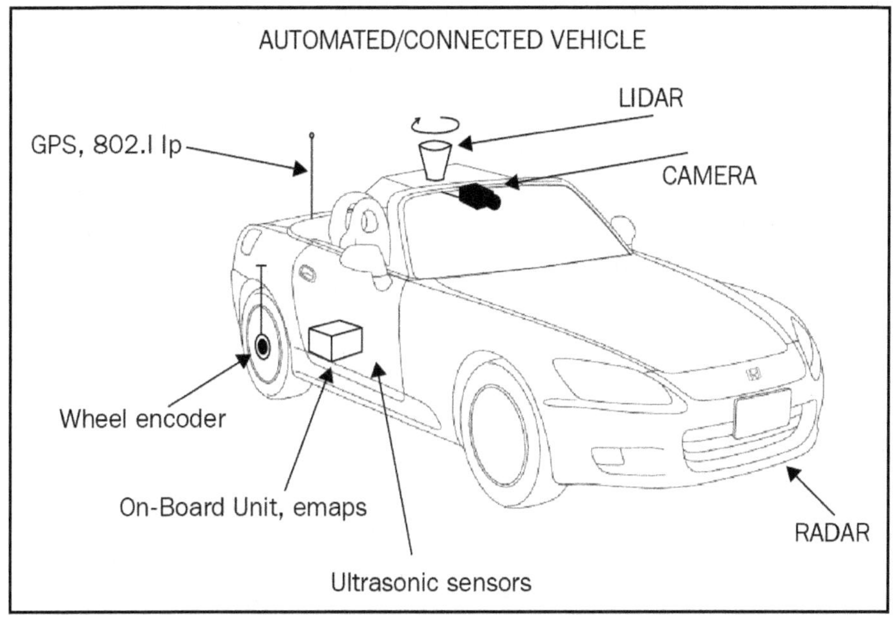

Important components of a self-driving car

Let's look at the important components starting with the GPS, IMU, and wheel encoders.

GPS, IMU, and wheel encoders

As you know, the **Global Positioning System** (**GPS**) helps us to determine the global position of a vehicle with the help of GPS satellites. The latitude and longitude of the vehicle can be calculated from the GPS data. The accuracy of GPS can vary with the type of sensor; some sensors have an error range of within a few meters, and some have one within less than a meter. We can find vehicle state by combining GPS, **Inertial Measurement Unit** (**IMU**), and wheel odometer data and by using the sensor fusion algorithms. This can give a better estimate of the vehicle. Let's look at the position estimation modules used for the DARPA Challenge 2007.

The POS LV modules from Applanix are the modules used in the Stanford autonomous car called Junior. They are a combination of GPS, IMU, and wheel encoders or **Distance Measurement Indicator** (**DMI**). You can find them at `http://www.applanix.com/products/poslv.htm`.

The OxTS module is another GPS/IMU combo module from **Oxford Technical Solution** (**OxTS**) (`http://www.oxts.com/`). This module was extensively used in the DARPA Challenge in 2007. The module is from the RT 3000 v2 family (`http://www.oxts.com/products/rt3000-family/`). The entire range of GPS modules from OxTS can be found at `http://www.oxts.com/industry/automotive-testing/`.

Xsens MTi IMU

The Xsens MTi series has independent IMU modules that can be used in autonomous cars. Here is the link to purchase this product: `http://www.xsens.com/products/mti-10-series/`.

Camera

Most autonomous vehicles are deployed with stereo or monocular cameras to detect various things, such as traffic signal status, pedestrians, cyclists, and vehicles. Companies such as Mobileye (`http://www.mobileye.com/`), which was acquired by Intel, built their **advanced driving assistance system** (**ADAS**) using a sensor fusion of cameras and LIDAR data to predict obstacles and path trajectory.

Other than ADAS, we can also use our own control algorithms by only using camera data. One of the cameras used by the Boss robot car in DARPA 2007 was **Point Grey Firefly** (**PGF**) (`https://www.ptgrey.com/firefly-mv-usb2-cameras`).

Ultrasonic sensors

In an ADAS system, ultrasonic sensors play an important role in the parking of vehicles, avoiding obstacles in blind spots, and detecting pedestrians. One of the companies providing ultrasound sensors for ADAS systems is Murata (`http://www.murata.com/`). They provide ultrasonic sensors for up to 10 meters, which are optimal for a **Parking Assistance System** (**PAS**). The following diagram shows where ultrasonic sensors are placed on a car:

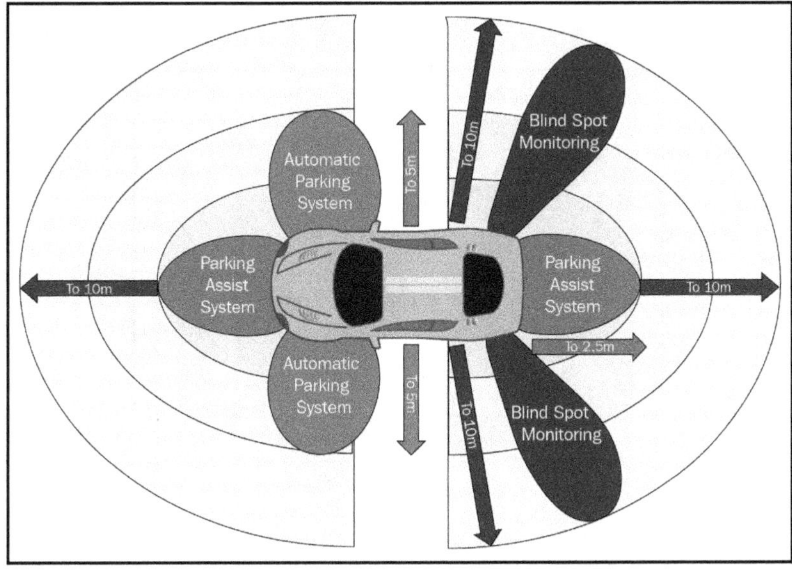

Placement of ultrasonic sensors for PAS

Let's now look at LIDAR and RADAR and learn how they are used in a self-driving car.

LIDAR and RADAR

LIDAR (short for **Light Detection and Ranging**) (`http://oceanservice.noaa.gov/facts/lidar.html`) sensors are the core sensors of a self-driving car. A LIDAR sensor basically measures the distance to an object by sending a laser signal and receiving its reflection. It can provide accurate 3D data of the environment, computed from each received laser signal. The main application of LIDAR in autonomous cars is mapping the environment from the 3D data, obstacle avoidance, object detection, and so on.

Some of the LIDARs used in the DARPA Challenge will be discussed now.

Velodyne HDL-64 LIDAR

The Velodyne HDL-64 sensor is designed for obstacle detection, mapping, and navigation for autonomous cars. It can give us 360-degree view laser-point cloud data with a high data rate. The range of this laser scan is 80 to 120 meters. This sensor is used for almost all self-driving cars available today.

Here are a few of them:

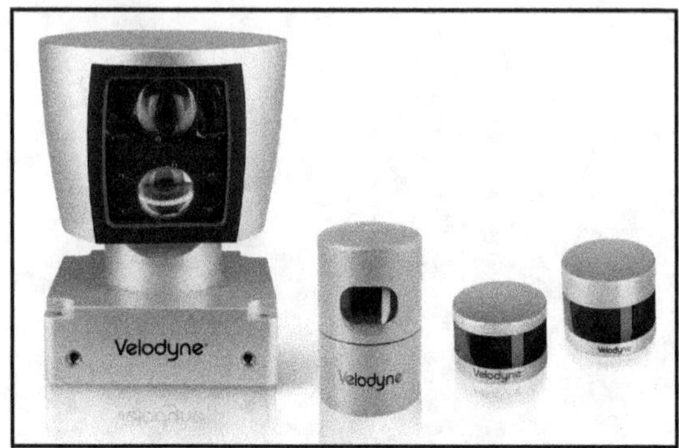

Some Velodyne sensors (source: https://commons.wikimedia.org/wiki/File:Velodyne_ProductFamily_BlueLens_32GreenLens.png. Image by APJarvis. Licensed under Creative Commons CC-BY-SA 4.0: https://creativecommons.org/licenses/by-sa/4.0/deed.en)

Let's now look at the SICK LMS 5xx/1xx and Hokuyo LIDAR.

SICK LMS 5xx/1xx and Hokuyo LIDAR

The company SICK provides a variety of laser scanners that can be used indoor or outdoor. The SICK **Laser Measurement System** (**LMS**) 5xx and 1xx models are commonly used in autonomous cars for obstacle detection. They provide a scanning range of 180 degrees and have high-resolution laser data. The list of SICK laser scanners available on the market can be found at `https://www.sick.com/in/en`. Check out the following example:

Mobile robot with a SICK laser (source: https://commons.wikimedia.org/wiki/File:LIDAR_equipped_mobile_robot.jpg)

Another company, called Hokuyo, also builds laser scanners for autonomous vehicles. Here is the link to a list of laser scanners provided by Hokuyo: `http://www.hokuyo-aut.jp/02sensor/`.

Some of the other LIDARs used in the DARPA Challenge are given here:

- **Continental**: `http://www.conti-online.com/www/industrial_sensors_de_en/`
- **Ibeo**: `https://www.ibeo-as.com/en/produkte`

Let's now look at the continental ARS 300 radar (ARS).

Continental ARS 300 radar (ARS)

Apart from LIDARs, self-driving cars are also deployed with long-range radars. One of the popular long-range radars is ARS 30X by Continental (`https://www.continental-automotive.com/getattachment/9b6de999-75d4-4786-bb18-8ab64fd0b181/ARS30X-Datasheet-EN.pdf.pdf`). It works using the Doppler principle and can measure up to 200 meters. Bosch also manufactures radars suitable for self-driving cars. The main application of radars is collision avoidance. Commonly, radars are deployed at the front of the vehicles.

The Delphi radar

Delphi has a new radar for autonomous cars. Here is the link to view the product: `https://autonomoustuff.com/product/delphi-esr-2-5-24v/`.

Onboard computer

The onboard computer is the heart of the self-driving car. It may have high-end processors such as Intel Xenon and GPUs to crunch data from various sensors. All sensors are connected to this computer, and it finally predicts the trajectory and sends control commands, such as steering angle, throttle, and braking for the self-driving car.

We will be able to learn more about self-driving cars by understanding the software block diagram. We will be covering that in the next section.

Software block diagram of self-driving cars

In this section, we will discuss a basic software block diagram of a self-driving car that was in the DARPA Challenge:

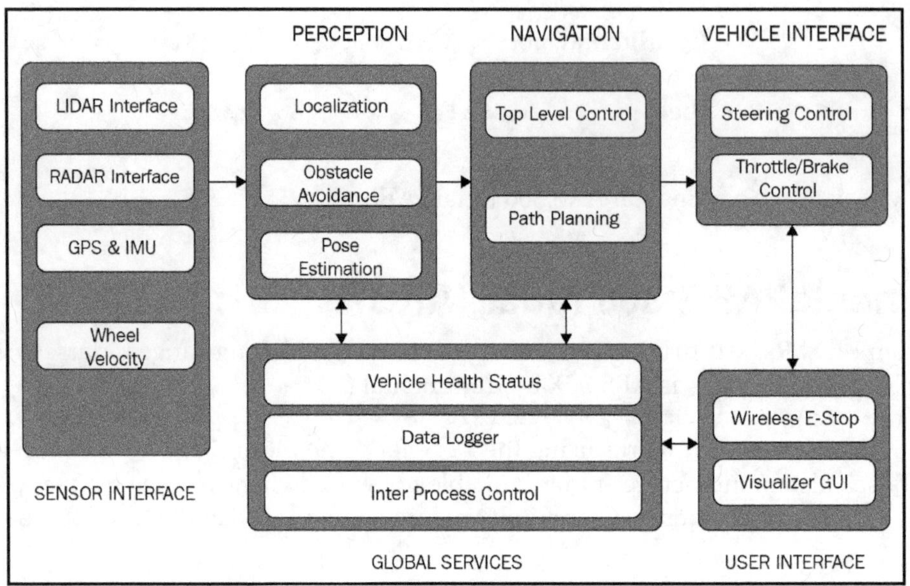

Software block diagram of a self-driving car

Each block can interact with others using **Inter-Process Communication** (**IPC**) or shared memory. ROS messaging middleware is a perfect fit in this scenario. In the DARPA Challenge, they implemented a publish/subscribe mechanism to do these tasks. One of the IPC library development by MIT for the 2006 DARPA Challenge was **Lightweight Communications and Marshalling** (**LCM**). You can learn more about LCM at the following link: `https://lcm-proj.github.io/`.

Let's learn what each block means:

- **Sensor interface modules**: As the name of the module indicates, all of the communication between the sensors and the vehicle is done in this block. The block enables us to provide the various kinds of sensor data to all other blocks. The main sensors include LIDAR, camera, radar, GPS, IMU, and wheel encoders.
- **Perception modules**: These modules perform processing on perception data from sensors such as LIDAR, camera, and radar and segment the data to find moving and static objects. They also help to localize the self-driving car relative to the digital map of the environment.

- **Navigation modules**: These modules determine the behavior of the autonomous car. It has motion planners and finite state machines for different behaviors in the robot.
- **Vehicle interface**: After the path planning, the control commands, such as steering, throttle, and brake control, are sent to the vehicle through a **Drive-By-Wire** (**DBW**) interface. DBW basically works through the CAN bus. Only some vehicles support the DBW interface. Some examples are the Lincoln MKZ, VW Passat Wagon, and some models from Nissan.
- **User interface**: The user interface section provides controls for the user. It can be a touch screen to view maps and set the destination. Also, there is an emergency stop button for the user.
- **Global services**: This set of modules helps to log the data and has time stamping and message-passing support to keep the software running reliably.

Now that we know the basics of self-driving cars, we can now start simulating and interfacing self-driving car sensors in ROS.

Simulating and interfacing self-driving car sensors in ROS

In the preceding section, we discussed the basic concepts of a self-driving car. That understanding will definitely help in this section too. In this section, we will simulate and interface some of the sensors that we are used in self-driving cars.

Here is the list of sensors that we are going to simulate and interface with ROS:

- Velodyne LIDAR
- Laser scanner
- Camera
- Stereo camera
- GPS
- IMU
- Ultrasonic sensor

We'll discuss how to set up the simulation using ROS and Gazebo and read the sensor values. This sensor interfacing will be useful when you build your own self-driving car simulation from scratch. So, if you know how to simulate and interface these sensors, it can definitely accelerate your self-driving car development. Let's start by simulating the Velodyne LIDAR.

Simulating the Velodyne LIDAR

The Velodyne LIDAR is becoming an integral part of a self-driving car. Because of high demand, there are enough software modules available to work with this sensor. We are going to simulate two popular models of Velodyne, called HDL-32E and VLP-16. Let's see how to do it in ROS and Gazebo.

In ROS Melodic, we can install from a binary package or compile from source code by following these steps:

1. Run the command to install Velodyne packages on ROS Melodic:

   ```
   $ sudo apt-get install ros-melodic-velodyne-simulator
   ```

2. To install it from source code, just clone the source package to the ROS workspace using the following command:

   ```
   $ git clone
   https://bitbucket.org/DataspeedInc/velodyne_simulator.git
   ```

3. After cloning the package, you can build it using the `catkin_make` command. Here is the ROS wiki page of the Velodyne imulator: `http://wiki.ros.org/velodyne_simulator`.
4. Now that we have installed the packages, it's time to start the simulation of the Velodyne sensor. You can start the simulation using the following command:

   ```
   $ roslaunch velodyne_description example.launch
   ```

This command will launch the sensor simulation in Gazebo.

Note that this simulation will consume a lot of the RAM of your system; your system should have at least 8 GB before you start the simulation.

You can add some obstacles around the sensor for testing, as shown in the following screenshot:

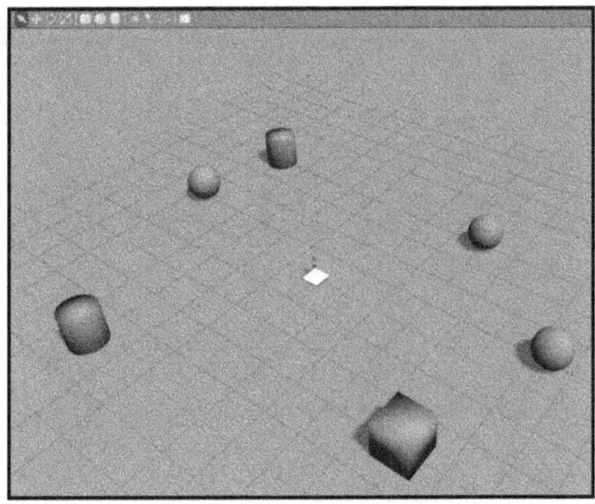

Simulation of Velodyne in Gazebo

You can visualize the sensor data in RViz by adding display types such as PointCloud2 and Robot Model to visualize sensor data and sensor models. You have to set the **Fixed Frame** to `velodyne`. You can clearly see the obstacles around the sensor in the following screenshot:

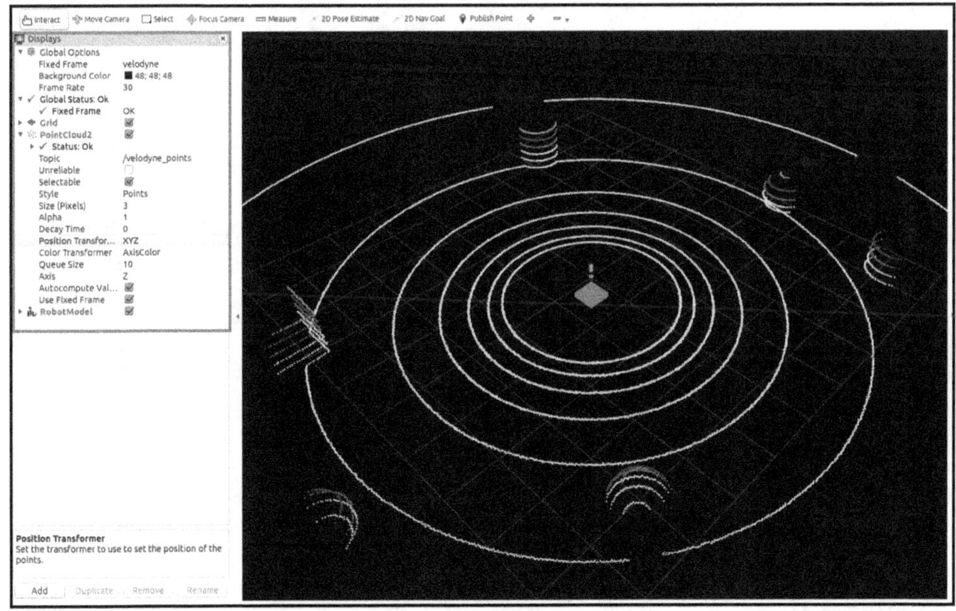

Visualization of a Velodyne sensor in RViz

We will now interface Velodyne sensors with ROS.

Interfacing Velodyne sensors with ROS

We have seen how to simulate a Velodyne sensor; now let's have a look at how we can interface a real Velodyne sensor with ROS.

The following commands are to install the Velodyne ROS driver package to convert Velodyne data into point cloud data.

For ROS Melodic, install the package using the following command:

```
$ sudo apt-get install ros-melodic-velodyne
```

Here is the command to start the driver nodelets:

```
$ roslaunch velodyne_driver nodelet_manager.launch model:=32E
```

In the preceding command, you need to mention the model name along with the launch file to start the driver for a specific model.

The following command will start the converter nodelets to convert Velodyne messages (`velodyne_msgs/VelodyneScan`) into a point cloud (`sensor_msgs/PointCloud2`). Here is the command to perform this conversion:

```
$ roslaunch velodyne_pointcloud cloud_nodelet.launch
calibration:=~/calibration_file.yaml
```

This will launch the calibration file for Velodyne, which is necessary for correcting noise from the sensor.

We can write all of these commands to a launch file, which is shown in the following code block. If you run this launch file, the driver node and point cloud convertor nodelets will start, and we can work with the sensor data:

```
<launch>
  <!-- start nodelet manager and driver nodelets -->
  <include file="$(find velodyne_driver)/launch/nodelet_manager.launch" />

  <!-- start transform nodelet -->
  <include file="$(find
velodyne_pointcloud)/launch/transform_nodelet.launch">
    <arg name="calibration"
         value="$(find velodyne_pointcloud)/params/64e_utexas.yaml"/>
  </include>
</launch>
```

The calibration files for each model are available in the velodyne_pointcloud package. We can now move on to simulating a laser scanner.

Note: The connection procedure for Velodyne to PC is given here: http://wiki.ros.org/velodyne/Tutorials/Getting%20Started%20with%20the%20HDL-32E.

Simulating a laser scanner

In this section, we will see how to simulate a laser scanner in Gazebo. We can simulate it by providing custom parameters according to our application. When you install ROS, you also automatically install several default Gazebo plugins, which includes the Gazebo laser scanner plugin.

We can simply use this plugin and apply our custom parameters. For demonstration, you can use a tutorial package inside chapter_10_ws called sensor_sim_gazebo (https://github.com/PacktPublishing/ROS-Robotics-Projects-SecondEdition/tree/master/chapter_10_ws/sensor_sim_gazebo). You can simply copy the package to the workspace and build it using the catkin_make command. This package contains a basic simulation of the laser scanner, camera, IMU, ultrasonic sensor, and GPS.

Before starting with this package, you should install a package called hector-gazebo-plugins using the following command:

```
$ sudo apt-get install ros-melodic-hector-gazebo-plugins
```

This package contains Gazebo plugins of several sensors that can be used in self-driving car simulations.

To start the laser scanner simulation, just use the following command:

```
$ roslaunch sensor_sim_gazebo laser.launch
```

We'll first look at the output of the laser scanner and then dig into the code.

When you launch the preceding command, you will see an empty world with an orange box. The orange box is our laser scanner. You can use any mesh file to replace this shape according to your application. To show laser scan data, we can place some objects in Gazebo, as shown here:

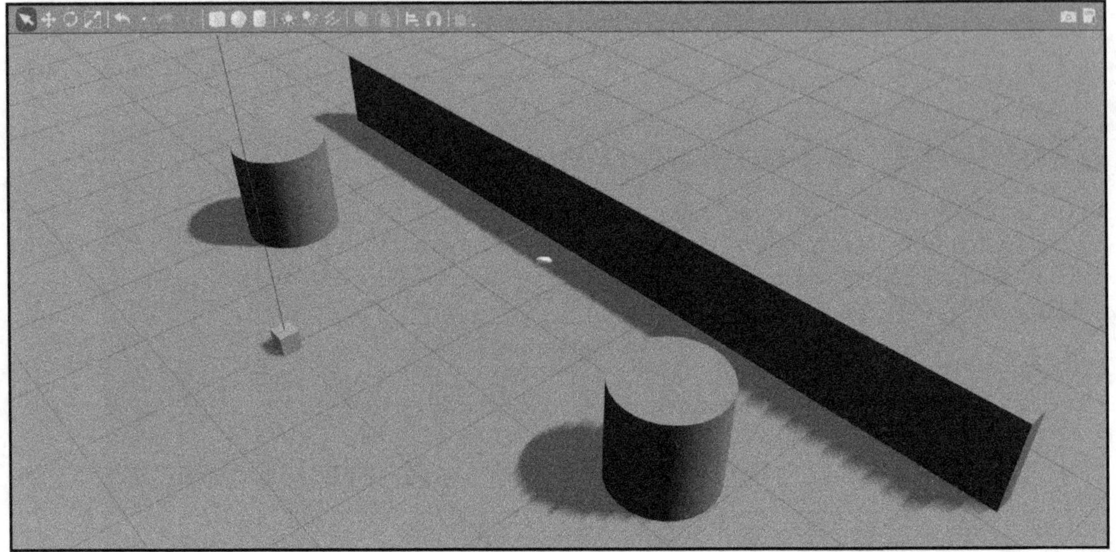

Simulation of a laser scanner in Gazebo

You can add models from Gazebo's top panel. You can visualize the laser data in RViz, as shown in the next screenshot. The topic to which the laser data is coming is `/laser/scan`. You can add a **LaserScan** display type to view this data:

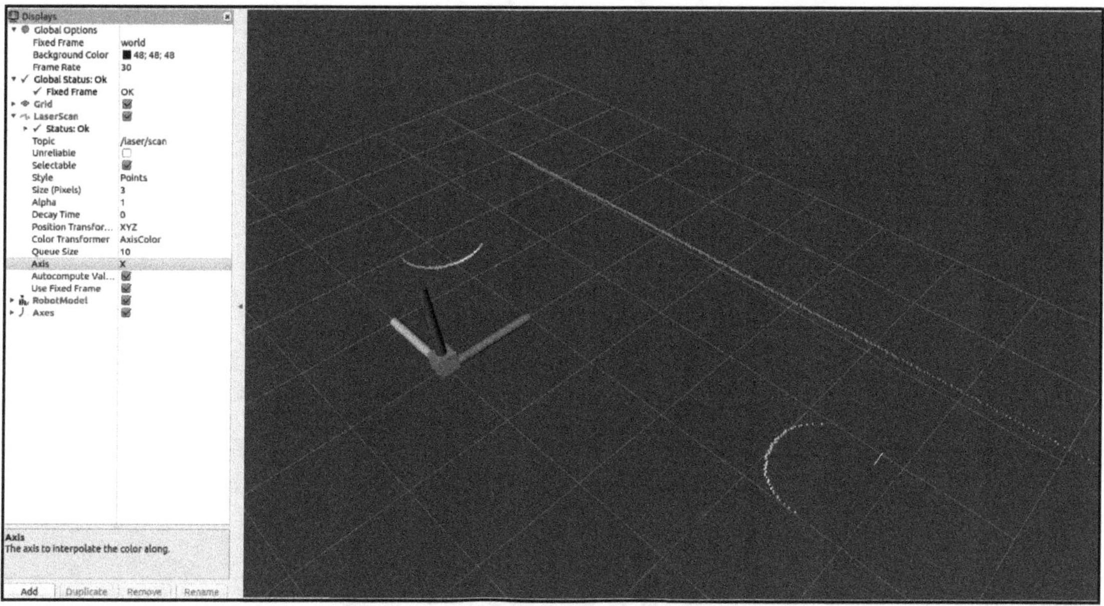

Visualization of laser scanner data in RViz

You have to set the **Fixed Frame** to a `world` frame and enable the **RobotModel** and **Axes** display types in RViz. The following is the list of topics generated while simulating this sensor:

```
robot@robot-pc:~$ rostopic list
/clicked_point
/clock
/gazebo/link_states
/gazebo/model_states
/gazebo/parameter_descriptions
/gazebo/parameter_updates
/gazebo/set_link_state
/gazebo/set_model_state
/initialpose
/joint_states
/laser/scan
/move_base_simple/goal
/rosout
/rosout_agg
/tf
/tf_static
```

List of topics from the laser scanner simulation

You should be able to see /laser/scan as highlighted in the preceding screenshot. Let's understand how the code works.

Explaining the simulation code

The sensor_sim_gazebo package has the following list of files for simulating all self-driving car sensors. Here is the directory structure of this package:

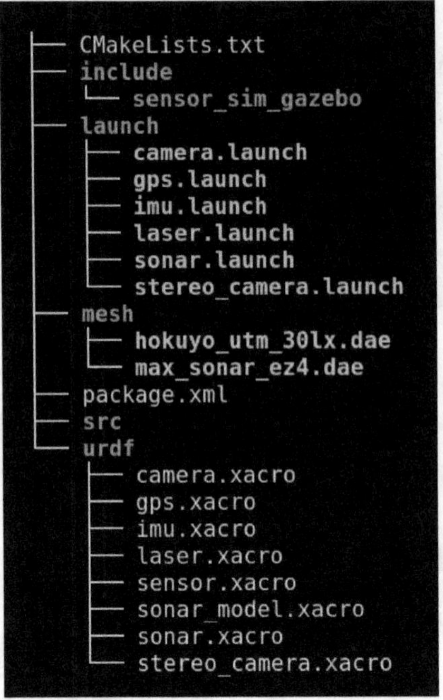

List of files in sensor_sim_gazebo

To simulate a laser, launch the laser.launch file; similarly, to start simulating the IMU, GPS, and camera, launch the corresponding launch files. Inside URDF, you can see the Gazebo plugin definition for each sensor.

The sensor.xacro file is the orange box definition that you saw in the preceding simulation. It is just a box for visualizing a sensor model. We are using this model to represent all of the sensors inside this package. You can use your own model instead of this, too.

The `laser.xacro` file has the Gazebo plugin definition of the laser, as shown in our repository here: `https://github.com/PacktPublishing/ROS-Robotics-Projects-SecondEdition/blob/master/chapter_10_ws/sensor_sim_gazebo/urdf/laser.xacro`.

Here, you can see various parameters of the laser scanner plugin. We can fine-tune these parameters for our custom applications. The plugin we've used here is `libgazebo_ros_laser.so`, and all of the parameters are passed to this plugin.

In the `laser.launch` file, we are creating an empty world and spawning the `laser.xacro` file. Here is the code snippet to spawn the model in Gazebo and start a joint-state publisher to start publishing TF data:

```
  <param name="robot_description" command="$(find xacro)/xacro --inorder
'$(find sensor_sim_gazebo)/urdf/laser.xacro'" />

  <node pkg="gazebo_ros" type="spawn_model" name="spawn_model" args="-urdf
-param /robot_description -model example"/>

  <node pkg="robot_state_publisher" type="robot_state_publisher"
name="robot_state_publisher">
    <param name="publish_frequency" type="double" value="30.0" />
  </node>
```

Now that we have seen how the laser plugin works, let's see how to interface real hardware in ROS in the next section.

Interfacing laser scanners with ROS

Now that we've discussed the simulation of the laser scanner, let's see how to interface real sensors with ROS.

Here are some links to guide you with setting up Hokuyo and SICK laser scanners in ROS. The complete installation instructions are available at the following links:

- **Hokuyo sensors:** `http://wiki.ros.org/hokuyo_node`
- **SICK lasers:** `http://wiki.ros.org/sick_tim`

You can install Hokuyo drivers from source packages by cloning the following package into your workspace:

- For 2D Hokuyo sensors, clone the following:

```
$ git clone https://github.com/ros-drivers/hokuyo_node.git
$ cd ~/workspace_ws
$ catkin_make
```

- For 3D Hokuyo sensors, clone the following:

```
$ git clone https://github.com/at-wat/hokuyo3d.git
$ cd ~/workspace_ws
$ catkin_make
```

For SICK laser scanners, installation can be done directly using binary packages:

```
$ sudo apt-get install ros-melodic-sick-tim ros-melodic-lms1xx
```

In the next section, we will simulate stereo and mono cameras in Gazebo.

Simulating stereo and mono cameras in Gazebo

In the previous section, we discussed laser scanner simulation. In this section, we will see how to simulate a camera. A camera is an important sensor for all kinds of robots. We will see how to launch both mono and stereo camera simulations.

You can use the following command to launch the simulations for a mono camera:

```
$ roslaunch sensor_sim_gazebo camera.launch
```

You can use the following command to launch the simulations for a stereo camera:

```
$ roslaunch sensor_sim_gazebo stereo_camera.launch
```

You can view the image from the camera either using RViz or using a tool called `image_view`.

You can look at the mono camera view using the following command:

```
$ rosrun image_view image_view image:=/sensor/camera1/image_raw
```

You should see the image window as shown here:

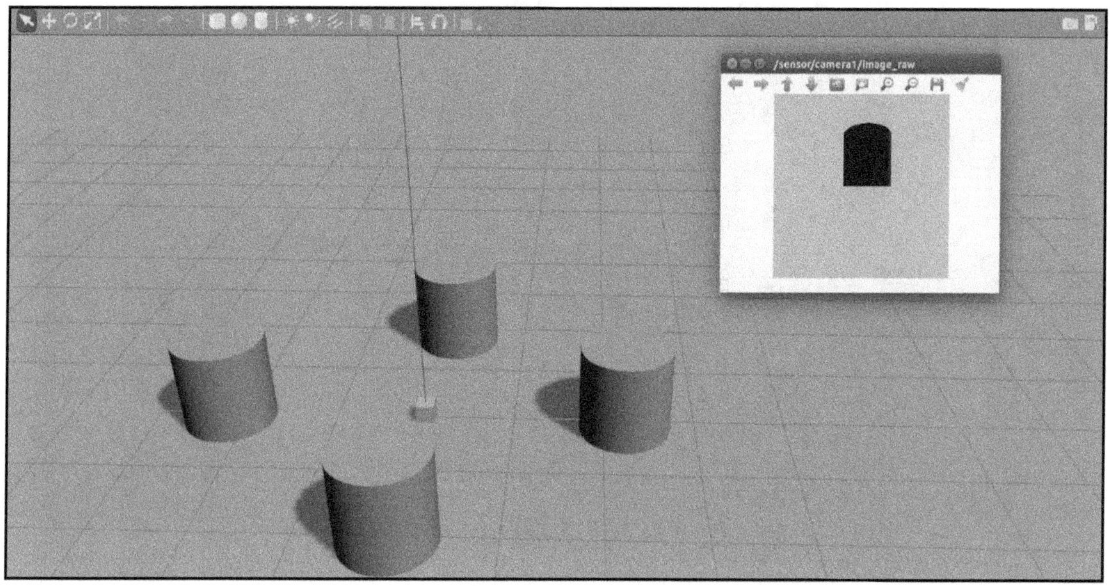

Image from simulated camera

To view images from a simulated stereo camera, use the following commands:

```
$ rosrun image_view image_view image:=/stereo/camera/right/image_raw
$ rosrun image_view image_view image:=/stereo/camera/left/image_raw
```

This will display two image windows from each camera of the stereo camera, as shown here:

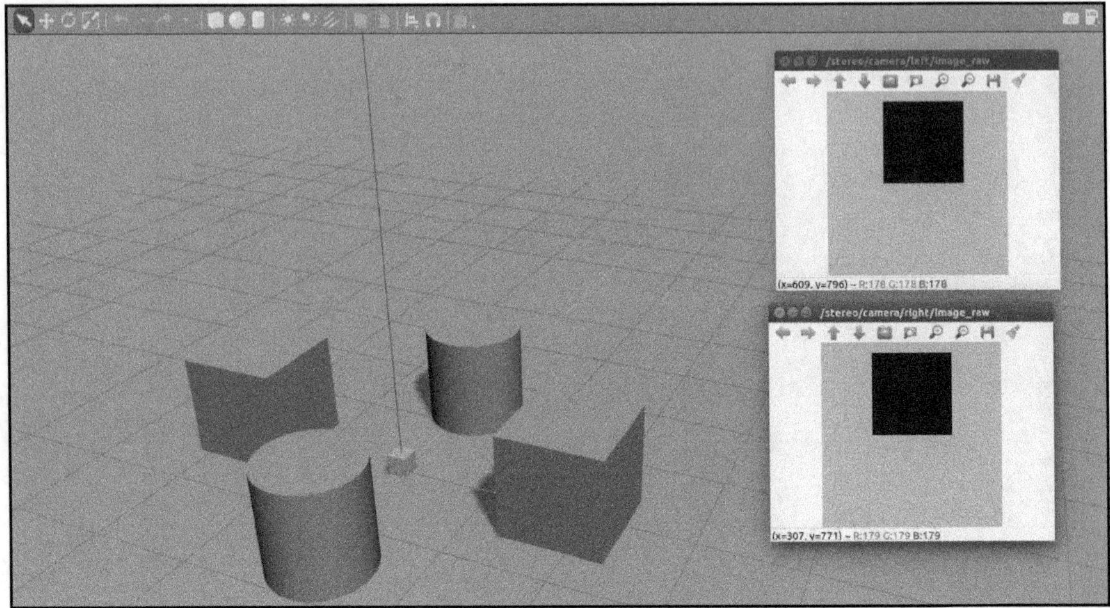

Image from simulated stereo camera

Similar to the laser scanner plugin, we are using a separate plugin for mono and stereo cameras. You can see the Gazebo plugin definition in `sensor_sim_gazebo/urdf/camera.xacro` and `stereo_camera.xacro` here, in our repository: `https://github.com/PacktPublishing/ROS-Robotics-Projects-SecondEdition/tree/master/chapter_10_ws/sensor_sim_gazebo/urdf`.

The `lib_gazebo_ros_camera.so` plugin is used to simulate a mono camera and `libgazebo_ros_multicamera.so` for a stereo camera. We will now learn how to interface the cameras with ROS.

Interfacing cameras with ROS

In this section, we will see how to interface an actual camera with ROS. There are a lot of cameras available on the market. We'll look at some of the commonly used cameras and their interfacing.

There are some links to guide you with setting up each driver in ROS:

- For the Point Gray camera, you can refer to the following link: `http://wiki.ros.org/pointgrey_camera_driver`.
- If you are working with a Mobileye sensor, you can get ROS drivers by contacting the company. All details of the drivers and their SDKs are available at the following link: `https://autonomoustuff.com/product/mobileye-camera-dev-kit`.
- If you are working on IEEE 1394 digital cameras, the following drivers can be used to interface with ROS: `http://wiki.ros.org/camera1394`.
- One of the latest stereo cameras available is the ZED camera (`https://www.stereolabs.com/`). The ROS drivers of this camera are available at the following link: `http://wiki.ros.org/zed-ros-wrapper`.
- If you are working with a normal USB web camera, the `usb_cam` driver package will be best for interfacing with ROS: `http://wiki.ros.org/usb_cam`.

In the next section, we will learn how to simulate GPS in Gazebo.

Simulating GPS in Gazebo

In this section, we will see how to simulate a GPS sensor in Gazebo. As you know, GPS is one of the essential sensors in a self-driving car. You can start a GPS simulation using the following command:

```
$ roslaunch sensor_sim_gazebo gps.launch
```

Now, you can list out the topic and find the GPS topics published from the Gazebo plugin. Here is a list of topics from the GPS plugin:

```
robot@robot-pc:~$ rostopic list
/clock
/gazebo/link_states
/gazebo/model_states
/gazebo/parameter_descriptions
/gazebo/parameter_updates
/gazebo/set_link_state
/gazebo/set_model_state
/gps/fix
/gps/fix/position/parameter_descriptions
/gps/fix/position/parameter_updates
/gps/fix/status/parameter_descriptions
/gps/fix/status/parameter_updates
/gps/fix/velocity/parameter_descriptions
/gps/fix/velocity/parameter_updates
/gps/fix_velocity
/joint_states
/rosout
/rosout_agg
/tf
/tf_static
```

List of topics from the Gazebo GPS plugin

You can echo the `/gps/fix` topic to confirm that the plugin is publishing the values correctly.

You can use the following command to echo this topic:

```
$ rostopic echo /gps/fix
```

You should see an output as shown here:

```
robot@robot-pc:~$ rostopic echo /gps/fix
header:
  seq: 161
  stamp:
    secs: 40
    nsecs: 500000000
  frame_id: sensor
status:
  status: 0
  service: 0
latitude: -30.0602249716
longitude: -51.17391374
altitude: 9.960587315
position_covariance: [0.0025010000000000006, 0.0, 0.0, 0.0, 0.00250100000
6, 0.0, 0.0, 0.0, 0.0025010000000000006]
position_covariance_type: 2
```

Values published to the /gps/fix topic

If you look at the code in `https://github.com/PacktPublishing/ROS-Robotics-Projects-SecondEdition/blob/master/chapter_10_ws/sensor_sim_gazebo/urdf/gps.xacro`, you will find `<plugin name="gazebo_ros_gps" filename="libhector_gazebo_ros_gps.so">`; these plugins belong to the `hector_gazebo_ros_plugins` package, which we installed at the beginning of the sensor interfacing. We can set all parameters related to GPS in this plugin description, and you can see the test parameters' values in the `gps.xacro` file. The GPS model is visualized as a box, and you can test the sensor values by moving this box in Gazebo. We will now interface GPS with ROS.

Interfacing GPS with ROS

In this section, we will see how to interface some popular GPS modules with ROS. One of the popular GPS modules is **Oxford Technical Solutions** (**OxTS**). You can find GPS/IMU modules at `http://www.oxts.com/products/`.

The ROS interface of this module can be found at `http://wiki.ros.org/oxford_gps_eth`. The Applanix GPS/IMU ROS module driver can be found at the following links:

- `applanix_driver`: `http://wiki.ros.org/applanix_driver`
- `applanix`: `http://wiki.ros.org/applanix`

Let's now simulate IMU on Gazebo.

Simulating IMU on Gazebo

Similar to GPS, we can start the IMU simulation using the following command:

```
$ roslaunch sensor_sim_gazebo imu.launch
```

You will get orientation values, linear acceleration, and angular velocity from this plugin. After launching this file, you can list out the topics published by the `imu` plugin. Here is the list of topics published by this plugin:

```
robot@robot-pc:~$ rostopic list
/clock
/gazebo/link_states
/gazebo/model_states
/gazebo/parameter_descriptions
/gazebo/parameter_updates
/gazebo/set_link_state
/gazebo/set_model_state
/imu
/joint_states
/rosout
/rosout_agg
/tf
/tf_static
```

List of topics published from the imu ROS plugin

We can check out the `/imu` topic by echoing the topic. You can find orientation, linear acceleration, and angular velocity data from this topic. The values are shown here:

```
robot@robot-pc:~$ rostopic echo /imu
header:
  seq: 0
  stamp:
    secs: 24
    nsecs:  95000000
  frame_id: sensor
orientation:
  x: -9.88131291682e-324
  y: -9.88131291682e-324
  z: 8.87671670196e-17
  w: 1.0
orientation_covariance: [0.0, 0.0, 0.0, 0.0, 0.0, 0.0, 0.0, 0.0, 0.0]
angular_velocity:
  x: 3.95252516673e-321
  y: 3.95252516673e-321
  z: 0.0
angular_velocity_covariance: [0.0, 0.0, 0.0, 0.0, 0.0, 0.0, 0.0, 0.0, 0.0]
linear_acceleration:
  x: -1.95719626798e-20
  y: 8.93613280022e-20
  z: 7.28456264068e-12
```

Data from the /imu topic

If you look at the IMU plugin definition code from `sensor_sim_gazebo/urdf/imu.xacro`, you can find the name of the plugin and its parameters.

The name of the plugin is mentioned in the following code snippet:

```
<gazebo>
  <plugin name="imu_plugin" filename="libgazebo_ros_imu.so">
    <alwaysOn>true</alwaysOn>
    <bodyName>sensor</bodyName>
    <topicName>imu</topicName>
    <serviceName>imu_service</serviceName>
    <gaussianNoise>0.0</gaussianNoise>
    <updateRate>20.0</updateRate>
  </plugin>
</gazebo>
```

The plugin's name is `libgazebo_ros_imu.so`, and it is installed along with a standard ROS installation.

You can also visualize IMU data in RViz. Choose the `imu` display type to view it. The IMU is visualized as a box itself, so if you move the box in Gazebo, you can see an arrow moving in the direction of movement. The Gazebo and RViz visualizations are shown here:

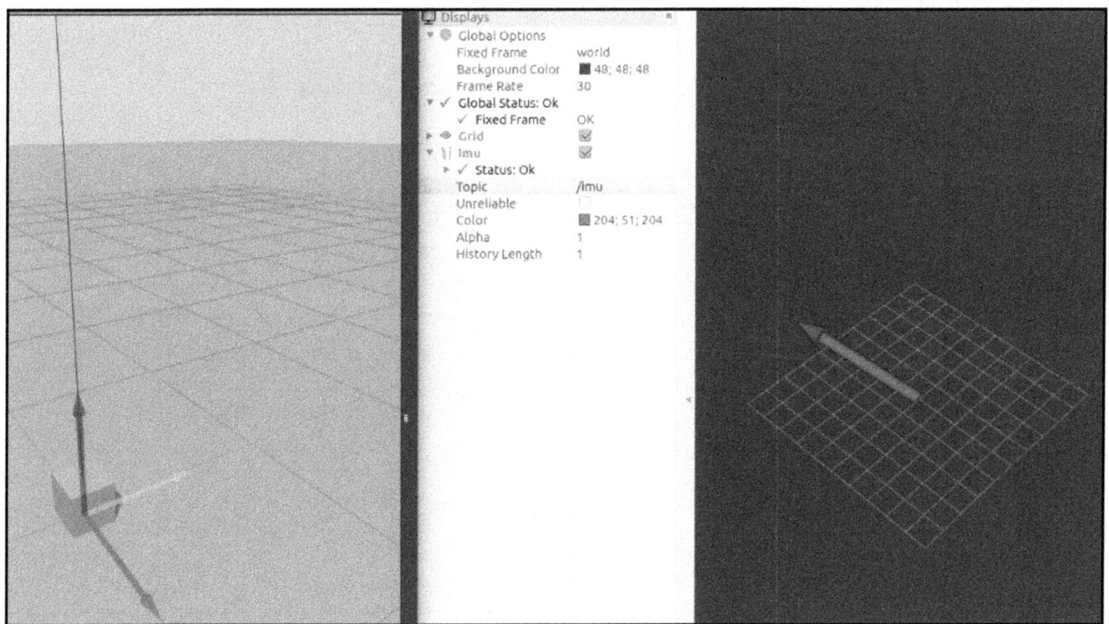

Visualization of the /imu topic

Now, let's see how to interface real hardware with ROS.

Interfacing IMUs with ROS

Most self-driving cars use integrated modules for GPS, IMU, and wheel encoders for accurate position prediction. In this section, we will look at some popular IMU modules that you can use if you want to use IMU alone.

I'll point you to a few links for ROS drivers used to interface with it. One of the popular IMUs is the MicroStrain 3DM-GX2 (`http://www.microstrain.com/inertial/3dm-gx2`).

Here are the ROS drivers for this IMU series:

- `microstrain_3dmgx2_imu`: `http://wiki.ros.org/microstrain_3dmgx2_imu`
- `microstrain_3dm_gx3_45`: `http://wiki.ros.org/microstrain_3dm_gx3_45`

Other than that, there are IMUs from Phidget (`http://wiki.ros.org/phidgets_imu`) and popular IMUs such as the InvenSense MPU 9250, 9150, and 6050 models (`https://github.com/jeskesen/i2c_imu`).

Another IMU sensor series called MTi from Xsens and its drivers can be found at `http://wiki.ros.org/xsens_driver`.

In the next section, we will simulate an ultrasonic sensor in Gazebo.

Simulating an ultrasonic sensor in Gazebo

Ultrasonic sensors also play a key role in self-driving cars. We've already seen that range sensors are widely used in parking assistant systems. In this section, we are going to see how to simulate a range sensor in Gazebo. The range sensor Gazebo plugin is already available in the hector Gazebo ROS plugin, so we can just use it in our code.

As we did in earlier demos, we will first see how to run the simulation and watch the output.

The following command will launch the range sensor simulation in Gazebo:

```
$ roslaunch sensor_sim_gazebo sonar.launch
```

In this simulation, we are taking the actual 3D model of the sonar, and it's very small. You may need to zoom into Gazebo to view the model. We can test the sensor by putting an obstacle in front of it. We can start RViz and can view the distance using the range display type. The topic name is `/distance` and the **Fixed Frame** is `world`.

Here is the range sensor value when the obstacle is far away:

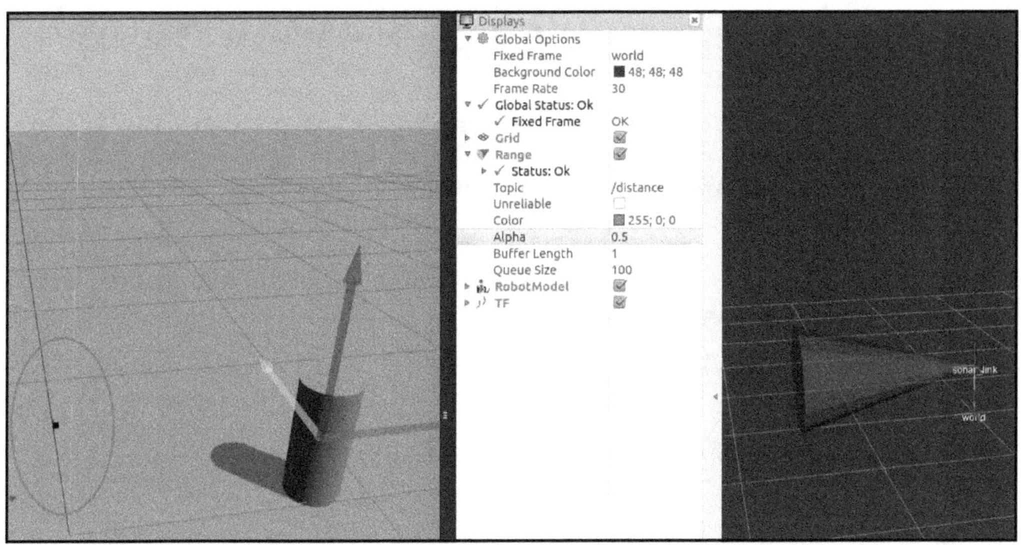

Range sensor value when the obstacle is far

You can see that the marked point is the ultrasonic sound sensor, and on the right, you can view the RViz range data as a cone-shaped structure. If we move the obstacle near the sensor, we can see what happens to the range sensor data:

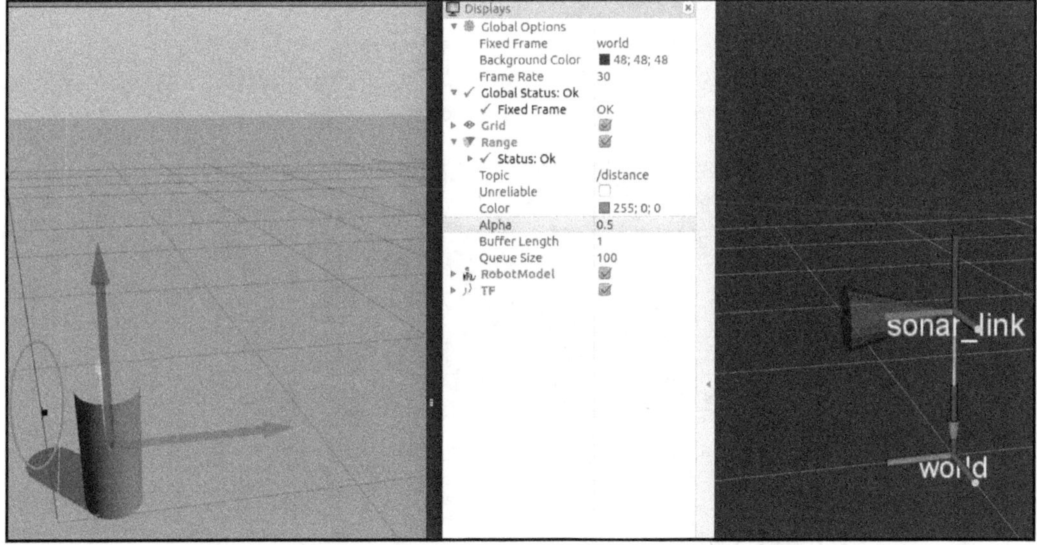

Range sensor value when the obstacle is near

When the obstacle is too near the sensor, the cone size is reduced, which means the distance to the obstacle is very short.

Open the Gazebo sonar plugin definition from `https://github.com/PacktPublishing/ROS-Robotics-Projects-SecondEdition/blob/master/chapter_10_ws/sensor_sim_gazebo/urdf/sonar.xacro`. This file includes a reference to another file called `sonar_model.xacro`, which has the complete sonar plugin definition.

We are using the `libhector_gazebo_ros_sonar` plugin to run this simulation, which is given in the following code snippet from `sonar_mode.xacro`:

```
<plugin name="gazebo_ros_sonar_controller"
filename="libhector_gazebo_ros_sonar.so">
```

Let's now look at a few popular low-cost LIDAR sensors.

Low-cost LIDAR sensors

This is an add-on section for hobbyists. If you are planning to build a miniature model of a self-driving car, you can use the following LIDAR sensors.

Sweep LIDAR

The Sweep 360-degree rotating LIDAR (`https://scanse.io/download/sweep-visualizer#r`) has a range of 40 meters. Compared to high-end LIDARs such as Velodyne, it is very cheap and good for research and hobby projects:

Sweep LIDAR (source: https://commons.wikimedia.org/wiki/File:Scanse_Sweep_LiDAR.jpg. Image by Simon Legner. Licensed under Creative Commons CC-BY-SA 4.0: https://creativecommons.org/licenses/by-sa/4.0/deed.en)

There is a good ROS interface available for this sensor. Here's the link to the Sweep sensor ROS package: `https://github.com/scanse/sweep-ros`. Before building the package, you need to install some dependencies:

```
$ sudo apt-get install ros-melodic-pcl-conversions ros-melodic-pointcloud-to-laserscan
```

Now, you can simply copy the `sweep-ros` package to your catkin workspace and build it using the `catkin_make` command.

After building the package, you can plug the LIDAR in to your PC through a serial-to-USB converter. If you plug this converter into a PC, Ubuntu will assign a device called `/dev/ttyUSB0`. First, you need to change the permission of the device using the following command:

```
$ sudo chmod 777 /dev/ttyUSB0
```

After changing the permission, we can start launching any of the launch files to view the laser's `/scan` point cloud data from the sensor.

The launch file will display the laser scan in RViz:

```
$ roslaunch sweep_ros view_sweep_laser_scan.launch
```

The launch file will display the point cloud in RViz:

```
$ roslaunch sweep_ros view_sweep_pc2.launch
```

Here is the visualization of the Sweep LIDAR:

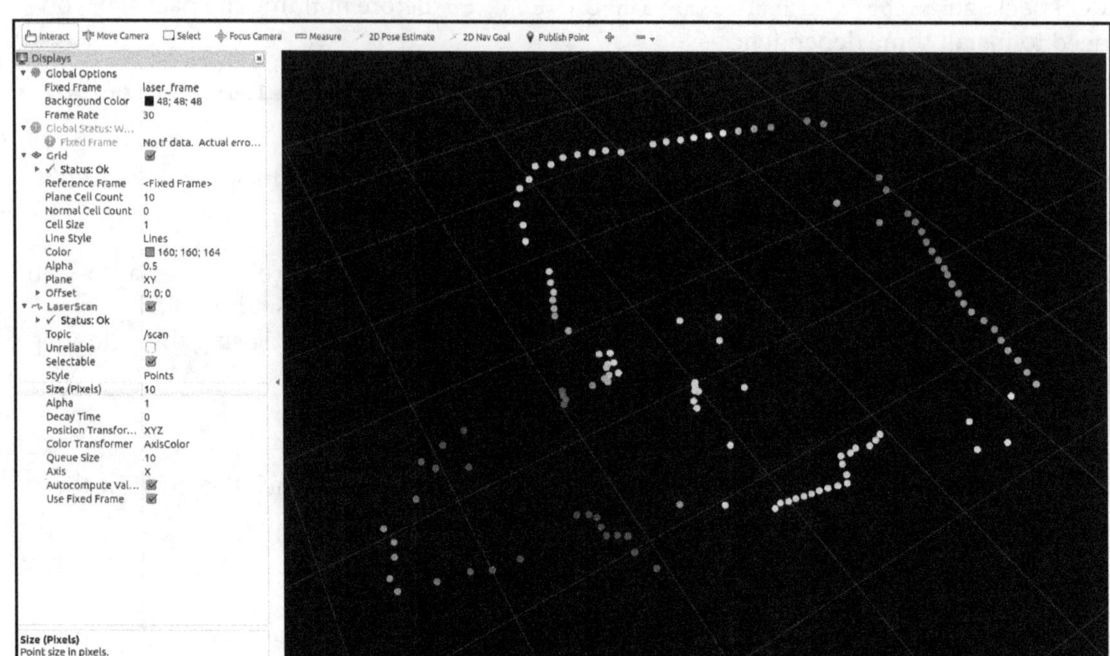

Sweep LIDAR visualization in RViz

The next section will give us some insight into RPLIDAR.

RPLIDAR

Similar to the Sweep LIDAR, RPLIDAR (`http://www.slamtec.com/en/lidar`) is another low-cost LIDAR for hobby projects. RPLIDAR and Sweep have the same applications, SLAM and autonomous navigation.

There is an ROS driver for interfacing the RPLIDAR with ROS. The ROS package can be found at `http://wiki.ros.org/rplidar`. The GitHub link of the package is `https://github.com/robopeak/rplidar_ros`.

Now that we have learned how to interface self-driving car sensors in ROS, we will simulate a self-driving car with sensors in Gazebo.

Simulating a self-driving car with sensors in Gazebo

In this section, we are going to discuss an open source self-driving car project done in Gazebo. In this project, we will learn how to implement a robot car model in Gazebo and how to integrate all sensors into it. Also, we will move the robot around the environment using a keyboard, and finally, we will build a map of the environment using SLAM.

Installing prerequisites

Let's take a look at the prerequisites for setting up packages in ROS Melodic.

The commands given here will install the ROS Gazebo controller manager:

```
$ sudo apt-get install ros-melodic-controller-manager
$ sudo apt-get install ros-melodic-ros-control ros-melodic-ros-controllers
$ sudo apt-get install ros-melodic-gazebo-ros-control
```

After installing this, we can install the Velodyne simulator packages in Melodic using the following command:

```
$ sudo apt-get install ros-melodic-velodyne
```

This project uses SICK laser scanners, so we have to install the SICK ROS toolbox packages. We do a source installation by cloning the package into the workspace and compiling:

```
$ git clone https://github.com/ros-drivers/sicktoolbox.git
$ cd ~/workspace_ws/
$ catkin_make
```

After installing all of these dependencies, we can clone the project files into a new ROS workspace. Use these commands:

```
$ cd ~
$ mkdir -p catvehicle_ws/src
$ cd catvehicle_ws/src
$ catkin_init_workspace
```

We have created a new ROS workspace, and now it's time to clone the project files to the workspace. The following commands will do this:

```
$ cd ~/catvehicle_ws/src
$ git clone https://github.com/sprinkjm/catvehicle.git
$ git clone https://github.com/sprinkjm/obstaclestopper.git
```

```
$ cd ../
$ catkin_make
```

If all packages have compiled successfully, you can add the following line to the `.bashrc` file:

```
$ source ~/catvehicle_ws/devel/setup.bash
```

You can launch the vehicle simulation using the following command:

```
$ roslaunch catvehicle catvehicle_skidpan.launch
```

This command will only start the simulation in the command line. And in another Terminal window, run the following command:

```
$ gzclient
```

We get the following robot car simulation in Gazebo:

Robot car simulation in Gazebo

You can see the Velodyne scan in front of the vehicle. We can list out all ROS topics from the simulation using the `rostopic` command.

Here are the main topics generated in the simulation:

```
/catvehicle/cmd_vel
/catvehicle/cmd_vel_safe
/catvehicle/distanceEstimator/angle
/catvehicle/distanceEstimator/dist
/catvehicle/front_laser_points
/catvehicle/front_left_steering_position_controller/command
/catvehicle/front_right_steering_position_controller/command
/catvehicle/joint1_velocity_controller/command
/catvehicle/joint2_velocity_controller/command
/catvehicle/joint_states
/catvehicle/lidar_points
/catvehicle/odom
/catvehicle/path
/catvehicle/steering
/catvehicle/vel
/clock
/gazebo/link_states
/gazebo/model_states
/gazebo/parameter_descriptions
/gazebo/parameter_updates
/gazebo/set_link_state
/gazebo/set_model_state
/rosout
/rosout_agg
/tf
/tf_static
```

Main topics generated by robotic car simulation

Let's now see how to visualize our robotic car sensor data.

Visualizing robotic car sensor data

We can view each type of sensor data from the robotic car in RViz. Just run RViz and open the `catvehicle.rviz` configuration from `chapter_10_ws`. You can see the Velodyne points and robot car model from RViz, as shown here:

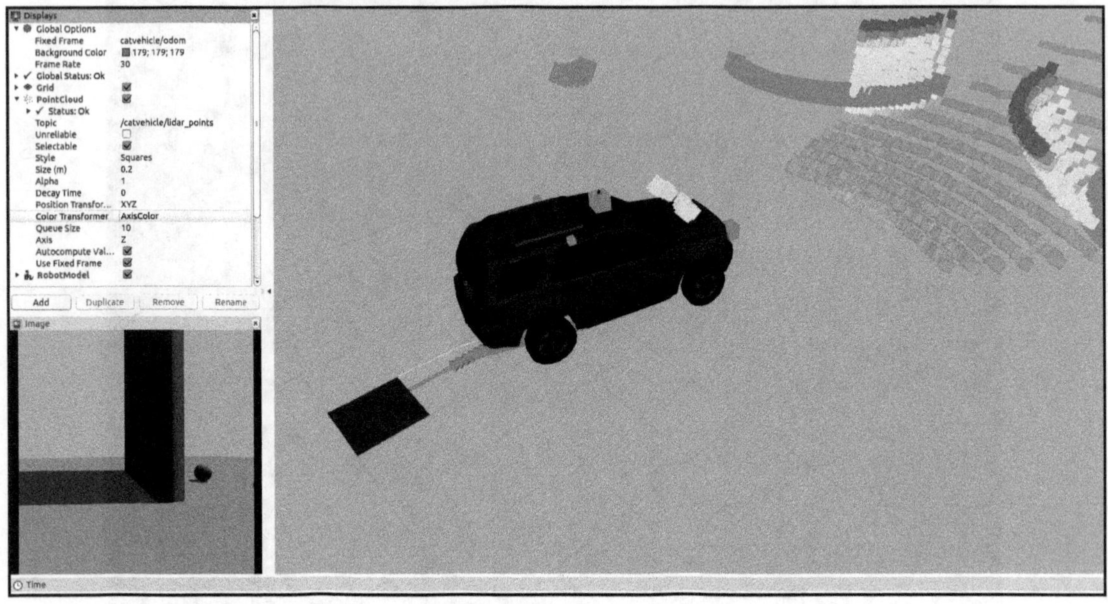

Complete robot car simulation in RViz

You can also add a camera view in RViz. There are two cameras, on the left and right side of the vehicle. We have added some obstacles in Gazebo to check whether the sensor is detecting obstacles. You can add more sensors, such as a SICK laser and IMU, to RViz. We will now move the self-driving car in Gazebo.

Moving a self-driving car in Gazebo

Okay, so we are done with simulating a complete robotic car in Gazebo; now, let's move the robot around the environment. We can do this using a `keyboard_teleop` node.

We can launch an existing TurtleBot teleop node using the following command:

```
$ roslaunch turtlebot_teleop keyboard_teleop.launch
```

The TurtleBot teleop node is publishing Twist messages to /cmd_vel_mux/input/teleop, and we need to convert them into /catvehicle/cmd_vel.

The following command can do this conversion:

```
$ rosrun topic_tools relay /cmd_vel_mux/input/teleop /catvehicle/cmd_vel
```

Now, you can move the car around the environment using the keyboard. This will be useful while we perform SLAM. Let's now run hector SLAM using a robotic car.

Running hector SLAM using a robotic car

After moving the robot around the world, let's do some mapping of the world. There are launch files present to start a new world in Gazebo and start mapping. Here is the command to start a new world in Gazebo:

```
$ roslaunch catvehicle catvehicle_canyonview.launch
```

This will launch the Gazebo simulation in a new world. You can enter the following command to view Gazebo:

```
$ gzclient
```

The Gazebo simulator with a new world is shown here:

Visualization of a robotic car in an urban environment

You can start the teleoperation node to move the robot, and the following command will start hector SLAM:

```
$ roslaunch catvehicle hectorslam.launch
```

To visualize the generated map, you can start RViz and open the configuration file called `catvehicle.rviz`.

You will get the following kind of visualization in RViz:

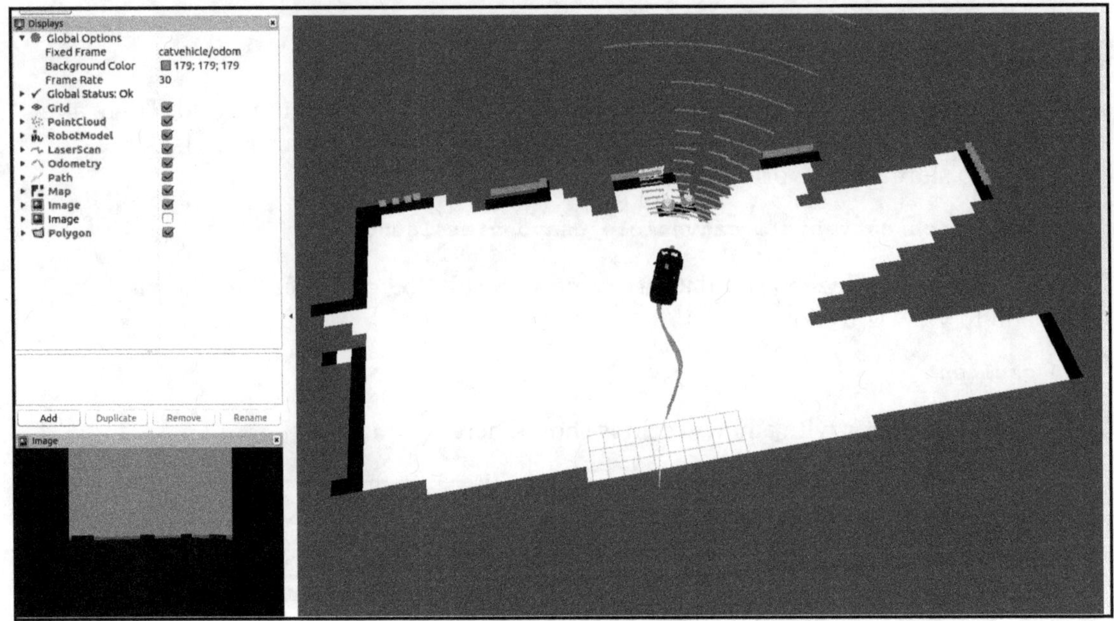

Visualization of a map in RViz using a robotic car

After completing the mapping process, we can save the map using the following command:

```
$ rosrun map_server map_saver -f map_name
```

The preceding command will save the current map as two files called `map_name.pgm` and `map_name.yaml`.

For more details of this project, you can check the following link: `http://cps-vo.org/group/CATVehicleTestbed`.

In the next section, we will interface a DBW car with ROS.

Interfacing a DBW car with ROS

In this section, we will see how to interface a real car with ROS and make it autonomous. As we discussed earlier, the DBW interface enables us to control a vehicle's throttle, brake, and steering using the CAN protocol.

There's an existing open source project that does this job. The project is owned by a company called Dataspeed Inc. (`http://dataspeedinc.com/`). Here is the list of projects related to self-driving cars from Dataspeed: `https://bitbucket.org/DataspeedInc/`.

We are going to discuss Dataspeed's ADAS vehicle development project. First, we will see how to install the ROS packages of this project and look at the functionality of each package and node.

Installing packages

Here are the complete instructions to install these packages. We only need a single command to install all of these packages.

We can install this on ROS Melodic using the following command:

```
bash <(wget -q -O -
https://bitbucket.org/DataspeedInc/dbw_mkz_ros/raw/default/dbw_mkz/scripts/
ros_install.bash)
```

You will get other methods of installation from the following link: `http://wiki.ros.org/dbw_mkz`.

Let's visualize the self-driving car and sensor data.

Visualizing the self-driving car and sensor data

The previous packages can help you to interface a DBW car with ROS. If we don't have a real car, we can work with ROS bag files, visualize data, and process it offline.

The following command helps you to visualize the URDF model of a self-driving car:

```
$ roslaunch dbw_mkz_description rviz.launch
```

You will get the following model when you execute it:

Visualization of a self-driving car

If you want to visualize the Velodyne sensor data, other sensors such as GPS and IMU, and control signals such as steering commands, braking, and acceleration, you can use the following commands:

1. Download the ROS bag file using the following command:

```
$ wget
https://bitbucket.org/DataspeedInc/dbw_mkz_ros/downloads/mkz_201512
07_extra.bag.tar.gz
```

   You will get a compressed file from the preceding command; extract it to your home folder.

2. Now you can run the following command to read the data from the bag file:

```
$ roslaunch dbw_mkz_can offline.launch
```

3. The following command will visualize the car model:

```
$ roslaunch dbw_mkz_description rviz.launch
```

4. And finally, we have to run the bag file:

```
$ rosbag play mkz_20151207.bag -clock
```

5. To view the sensor data in RViz, we have to publish a static transform:

```
$ rosrun tf static_transform_publisher 0.94 0 1.5 0.07 -0.02 0
base_footprint velodyne 50
```

This is the result:

Visualization of a self-driving car

You can set **Fixed Frame** as `base_footprint` and view the car model and Velodyne data.

This data is provided by Dataspeed Inc. located in Rochester Hills, Michigan.

Communicating with DBW from ROS

In this section, we will see how we can communicate with DBW-based cars from ROS.

This is the command to do so:

```
$ roslaunch dbw_mkz_can dbw.launch
```

Now, you can test the car using a joystick. Here is the command to launch its nodes:

```
$ roslaunch dbw_mkz_joystick_demo joystick_demo.launch sys:=true
```

In the next section, we will cover the Udacity open source self-driving car project.

Introducing the Udacity open source self-driving car project

There is another open source self-driving car project by Udacity (`https://github.com/udacity/self-driving-car`) that was created for teaching their Nanodegree self-driving car program. This project aims to create a completely autonomous self-driving car using deep learning and using ROS as middleware for communication.

The project is split into a series of challenges, and anyone can contribute to the project and win a prize. The project is trying to train a **Convolution Neural Network** (**CNN**) from a vehicle camera dataset to predict steering angles. This approach is a replication of end-to-end deep learning from NVIDIA (`https://devblogs.nvidia.com/parallelforall/deep-learning-self-driving-cars/`), used in their self-driving car project called DAVE-2.

The following is the block diagram of DAVE-2. DAVE-2 stands for DARPA Autonomous Vehicle-2, which is inspired by the DAVE project by DARPA:

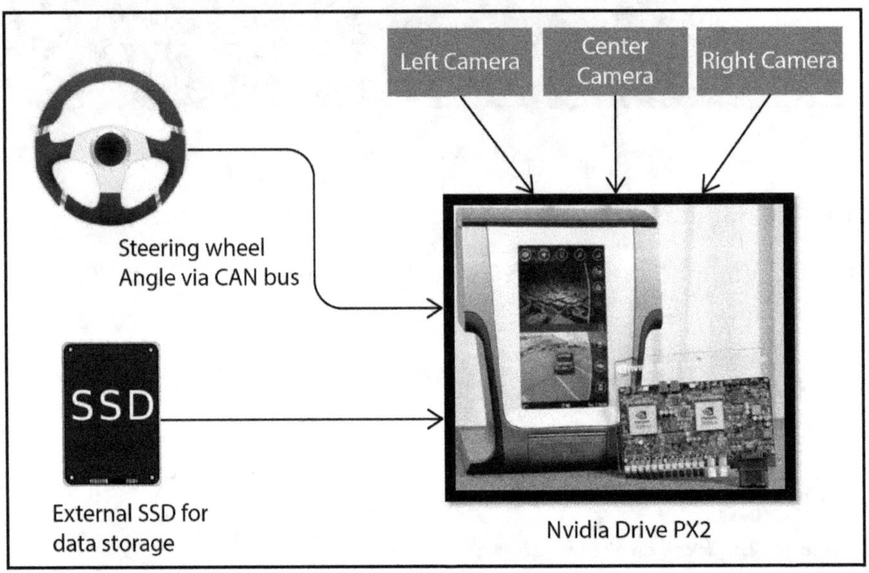

DAVE-2 block diagram (source: https://en.wikipedia.org/wiki/Nvidia_Drive#/media/File:NVIDIA_Drive_PX,_Computex_Taipei_20150601.jpg. Image by: NVIDIA Taiwan. Licensed under Creative Commons CC-BY-SA 2.0: https://creativecommons.org/licenses/by/2.0/legalcode)

This system basically consists of three cameras and an NVIDIA supercomputer called NVIDIA PX. This computer can train images from this camera and predict the steering angle of the car. The steering angle is fed to the CAN bus and controls the car.

The following are the sensors and components used in the Udacity self-driving car:

- 2016 Lincoln MKZ—this is the car that is going to be made autonomous. In the previous section, we saw the ROS interfacing of this car. We are using that project here too.
- Two Velodyne VLP-16 LIDARs
- Delphi radar
- Point Grey Blackfly cameras
- Xsens IMU
- **Engine control unit** (ECU)

This project uses the `dbw_mkz_ros` package to communicate from ROS to the Lincoln MKZ. In the previous section, we set up and worked with the `dbw_mkz_ros` package. Here is the link to obtain a dataset to train the steering model: `https://github.com/udacity/self-driving-car/tree/master/datasets`. You will get an ROS launch file from this link to play with these bag files too.

Here is the link to get an already trained model that can only be used for research purposes: `https://github.com/udacity/self-driving-car/tree/master/steering-models`. There is an ROS node for sending steering commands from the trained model to the Lincoln MKZ. Here, the `dbw_mkz_ros` packages act as an intermediate layer between the trained model commands and the actual car.

Open source self-driving car simulator from Udacity

Udacity also provides an open source simulator for training and testing self-driving deep learning algorithms. The simulator project is available at `https://github.com/udacity/self-driving-car-sim`. You can also download the pre-compiled version of a simulator for Linux, Windows, and macOS from the same link.

We can discuss the workings of the simulator along with it. Here is a screenshot of this simulator:

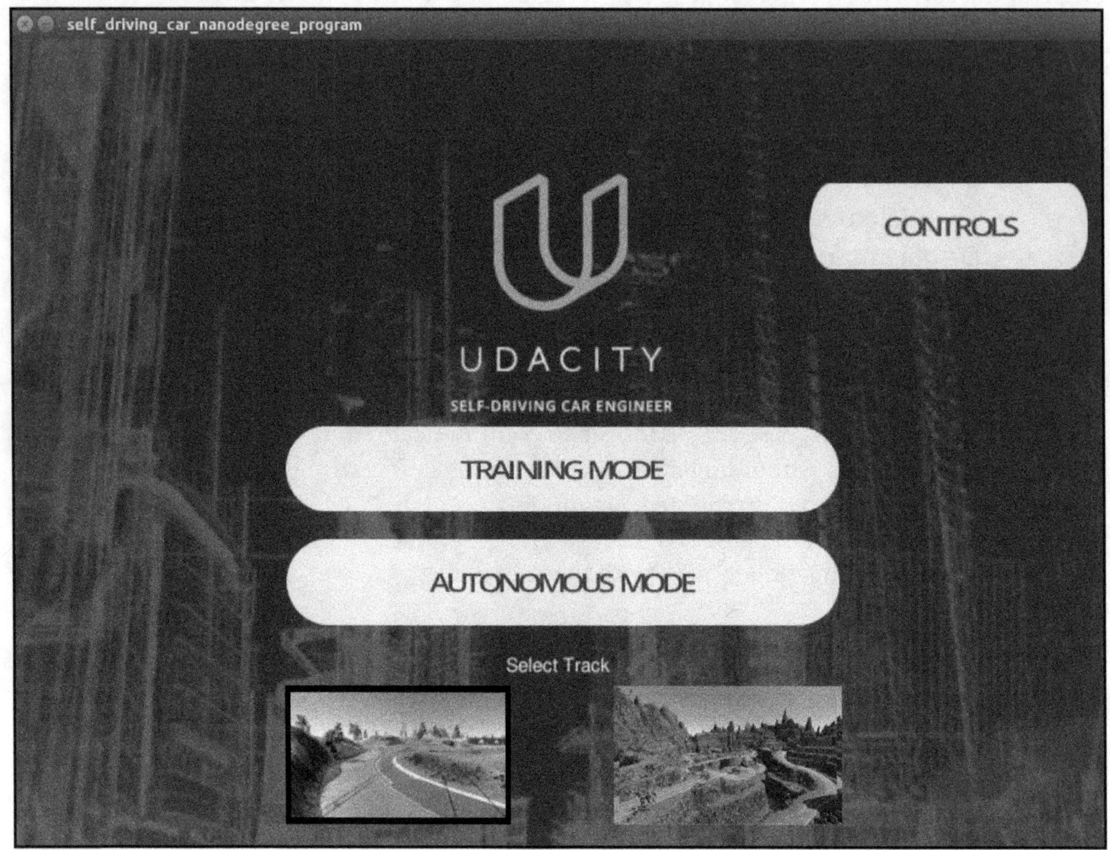

Udacity self-driving car simulator

You can see two options in the simulator; the first is for training and the second is for testing autonomous algorithms. We can also select the track in which we have to drive the vehicle. When you click on the **TRAINING MODE** button, you will get a racing car on the selected track. You can move the car using the WASD key combination, like a game. Here is a screenshot of the training mode.

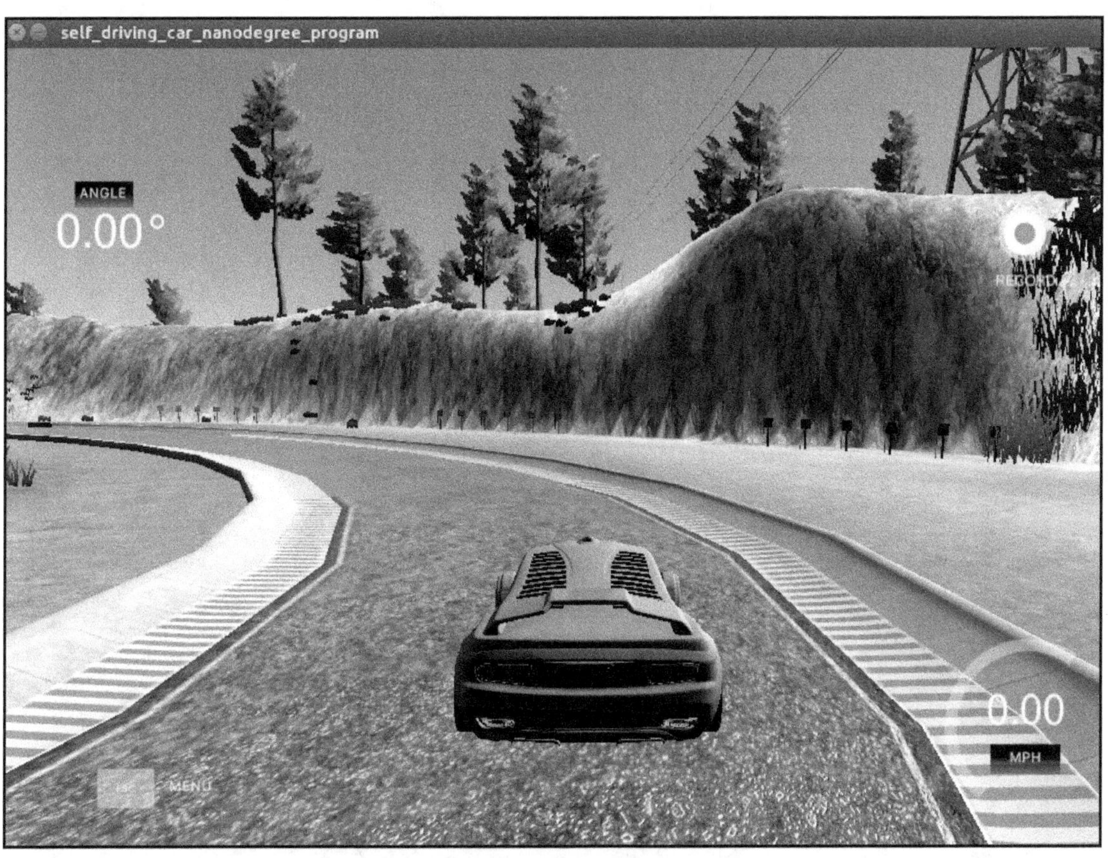

Udacity self-driving car simulator in training mode

You can see a **RECORD** button in the top-right corner, which is used to capture the front camera images of the car. We can browse to a location, and those captured images will be stored in that location.

After capturing the images, we have to train the car using deep learning algorithms to predict steering angle, acceleration, and braking. We are not discussing the code, but I'll provide a reference for you to write it. The complete code reference to implement the driving model using deep learning and the entire explanation for it are at `https://github.com/thomasantony/sdc-live-trainer`. The `live_trainer.py` code helps us to train the model from captured images.

After training the model, we can run `hybrid_driver.py` for autonomous driving. For this mode, we need to select the autonomous mode in the simulator and execute the `hybrid_driver.py` code:

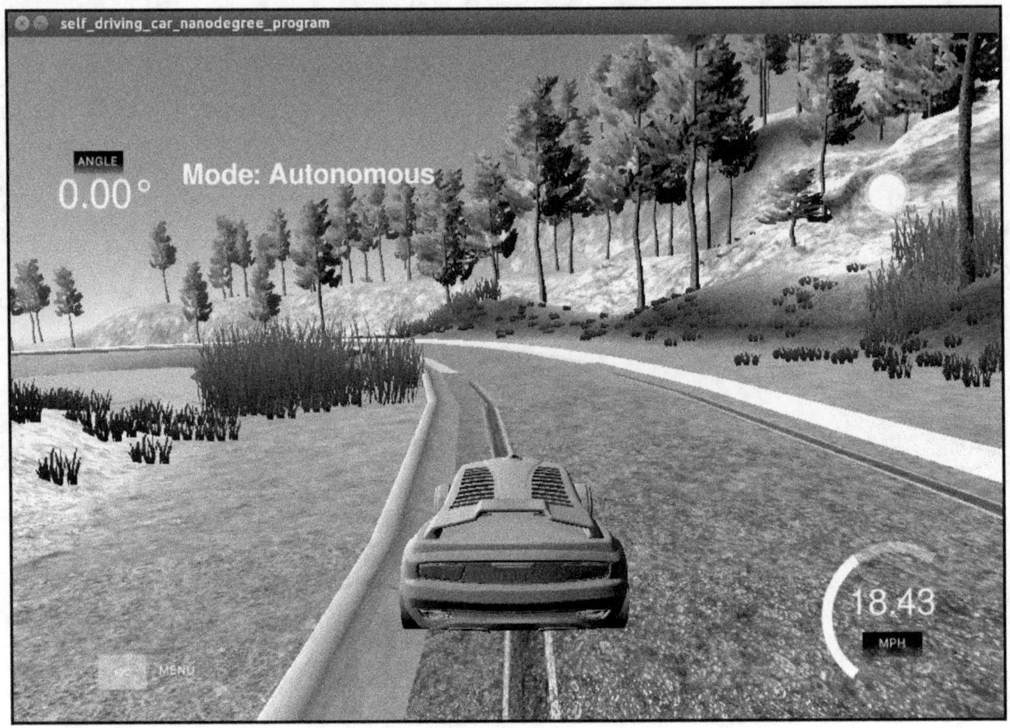

Udacity self-driving car simulator in autonomous mode

You can see the car moving autonomously and manually override the steering control at any time. This simulator can be used to test the accuracy of the deep learning algorithm we are going to use in a real self-driving car.

MATLAB ADAS Toolbox

MATLAB also provides a toolbox for working with ADAS and autonomous systems. You can design, simulate, and test ADAS and autonomous driving systems using this toolbox. Here is the link to check the new toolbox: `https://in.mathworks.com/products/automated-driving.html`

Summary

This chapter was an in-depth discussion of self-driving cars and their implementation. This chapter started by discussing the basics of self-driving car technology and its history. After that, we discussed the core blocks of a typical self-driving car. We also discussed the concept of autonomy levels in self-driving cars. Then, we took a look at different sensors and components commonly used in a self-driving car.

We discussed how to simulate such a car in Gazebo and interfacing it with ROS. After discussing all sensors, we looked at an open source self-driving car project that incorporated all sensors and simulated the car model itself in Gazebo. We visualized its sensor data and moved the robot using a teleoperation node. We also mapped the environment using hector SLAM. The next project was from Dataspeed Inc., in which we saw how to interface a real DBW-compatible vehicle with ROS. We visualized the offline data of the vehicle using RViz. Finally, we took a look at the Udacity self-driving car project and its simulator. This chapter helped us to acquire the skills required to simulate self-driving cars.

In the next chapter, we will see how to teleoperate a robot using a VR headset and Leap.

11
Teleoperating Robots Using a VR Headset and Leap Motion

The term **virtual reality** (**VR**) is gaining popularity nowadays even though it is an old invention. The concept of VR first emerged in the 1950s as science fiction, but it was another 60 years before it gained popularity and became widely accepted.

So, why is it more popular now? Well, the answer is the availability of cheap computing. Before, a VR headset was very expensive; now we can build one for $5. You may have heard of Google Cardboard, which is the cheapest VR headset that is currently available, and there are many upcoming models based on it. Now, we only need a good smartphone and cheap VR headset to get the VR experience. There are also high-end VR headsets such as the Oculus Rift and HTC Vive, which have a high frame rate and response time.

In this chapter, we will discuss a ROS project in which we can control a robot using a Leap Motion sensor and experience the robot environment using a VR headset. We will demonstrate this project using a TurtleBot simulation in Gazebo and control the robot using Leap Motion. To visualize the robot environment, we will use a cheap VR headset along with an Android smartphone.

Here are the main topics that we will discuss in this chapter:

- Getting started with a VR headset and Leap Motion
- Designing and working on the project
- Installing the Leap Motion SDK on Ubuntu
- Playing with the Leap Motion Visualizer tool
- Installing ROS packages for Leap Motion
- Visualizing Leap Motion data in RViz
- Creating a teleoperation node for Leap Motion

- Building and installing the ROS-VR Android application
- Working with the ROS-VR application and interfacing with Gazebo
- TurtleBot simulation in VR
- Integrating the ROS-VR application and Leap Motion teleoperation
- Troubleshooting the ROS-VR application

Technical requirements

The following are the software and hardware prerequisites of this project:

- **Low-cost VR headset**: `https://vr.google.com/cardboard/get-cardboard/`
- **Leap Motion controller**: `https://www.leapmotion.com/`
- **Wi-Fi router**: Any router that can connect to a PC or Android phone
- **Ubuntu 14.04.5 LTS**: `http://releases.ubuntu.com/14.04/`
- **ROS Indigo**: `http://wiki.ros.org/indigo/Installation/Ubuntu`
- **Leap Motion SDK**: `https://www.leapmotion.com/setup/linux`

This project has been tested on ROS Indigo, and the code is compatible with ROS Kinetic too; however, the Leap Motion SDK is still in development for Ubuntu 16.04 LTS. So, here, the code is tested using Ubuntu 14.04.5 and ROS Indigo.

Here are the estimated timelines and testing platform that we will use:

- **Estimated learning time**: 90 minutes (on average)
- **Project build time (inclusive of compile and runtime)**: 60 minutes (on average)

The code for this chapter is available at `https://github.com/PacktPublishing/ROS-Robotics-Projects-SecondEdition/tree/master/chapter_11_ws`.

Let's begin this chapter by learning about VR headsets and Leap Motion sensors in more detail.

Getting started with a VR headset and Leap Motion

This section is for beginners who haven't worked with VR headsets or Leap Motion yet. A VR headset is a head-mounted display, in which we can put a smartphone, or which has a built-in display that can be connected to HDMI or some other display port. A VR headset can create a virtual 3D environment by mimicking human vision, that is, stereo vision.

Human vision works like this: we have two eyes and get two separate and slightly different images in each eye. The brain then combines these two images and generates a 3D image of the surroundings. Similarly, VR headsets have two lenses and a display. The display can be inbuilt or a smartphone. This screen will show a separate view of the left and right images, and when we put the smartphone or built-in display into the headset, it will focus and reshape using two lenses and will simulate 3D stereoscopic vision.

In effect, we can explore a 3D world inside this headset. In addition to visualizing the world, we can also control the events in the 3D world and hear sounds too. Cool, right?

Here is the internal structure of a Google Cardboard VR headset:

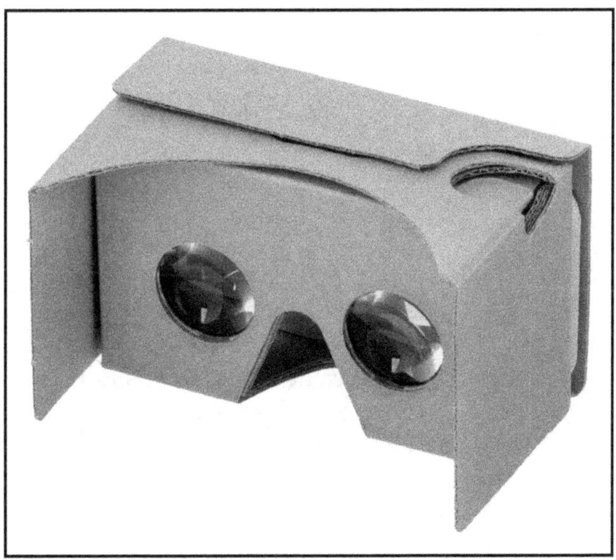

Google Cardboard VR headset (source: https://commons.wikimedia.org/wiki/File:Google-Cardboard.jpg)

There are a variety of VR headsets models available, including high-end models such as Oculus Rift, HTC Vive, and more. The following photograph shows one of these VR headsets, which we will use in this chapter. It follows the principle of Google Cardboard, but rather than cardboard, it uses a plastic body:

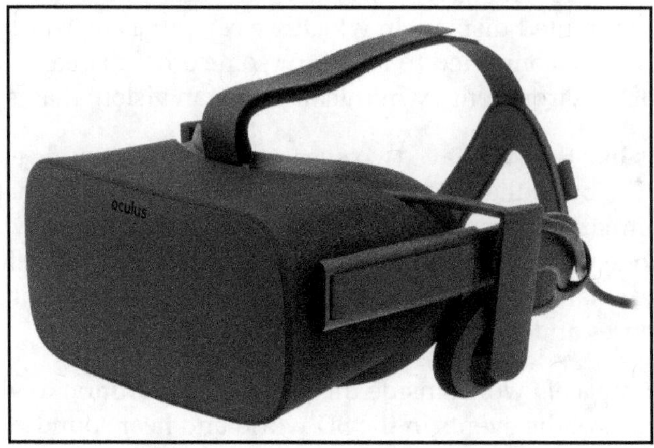

Oculus Rift (source: https://commons.wikimedia.org/wiki/File:Oculus-Rift-CV1-Headset-Front.jpg)

You can test the VR feature of the headset by downloading Android VR applications from Google Play Store.

You can search for `Cardboard` in Google Play Store to get the Google VR application. You can use it for testing VR on your smartphone.

The next device we will use in this project is the Leap Motion controller (`https://www.leapmotion.com/`). The Leap Motion controller is essentially an input device like a PC mouse, in which we can control everything using hand gestures. The controller can accurately track the two hands of a user and map the position and orientation of each finger joint accurately. It has two IR cameras and several IR projectors facing upward. The user can position their hand above the device and then move their hand. The position and orientation of the hands and fingers can be accurately retrieved from their SDK.

Here is an example of the Leap Motion controller and how we can interact with it:

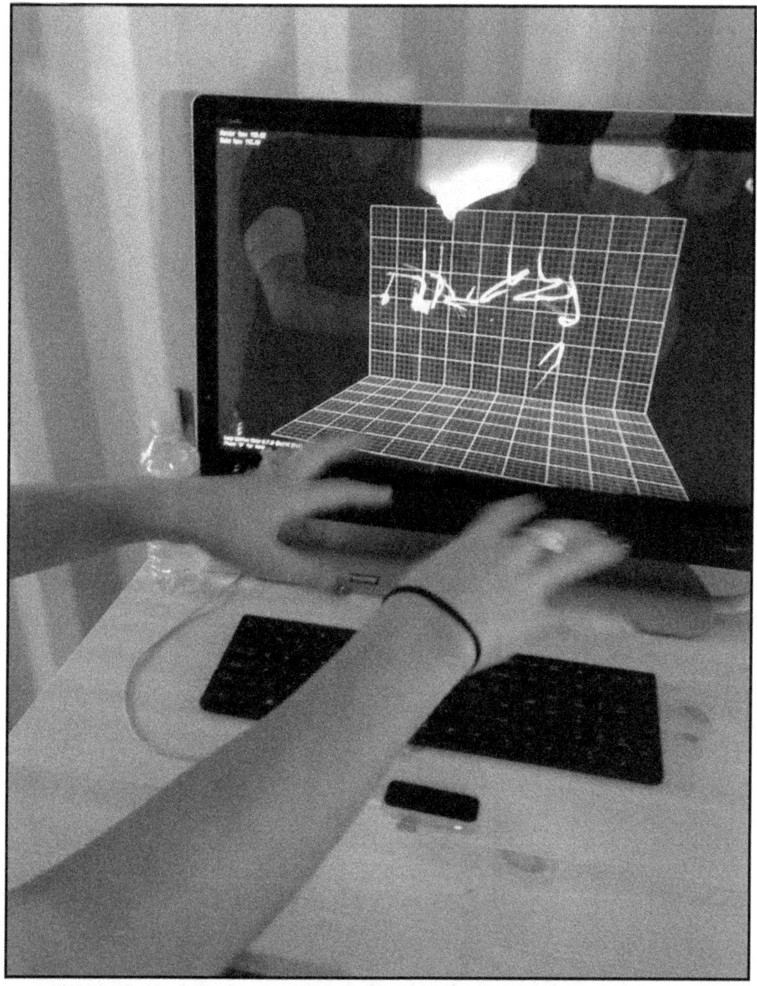

Interacting with the Leap Motion controller (source: https://www.flickr.com/photos/davidberkowitz/8598269932. Image by David Berkowitz. Licensed under Creative Commons CC-BY 2.0: https://creativecommons.org/licenses/by/2.0/legalcode)

Let's now look at the design of our project.

Designing and working on the project

This project can be divided into two sections: teleoperation using **Leap Motion** and streaming images to an Android phone to get a VR experience inside a VR headset. Before going on to discuss each design aspect, let's look at how we can interconnect these devices. The following diagram shows how the components are interconnected for this project:

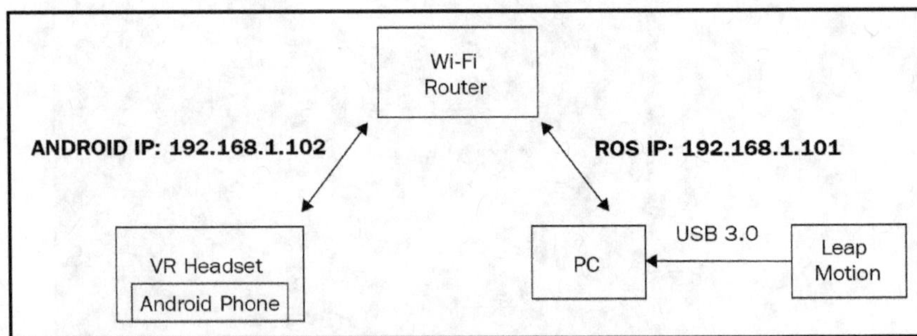

Hardware components and connection

You can see that each device (that is, the PC and Android phone) is connected to a Wi-Fi router, and the router has assigned an IP address to each device. Each device communicates using these IP addresses. You will see the importance of these addresses in the upcoming sections.

Next, we will examine how we can teleoperate a robot in ROS using Leap Motion. We will be controlling it while wearing the VR headset. This means that we don't need to press any keys to move the robot; rather, we can just move it with our hands.

The basic operation involved here is converting the Leap Motion data into ROS Twist messages. Here, we are only interested in reading the orientation of the hand. We are taking roll, pitch, and yaw and mapping them into ROS Twist messages, as shown here:

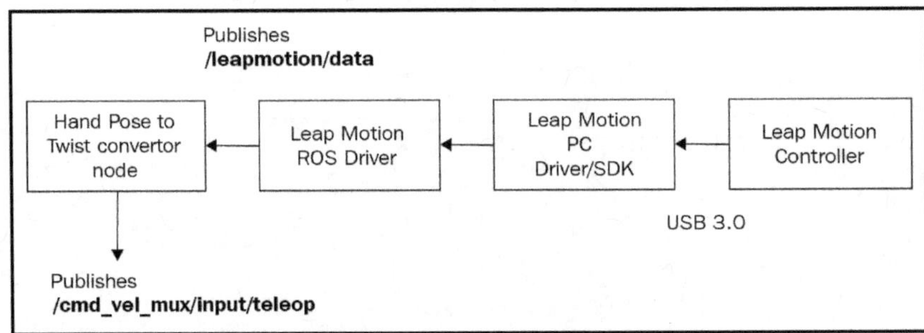

Leap Motion data to ROS command velocity

The preceding diagram shows how Leap Motion data is manipulated into ROS Twist messages. The **Leap Motion PC Driver/SDK** interfaces the controller with Ubuntu, and the **Leap Motion ROS Driver**, which works on top of this driver/SDK, fetches the hand and finger position and publishes it as ROS topics. The node we are going to write can convert the hand position to twist data, which will subscribe to the Leap Motion data topic called `/leapmotion/data`, convert it into corresponding command velocities, and then publish it to the topic called `/cmd_vel_mux/input/teleop`. The conversion algorithm is based on comparing the hand orientation value. If the value is in a particular range, we will publish a particular Twist value.

Here is a representation of the workings of a simple algorithm that converts Leap Motion orientation data into Twist messages:

1. Take the orientation values of a hand, such as yaw, pitch, and roll, from the Leap Motion ROS driver.
2. The roll movement of the hand corresponds to robot rotation. If the hand rotates anticlockwise, the robot will be triggered to rotate anticlockwise by sending a command velocity and the roll movement of the hand in the clockwise direction is the opposite.
3. If the hand is pitched down, the robot will move forward, and if the hand is pitched up, the robot will move backward.
4. If there is no hand movement, the robot will stop.

This is a simple algorithm to move a robot using Leap Motion. Okay, let's start by setting up a Leap Motion controller in Ubuntu and working with its ROS interface.

Installing the Leap Motion SDK on Ubuntu 14.04.5

In this project, we have chosen Ubuntu 14.04.5 LTS and ROS Indigo because the Leap Motion SDK will work smoothly with this combination. The Leap Motion SDK is not fully supported by Ubuntu 16.04 LTS; if there are any further fixes from the company, this code will work on Ubuntu 16.04 LTS with ROS Kinetic.

The Leap Motion SDK is the core of the Leap Motion controller. The Leap Motion controller has two IR cameras facing upward, and it also has several IR projectors. This is interfaced with a PC, and the Leap Motion SDK runs on the PC, which has drivers for the controller. It also has algorithms to process the hand image to produce the joint values of each finger joint.

Here is the procedure to install the Leap Motion SDK in Ubuntu:

1. Download the SDK from `https://www.leapmotion.com/setup/linux`; you can extract this package and you will find two DEB files that can be installed on Ubuntu.
2. Open Terminal on the extracted location, and install the DEB file using the following command (for 64-bit PCs):

    ```
    $ sudo dpkg -install Leap-*-x64.deb
    ```

    If you are installing it on a 32-bit PC, you can use the following command:

    ```
    $ sudo dpkg -install Leap-*-x86.deb
    ```

If you can install this package without any errors, then you are done with installing the Leap Motion SDK and driver.

More detailed installation and debugging tips are given here: `https://support.leapmotion.com/hc/en-us/articles/223782608-Linux-Installation`.

Once we have installed the Leap Motion SDK and driver, we can start visualizing the Leap Motion controller data.

Visualizing the Leap Motion controller data

If you have successfully installed the Leap Motion driver/SDK, we can start the device by following these steps:

1. Plug the Leap Motion controller into a USB port; you can plug it into USB 3.0, but 2.0 is fine too.
2. Open Terminal and execute the `dmesg` command to verify that the device is properly detected on Ubuntu:

    ```
    $ dmesg
    ```

It will give you the following result if it's detected properly:

```
[10010.420978] usb 2-1.2: new high-speed USB device number 8 using ehci-pci
[10010.513671] usb 2-1.2: New USB device found, idVendor=f182, idProduct=0003
[10010.513682] usb 2-1.2: New USB device strings: Mfr=1, Product=2, SerialNumber=0
[10010.513688] usb 2-1.2: Product: Leap Dev Kit
[10010.513692] usb 2-1.2: Manufacturer: Leap Motion
[10010.514270] uvcvideo: Found UVC 1.00 device Leap Dev Kit (f182:0003)
lentin@lentin-Aspire-4755:~$
```

Kernel message when plugging in Leap Motion

If you see this message, you're ready to start the Leap Motion controller manager.

Playing with the Leap Motion Visualizer tool

You could visualize the motion tracking data from the Leap Motion sensor using a Visualizer called the Leap Motion Visualizer tool and the following command.

You can invoke the Leap Motion controller manager by executing the following:

```
$ sudo LeapControlPanel
```

If you only want to start the driver, you can use the following command:

```
$ sudo leapd
```

Additionally, you can use this command to restart the driver:

```
$ sudo service leapd stop
```

If you are running the Leap Motion control panel, you will see an additional menu on the left-hand side of the screen. Select **Diagnostic Visualizer...** to view the data from Leap Motion:

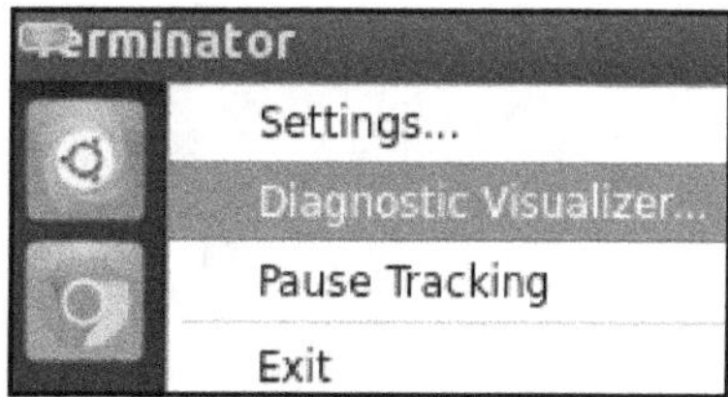

The Leap Motion control panel

When you click on this option, a window will pop up, where you can see your hand and figures get tracked when you put your hand over the device. You can also see the two IR camera views from the device. Here is a screenshot of the Visualizer application. You can quit the driver from the same drop-down menu, too:

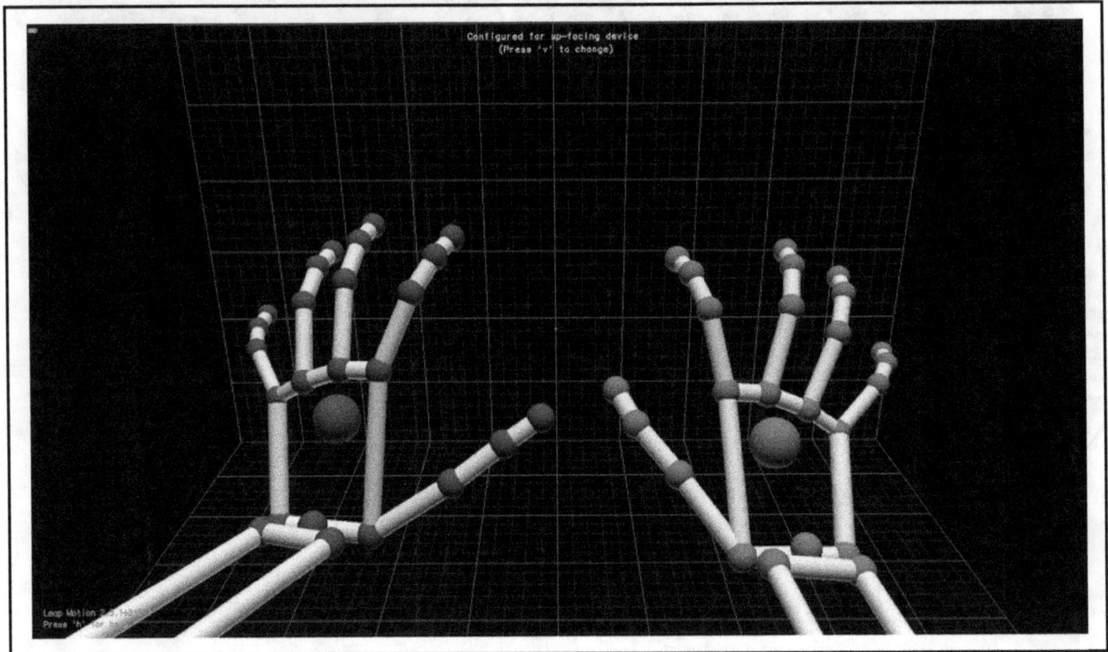

The Leap Motion controller Visualizer application

You can interact with the device and visualize the data here. If everything is working well, we can proceed to the next stage: installing the ROS driver for the Leap Motion controller.

You can find more shortcuts for Visualizer at `https://developer.leapmotion.com/documentation/cpp/supplements/Leap_Visualizer.html`.

Installing the ROS driver for the Leap Motion controller

To interface the Leap Motion controller with ROS, we will need the ROS driver for it. Here is the link to get the ROS driver for Leap Motion; you can clone it using the following command:

```
$ git clone https://github.com/ros-drivers/leap_motion
```

Before installing the `leap_motion` driver package, we have to do a few things to ensure that it is properly compiled:

1. The first step is to set the path of the Leap Motion SDK in the `.bashrc` file.

   Assuming that the Leap SDK is in the user's `home` folder with the name `LeapSDK`, we have to set the path variable in `.bashrc`, as follows:

   ```
   $ export LEAP_SDK=$LEAP_SDK:$HOME/LeapSDK
   ```

   This environment variable is needed for compiling the code of the ROS driver, which has Leap SDK APIs.

2. We also have to add the path of the Python extension of the Leap Motion SDK to `.bashrc`.

   Here is the command used to do it:

   ```
   export
   PYTHONPATH=$PYTHONPATH:$HOME/LeapSDK/lib:$HOME/LeapSDK/lib/x64
   ```

   This will enable Leap Motion SDK APIs in Python.

3. After performing the preceding steps, you can save `.bashrc` and open a new Terminal so that we will get the preceding variables in the new Terminal.
4. The final step is to copy the `libLeap.so` file to `/usr/local/lib`.

   Here is how to do it:

   ```
   $ sudo cp $LEAP_SDK/lib/x64/libLeap.so /usr/local/lib
   ```

5. After copying the file, execute `ldconfig`:

```
$ sudo ldconfig
```

Okay, you are now finished with setting the environment variables.

6. Now you can compile the `leap_motion` ROS driver package. You can create a ROS workspace or copy the `leap_motion` package to an existing ROS workspace and use `catkin_make`.

You can use the following command to install the `leap_motion` package:

```
$ catkin_make install --pkg leap_motion
```

The preceding will install the `leap_motion` driver. Now, let's test the device in ROS.

Testing the Leap Motion ROS driver

If everything has been installed properly, we can test it using a couple of commands:

1. First, launch the Leap Motion driver or control panel using the following command:

```
$ sudo LeapControlPanel
```

2. After launching the command, you can verify that the device is working by opening the Visualizer application. If it's working well, you can launch the ROS driver using the following command:

```
$ roslaunch leap_motion sensor_sender.launch
```

If it's working properly, you will get topics with this:

```
$ rostopic list
```

The list of available topics is shown here:

```
lentin@lentin-Aspire-4755:~$ rostopic list
/leapmotion/data
/rosout
/rosout agg
```

Leap Motion ROS driver topics

If you can see `rostopic/leapmotion/data` in the list, you can confirm that the driver is working. You can just echo the topic and see that the hand and finger values are coming in, as shown in the following screenshot:

```
header:
  seq: 847
  stamp:
    secs: 0
    nsecs:
  frame_id: ''
direction:
  x: 0.24784040451
  y: 0.227308988571
  z: -0.941756725311
normal:
  x: 0.0999223664403
  y: -0.972898304462
  z: -0.208529144526
palmpos:
  x: -52.5600471497
  y: 173.553512573
  z: 66.0648040771
ypr:
  x: 25.602668997
  y: 13.5697675013
  z: 132.525765862
```

Data from the Leap ROS driver topic

Let's now visualize our Leap Motion data in RViz.

Visualizing Leap Motion data in RViz

We can also visualize Leap Motion data in RViz. There is a ROS package called `leap_client` (`https://github.com/qboticslabs/leap_client`).

You can install this package by setting the following environment variable in `~/.bashrc`:

```
export LEAPSDK=$LEAPSDK:$HOME/LeapSDK
```

Note that when we add new variables in `~/.bashrc`, you may need to open a new Terminal or type `bash` into the existing Terminal.

Now, we can clone the code in a ROS workspace and build the package using `catkin_make`. Let's play around with this package:

1. To launch the nodes, we have to start `LeapControlPanel`:

   ```
   $ sudo LeapControlPanel
   ```

2. Then, start the ROS Leap Motion driver launch file:

   ```
   $ roslaunch leap_motion sensor_sender.launch
   ```

3. Now launch the `leap_client` launch file to start the visualization nodes. This node will subscribe to the `leap_motion` driver and convert it into visualization markers in RViz:

   ```
   $ roslaunch leap_client leap_client.launch
   ```

4. Now, you can open RViz using the following command, and select the `leap_client/launch/leap_client.rviz` configuration file to visualize the markers properly:

   ```
   $ rosrun rviz rviz
   ```

If you load the `leap_client.rviz` configuration, you may get hand data like the following (you have to put your hand over the Leap Motion controller):

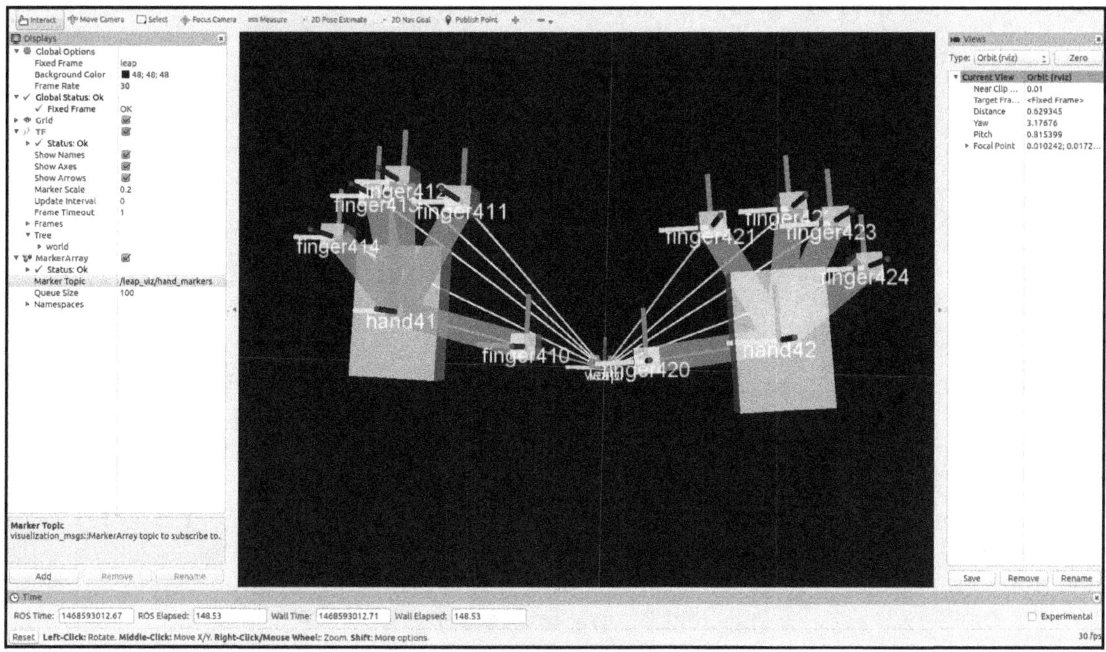

Data from the Leap ROS driver topic

Now that we have learned how to visualize Leap Motion data in RViz, we can now create a teleoperation node using the Leap Motion controller.

Creating a teleoperation node using the Leap Motion controller

In this section, we will demonstrate how to create a teleoperation node for a robot using Leap Motion data. The procedure is very simple. We can create a ROS package for this node using the following steps:

1. The following is the command to create a new package. You can also find this package in `chapter_11_ws/vr_leap_teleop`:

   ```
   $ catkin_create_pkg vr_leap_teleop roscpp rospy std_msgs
   visualization_msgs geometry_msgs message_generation
   visualization_msgs
   ```

   After creation, you could build the workspace using `catkin_make` command.

2. Now, let's create the node to convert Leap Motion data to Twist. You can create a folder called `scripts` inside the `vr_leap_teleop` package.
3. Now you can copy the node called `vr_leap_teleop.py` from our repository at `https://github.com/PacktPublishing/ROS-Robotics-Projects-SecondEdition/blob/master/chapter_11_ws/vr_leap_teleop/scripts/vr_leap_teleop.py`. Let's look at how this code works.

   We need the following Python modules in this node. Here, we require message definitions from the `leap_motion` package, which is the driver package:

   ```
   import rospy
   from leap_motion.msg import leap
   from leap_motion.msg import leapros
   from geometry_msgs.msg import Twist
   ```

4. Now, we have to set some range values in which we can check whether the current hand value is within range. We are also defining the `teleop_topic` name here:

   ```
   teleop_topic = '/cmd_vel_mux/input/teleop'

   low_speed = -0.5
   stop_speed = 0
   high_speed = 0.5

   low_turn = -0.5
   stop_turn = 0
   high_turn = 0.5
   ```

```
        pitch_low_range = -30
        pitch_high_range = 30

        roll_low_range = -150
        roll_high_range = 150
```

Here is the main code of this node. In this code, you can see that the topic from the Leap Motion driver is being subscribed to. When a topic is received, it will call the `callback_ros()` function:

```
def listener():
    global pub
    rospy.init_node('leap_sub', anonymous=True)
    rospy.Subscriber("leapmotion/data", leapros, callback_ros)
    pub = rospy.Publisher(teleop_topic, Twist, queue_size=1)

    rospy.spin()

if __name__ == '__main__':
    listener()
```

The following is the definition of the `callback_ros()` function. Essentially, it will receive the Leap Motion data and extract the orientation components of the palm only. Therefore, we will get yaw, pitch, and roll from this function.

We are also creating a `Twist()` message to send the velocity values to the robot:

```
def callback_ros(data):
    global pub

    msg = leapros()
    msg = data
    yaw = msg.ypr.x
    pitch = msg.ypr.y
    roll = msg.ypr.z

    twist = Twist()

    twist.linear.x = 0; twist.linear.y = 0; twist.linear.z = 0
    twist.angular.x = 0; twist.angular.y = 0; twist.angular.z = 0
```

We are performing a basic comparison with the current roll and pitch values again within the following ranges. Here are the actions we've assigned for each movement of the robot:

Action	**Description**
Hand gesture	Robot movement
Hand pitch low	Move forward
Hand pitch high	Move backward
Hand roll anticlockwise	Rotate anticlockwise
Hand roll clockwise	Rotate clockwise

Here is a code snippet that takes care of one condition. In this case, if the pitch is low, then we are providing a high value for linear velocity in the `x` direction for moving forward:

```
if(pitch > pitch_low_range and pitch < pitch_low_range + 30):
    twist.linear.x = high_speed; twist.linear.y = 0;
      twist.linear.z = 0   twist.angular.x = 0; twist.angular.y = 0;
twist.angular.z = 0
```

Okay, so we have built the node, and we can test it at the end of the project. In the next section, we will look at how to implement VR in ROS.

Building a ROS-VR Android application

In this section, we will look at how to create a VR experience in ROS, especially in robotics simulators such as Gazebo. Luckily, we have an open source Android project called ROS Cardboard (`https://github.com/cloudspace/ros_cardboard`). This project is exactly what we want we want for this application. This application is based on ROS-Android APIs, which help us to visualize compressed images from a ROS PC. It also does the splitting of the view for the left and right eye, and when we put this on a VR headset, it will feel like 3D.

Here is a diagram that shows how this application works:

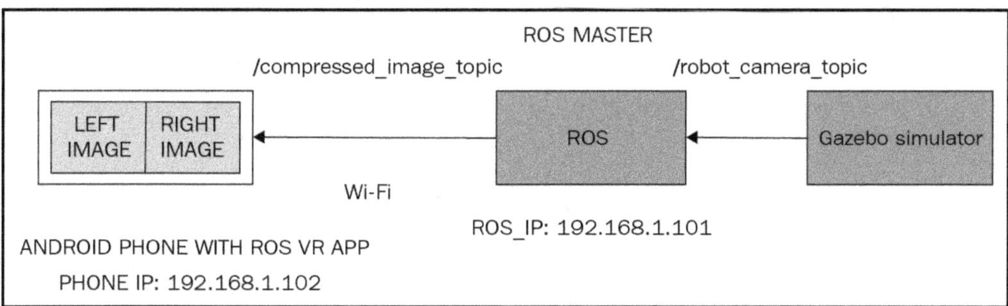

Communication between a ROS PC and Android phone

From the preceding diagram, you can see that the image topic from Gazebo can be accessed from a ROS environment, and the compressed version of that image is sent to the ROS-VR app, which will split the view into left and right to provide 3D vision. Setting the `ROS_IP` variable on the PC is important for the proper working of the VR application. The communication between the PC and the phone happens over Wi-Fi, both on the same network.

Building this application is not very tough; first, you can clone this app into a folder. You need to have the Android development environment and SDK installed. Just clone it and you can simply build it using the following instructions:

1. Install the `rosjava` packages from the source, as shown here:

```
$ mkdir -p ~/rosjava/src
$ wstool init -j4 ~/rosjava/src
https://raw.githubusercontent.com/rosjava/rosjava/melodic/rosjava.r
osinstall
$ initros1
$ cd ~/rosjava
$ rosdep update
$ rosdep install --from-paths src -i -y
$ catkin_make
```

Then, install `android-sdk` using the following command:

```
$ sudo apt-get install android-sdk
```

2. Plug your Android device into Ubuntu and execute the following command to check whether the device is detected on your PC:

   ```
   $ adb devices
   ```

   The `adb` command, which stands for **Android Debug Bridge**, will help you communicate with an Android device and emulator. If this command lists out the devices, then you are done; otherwise, do a Google search to find out how to make it work. It won't be too difficult.

3. After getting the device list, clone the ROS Cardboard project using the following command. You can clone it to `home` or `desktop`:

   ```
   $ git clone https://github.com/cloudspace/ros_cardboard.git
   ```

4. After cloning, enter the folder and execute the following command to build the entire package and install it on the device:

   ```
   $ ./gradlew installDebug
   ```

You may get an error saying the required Android platform is not available; what you need is to simply install it using the Android SDK GUI. If everything works fine, you will be able to install the APK on an Android device. If you are unable to build the APK, you can also find it at `https://github.com/PacktPublishing/ROS-Robotics-Projects-SecondEdition/tree/master/chapter_11_ws/ros_cardboard`.

If directly installing the APK to the device failed, you can find the generated APK from `ros_cardboard/ros_cardboard_module/build/outputs/apk`. You can copy this APK to the device and try to install it. Let's now work with the ROS-VR application and learn to interface it with Gazebo.

Working with the ROS-VR application and interfacing with Gazebo

Let's demonstrate how to use the ROS-VR application and interface it with Gazebo. The new APK will be installed with a name such as `ROSSerial`.

Before starting the app, we need to set up a few steps on the ROS PC. Let's follow these steps:

1. First, set the `ROS_IP` variable in the `~/.bashrc` file. Execute the `ifconfig` command and retrieve the Wi-Fi IP address of the PC, as shown here:

```
wlan0     Link encap:Ethernet  HWaddr 94:39:e5:4d:7d:da
          inet addr:192.168.1.101  Bcast:192.168.1.255  Mask:255.255.255.0
          inet6 addr: fe80::9639:e5ff:fe4d:7dda/64 Scope:Link
          UP BROADCAST RUNNING MULTICAST  MTU:1500  Metric:1
          RX packets:1303 errors:0 dropped:0 overruns:0 frame:0
          TX packets:1127 errors:0 dropped:0 overruns:0 carrier:0
          collisions:0 txqueuelen:1000
          RX bytes:1136655 (1.1 MB)  TX bytes:243000 (243.0 KB)
```

PC Wi-Fi adapter IP address

2. For this project, the IP address was `192.168.1.101`, so we have to set the `ROS_IP` variable as the current IP in `~/.bashrc`. You can simply copy the following line to the `~/.bashrc` file:

```
$ export ROS_IP=192.168.1.101
```

We need to set this; only then will the Android VR app work.

3. Now start the `roscore` command on the ROS PC:

```
$ roscore
```

4. The next step is to open the Android app, and you will get a window like the following. Enter `ROS_IP` in the edit box and click on the **CONNECT** button:

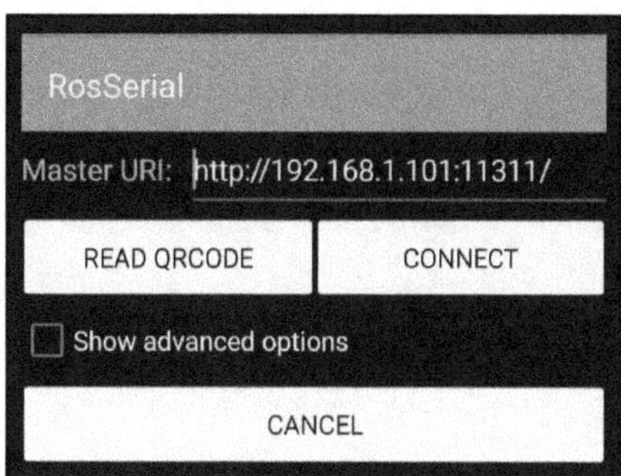

The ROS-VR application

If the app is connected to the ROS master on the PC, it will show up as connected and show a blank screen with a split view. Now list out the topics on the ROS PC:

```
lentin@lentin-Aspire-4755:~$ rostopic list
/rosout
/rosout_agg
/usb_cam/image_raw/compressed
lentin@lentin-Aspire-4755:~$
```

Listing the ROS-VR topics on the PC

Here, you can see topics such as `/usb_cam/image_raw/compressed` and `/camera/image/compressed` in the list. What we want to do is feed a compressed image to whatever image topic the app is going to subscribe to.

5. If you've installed the `usb_cam` (`https://github.com/bosch-ros-pkg/usb_cam`) ROS package already, you can launch the webcam driver using the following command:

   ```
   $ roslaunch usb_cam usb_cam-test.launch
   ```

   This driver will publish the camera image in compressed form to the `/usb_cam/image_raw/compressed` topic, and, when there is a publisher for this topic, it will display it on the app as well.

If you are getting some other topics from the app, say, `/camera/image/compressed`, you can use `topic_tools` (`http://wiki.ros.org/topic_tools`) for remapping the topic to the app topic. You can use the following command:

```
$ rosrun topic_tools relay /usb_cam/image_raw/compressed
/camera/image/compressed
```

Now you can see the camera view in the VR app like this:

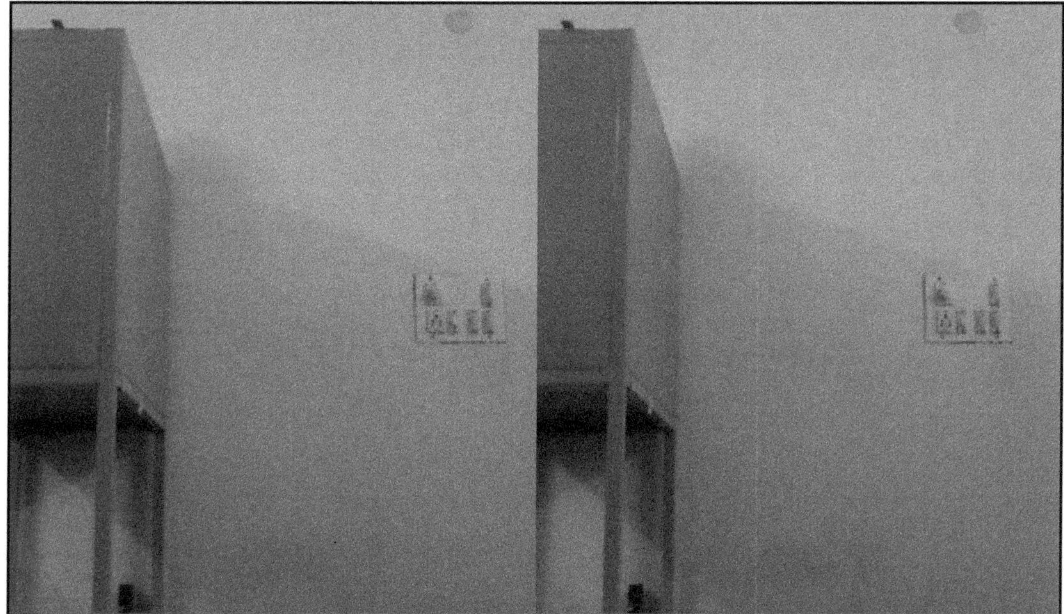

The ROS-VR app

This is the split view that we are getting in the application. We can also display images from Gazebo in a similar manner. Simple, right? Just remap the robot camera compressed image to the app topic. In the next section, we will learn how to view Gazebo images in the VR app.

TurtleBot simulation in VR

Let's look at how to install and use the TurtleBot simulator, followed by using the TurtleBot simulation with VR-App.

Installing the Turtlebot simulator

This is a prerequisite for our testing; therefore, let's go ahead with installing the TurtleBot packages.

Since we're using ROS Melodic, there isn't any Debian package of the TurtleBot simulator. Therefore, we shall source the workspace from `https://github.com/turtlebot/turtlebot_simulator`. Let's begin installing the Turtlebot simulator as follows:

1. Clone the package using the following command in your `workspace/src` folder:

   ```
   $ git clone https://github.com/turtlebot/turtlebot_simulator.git'
   ```

2. Once cloned, you could install the dependencies using the `rosinstall` command:

   ```
   $ rosinstall . turtlebot_simulator.rosinstall
   ```

   Also, install the following dependencies:

   ```
   $ sudo apt-get install ros-melodic-ecl ros-melodic-joy ros-melodic-kobuki-* ros-melodic-yocs-controllers ros-melodic-yocs-cmd-vel-mux ros-melodic-depthimage-to-laserscan
   ```

3. Once done, remove the following two packages, `kobuki_desktop` and `turtlebot_create_desktop`, as they cause a Gazebo library error while compiling. However, we do not want them for our simulation.
4. Now compile the package using `catkin_make` and you should be all set.

If everything was successful, you should have a proper Turtlebot simulation workspace ready for use. Let's run our web teleoperation application now.

Working with TurtleBot in VR

We can start a TurtleBot simulation using the following command:

```
$ roslaunch turtlebot_gazebo turtlebot_playground.launch
```

You will get the TurtleBot simulation in Gazebo like this:

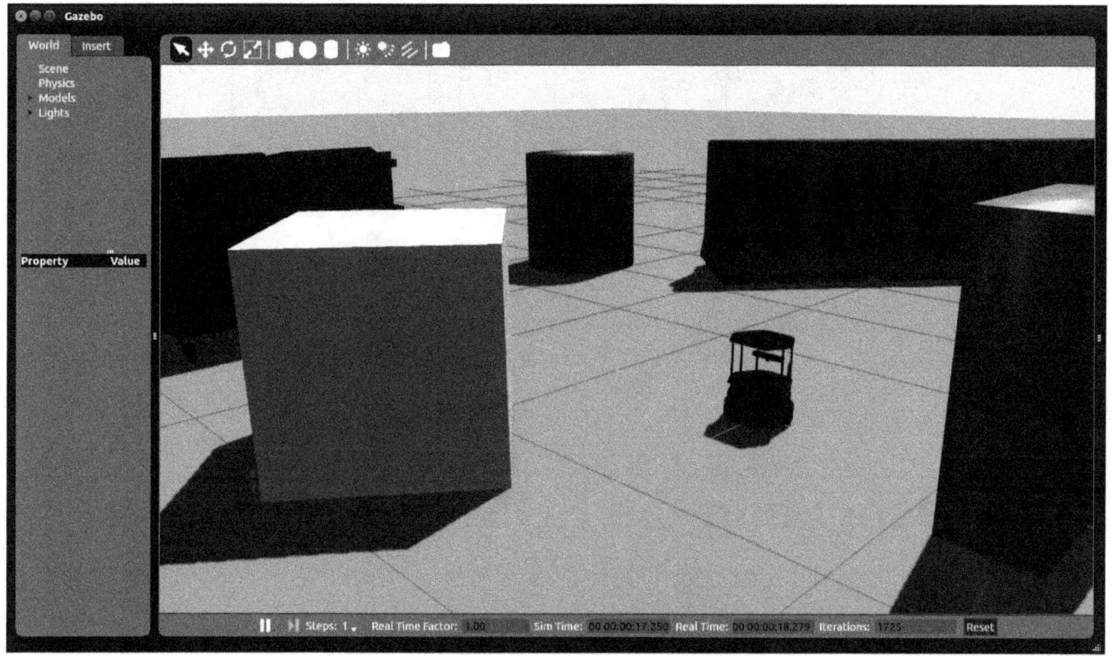

TurtleBot simulation in Gazebo

You can move the robot by launching the teleoperation node with the following command:

```
$ roslaunch turtlebot_teleop keyboard_teleop.launch
```

You can now move the robot using the keyboard. Launch the app again and connect to the ROS master running on the PC. Then, you can remap the compressed Gazebo RGB image data into an app image topic, like this:

```
$ rosrun topic_tools relay /camera/rgb/image_raw/compressed
/usb_cam/image_raw/compressed
```

Now, what happens is that the robot camera image is visualized in the app, and if you put the phone into a VR headset, it will simulate a 3D environment. The following screenshot shows the split view of the images from Gazebo:

Gazebo image view in the ROS-VR app

You can now move the robot using a keyboard. In the next section, we will examine possible issues that you may encounter when you work with the application and their solutions.

Troubleshooting the ROS-VR application

You may get issues working with ROS-VR applications. One of the issues might be the size of the image. The left and right image size can vary according to the device screen size and resolution. This project was tested on a full-HD 5-inch screen, and if you have a different screen size or resolution, you may need to hack the application code.

You can go to the app's project folder and open the code at `ros_cardboard/ros_cardboard_module/src/main/java/com/cloudspace/cardboard/CardboardOverlayEyeView.java`.

You can change the `final float imageSize = 1.0f` value to `1.8f` or `2f`; this will stretch the image and fill the screen, but we might lose some part of the image. After this change, build it again and install it.

One of the other issues associated with the working of this app is that the app will not work until we set the `ROS_IP` value on the PC. So, you should check whether `ROS_IP` is set.

If you want to change the topic name of the app, then go to `ros_cardboard/ros_cardboard_module/src/main/java/com/cloudspace/cardboard/CardboardViewerActivity.java` and change this line:

```
mOverlayView.setTopicInformation("/camera/image/compressed",
CompressedImage._TYPE);
```

If you want to work with other high-end VR headsets such as Oculus and HTC Vive, you can follow these links:

- `ros_ovr_sdk`: `https://github.com/OSUrobotics/ros_ovr_sdk`
- `vive_ros`: `https://github.com/robosavvy/vive_ros`
- `oculus_rviz_plugins`: `http://wiki.ros.org/oculus_rviz_plugins`

In the next section, we will combine the power of the VR headset and the Leap Motion robot controller node.

Integrating the ROS-VR application and Leap Motion teleoperation

In this section, we are going to replace the keyboard teleoperation with Leap Motion-based teleoperation. When we roll our hand in the anticlockwise direction, the robot also rotates anticlockwise, and vice versa. If we pitch our hand down, the robot will move forward, and if we pitch it up, it will move backward. So, we can start the VR application and TurtleBot simulation like in the previous section, and, instead of the keyboard teleoperation, run the Leap Motion teleoperation node.

So, before starting the Leap Motion teleoperation node:

1. Launch the PC driver and ROS driver using the following commands:

   ```
   $ sudo LeapControlPanel
   ```

2. Start the ROS driver using the following command:

   ```
   $ roslaunch leap_motion sensor_sender.launch
   ```

3. Launch Leap Motion on the Twist node using the following command:

```
$ rosrun vr_leap_teleop vr_leap_teleop.py
```

You can now put the VR headset on your head and control the robot using your hand.

Summary

In this chapter, we looked at how to make use of Leap Motion sensors and VR headsets in Ubuntu with ROS. Additionally, we looked at how to set up the Leap Motion sensor in ROS and make use of a VR headset to visualize data in RViz. Later, we created a custom teleoperation node to help teleoperate a mobile robot in Gazebo using hand gestures that were recognized by the Leap Motion sensor. We also examined how to visualize the Gazebo environment using the VR headset. Using these skills, you could also manually control a real mobile robot, by only using hand gestures and especially those that are not in direct human vicinity.

In the next chapter, we will demonstrate how to detect and track a face in ROS.

12
Face Detection and Tracking Using ROS, OpenCV, and Dynamixel Servos

One of the capabilities of most service and social robots is face detection and tracking. These robots can identify faces and track head movements. There are numerous implementations of face detection and tracking systems on the web. Most trackers have a pan-and-tilt mechanism, and a camera is mounted on the top of the servos. In this chapter, we will see a simple tracker that only has a pan mechanism. We are going to use a USB webcam mounted on an AX-12 Dynamixel servo. Controlling the Dynamixel servo and image processing are done in ROS.

In this chapter, we will begin by first configuring Dynamixel AX-12 servos and then interfacing Dynamixel with ROS. We will then create face-tracker ROS packages. By the end of this chapter, we will have learned how to work with the face-tracking ROS package.

The following topics will be covered in this chapter:

- Overview of the project
- Hardware and software prerequisites
- Configuring Dynamixel AX-12 servos
- Interfacing Dynamixel with ROS
- Creating face tracker ROS packages
- Working with the face-tracking ROS package

Technical requirements

Let's look into the technical requirements for this chapter:

- ROS Melodic Morenia on Ubuntu 18.04 (Bionic)
- Timelines and test platform:
    - **Estimated learning time**: On average, 100 minutes
    - **Project build time (inclusive of compile and runtime)**: On average, 45-90 minutes (depending on setting up hardware boards with the indicated requirements)
    - **Project test platform**: HP Pavilion laptop (Intel® Core™ i7-4510U CPU @ 2.00 GHz × 4 with 8 GB Memory and 64-bit OS, GNOME-3.28.2)

The code for this chapter is available at `https://github.com/PacktPublishing/ROS-Robotics-Projects-SecondEdition/tree/master/chapter_12_ws`.

Let's begin by looking at the overview of the project.

Overview of the project

The project aims to build a simple face tracker that can track a face only along the horizontal axis of the camera. The face tracker hardware consists of a webcam, a Dynamixel servo called AX-12, and a supporting bracket to mount the camera on the servo. The servo tracker will follow a face until it aligns with the center of the image from the webcam. Once it reaches the center, it will stop and wait for face movement. The face detection is done using an OpenCV and ROS interface, and the controlling of the servo is done using a Dynamixel motor driver in ROS.

We are going to create two ROS packages for this complete tracking system; one will be for face detection and finding the centroid of the face, and the other will be for sending commands to the servo to track the face using the centroid values.

Okay! Let's start discussing the hardware and software prerequisites of this project.

The complete source code of this project can be cloned from the following Git repository. The following command will clone the project repository: `$ git clone https://github.com/PacktPublishing/ROS-Robotics-Projects-SecondEdition.git`.

Hardware and software prerequisites

The following lists the hardware components that are required for building this project:

- Webcam
- Dynamixel AX-12A servo with mounting bracket
- USB-to-Dynamixel adapter
- Extra 3-pin cables for AX-12 servos
- Power adapter
- 6-port AX/MX power hub
- USB extension cable

If you are thinking that the total cost is not affordable, then there are cheap alternatives to do this project too. The main heart of this project is the Dynamixel servo. We can replace this servo with RC servos, which only cost around $10, and an Arduino board costing around $20 can be used to control the servo too. The ROS and Arduino interfacing will be discussed in the upcoming chapters, so you can think about porting the face tracker project using an Arduino and RC servo.

Okay, let's look at the software prerequisites of the project. The prerequisites include the ROS framework, OS version, and ROS packages mentioned here:

- Ubuntu 18.04 LTS
- ROS Melodic LTS
- ROS `usb_cam` package
- ROS `cv_bridge` package
- ROS Dynamixel controller
- Windows 7 or higher
- RoboPlus (Windows application)

This gives you an idea of the software we are going to be using for this project. We may need both Windows and Ubuntu for this project. It would be great if you have dual operating systems on your computer.

Let's see how to install the software first.

Installing the usb_cam ROS package

Let's look at the use of the `usb_cam` package in ROS first. The `usb_cam` package is the ROS driver for **Video4Linux (V4L)** USB cameras. V4L is a collection of device drivers in Linux for real-time video capture from webcams. The `usb_cam` ROS package works using V4L devices and publishes the video stream from devices as ROS image messages. We can subscribe to it and perform our own processing using it. The official ROS page of this package is given in the previous table. You can check out that page for different settings and configurations this package offers.

Creating an ROS workspace for dependencies

Before starting to install the `usb_cam` package, let's create an ROS workspace for storing the dependencies of all of the projects mentioned in this book. We can create another workspace for keeping the project code as shown in the following steps:

1. Create an ROS workspace called `ros_project_dependencies_ws` in the home folder. Clone the `usb_cam` package into the `src` folder:

```
$ git clone https://github.com/bosch-ros-pkg/usb_cam.git
```

2. Build the workspace using `catkin_make`.
3. After building the package, install the `v4l-util` Ubuntu package. It is a collection of command-line V4L utilities used by the `usb_cam` package:

```
$ sudo apt-get install v4l-utils
```

Let's now look at configuring a webcam on Ubuntu.

Configuring a webcam on Ubuntu 18.04

After installing the preceding dependencies, we can connect the webcam to the PC to check whether it is properly detected by our PC by following these steps:

1. Open a Terminal and execute the `dmesg` command to check the kernel logs:

```
$ dmesg
```

The following screenshot shows the kernel logs of the webcam device:

```
[   86.483102] usb 1-1.5: new high-speed USB device number 6 using ehci-pci
[   86.620403] usb 1-1.5: New USB device found, idVendor=0c45, idProduct=6340
[   86.620409] usb 1-1.5: New USB device strings: Mfr=2, Product=1, SerialNumber=3
[   86.620412] usb 1-1.5: Product: iBall Face2Face Webcam C12.0
[   86.620414] usb 1-1.5: Manufacturer: iBall Face2Face Webcam C12.0
[   86.620416] usb 1-1.5: SerialNumber: iBall Face2Face Webcam C12.0
[   86.657389] media: Linux media interface: v0.10
[   86.677503] Linux video capture interface: v2.00
[   86.703833] usb 1-1.5: 3:1: cannot get freq at ep 0x84
[   86.722072] usbcore: registered new interface driver snd-usb-audio
[   86.722096] uvcvideo: Found UVC 1.00 device iBall Face2Face Webcam C12.0 (0c45:6340)
[   86.735670] input: iBall Face2Face Webcam C12.0 as /devices/pci0000:00/0000:00:1a.0/
t/input16
[   86.735747] usbcore: registered new interface driver uvcvideo
[   86.735749] USB Video Class driver (1.1.1)
```

Kernel logs of the webcam device

You can use any webcam that has driver support in Linux. In this project, an iBall Face2Face webcam is used for tracking. You can also go for the popular Logitech C310 webcam mentioned as a hardware prerequisite. You can opt for that for better performance and tracking.

2. If our webcam has support in Ubuntu, we can open the video device using a tool called Cheese. Cheese is simply a webcam viewer.
3. Enter the `cheese` command in the Terminal. If it is not installed, you can install it using the following command:

```
$ sudo apt-get install cheese
```

You could open `cheese` using the following command on the Terminal:

```
$ cheese
```

If the driver and device are proper, you will get a video stream from the webcam, like this:

Webcam video streaming using Cheese

Congratulations! Your webcam is working well in Ubuntu, but are we done with everything? No. The next thing is to test the ROS `usb_cam` package. We have to make sure that it's working well in ROS!

Interfacing the webcam with ROS

Let's test the webcam using the `usb_cam` package. The following command is used to launch the `usb_cam` nodes to display images from a webcam and publish ROS image topics at the same time:

```
$ roslaunch usb_cam usb_cam-test.launch
```

If everything works fine, you will get the image stream and logs in the Terminal, as shown here:

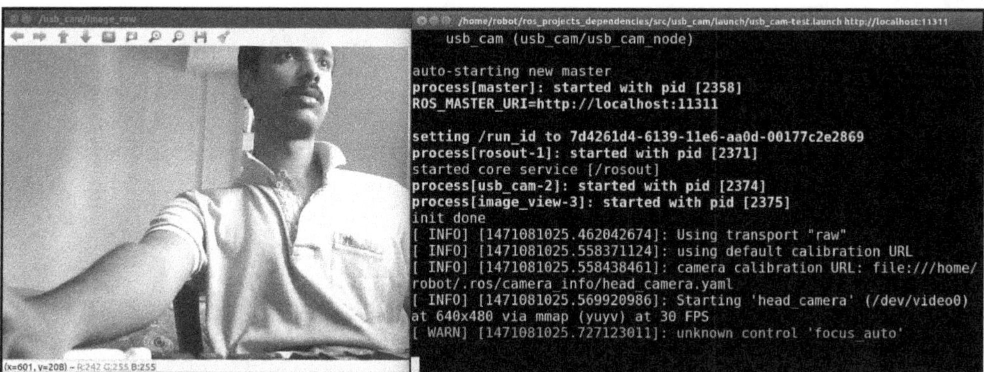

The workings of the usb_cam package in ROS

The image is displayed using the `image_view` package in ROS, which is subscribed to the topic called `/usb_cam/image_raw`.

Here are the topics that the `usb_cam` node is publishing:

```
robot@robot-pc:~$ rostopic list
/image_view/parameter_descriptions
/image_view/parameter_updates
/rosout
/rosout_agg
/usb_cam/camera_info
/usb_cam/image_raw
/usb_cam/image_raw/compressed
/usb_cam/image_raw/compressed/parameter_descriptions
/usb_cam/image_raw/compressed/parameter_updates
/usb_cam/image_raw/compressedDepth
/usb_cam/image_raw/compressedDepth/parameter_descriptions
/usb_cam/image_raw/compressedDepth/parameter_updates
/usb_cam/image_raw/theora
/usb_cam/image_raw/theora/parameter_descriptions
/usb_cam/image_raw/theora/parameter_updates
robot@robot-pc:~$
```

The topics being published by the usb_cam node

We've finished interfacing a webcam with ROS. So, what's next? We have to interface an AX-12 Dynamixel servo with ROS. Before we start interfacing, we have to do something to configure this servo.

Next, we are going to see how to configure a Dynamixel AX-12A servo.

Configuring a Dynamixel servo using RoboPlus

The Dynamixel servo can be configured using a program called **RoboPlus**, provided by **ROBOTIS, INC.** (`http://en.robotis.com/`), the manufacturer of Dynamixel servos.

To configure Dynamixel, you have to switch your operating system to Windows as the RoboPlus tool works on Windows. In this project, we are going to configure the servo in Windows 7.

Here is the link to download RoboPlus: `http://www.robotis.com/download/software/RoboPlusWeb%28v1.1.3.0%29.exe`.

If the link is not working, you can just search in Google for RoboPlus 1.1.3. After installing the software, you will get the following window. Navigate to the **Expert** tab in the software to get the application for configuring Dynamixel:

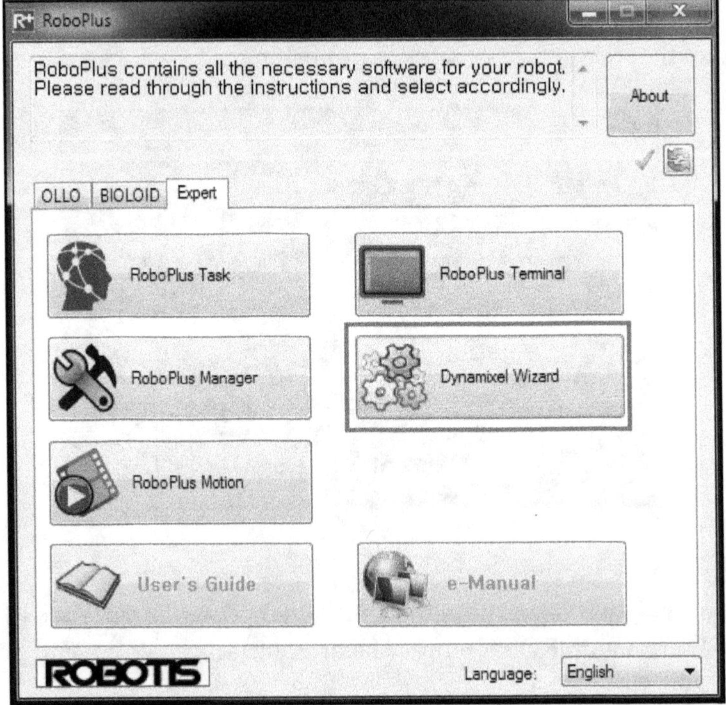

Dynamixel manager in RoboPlus

Before starting the Dynamixel wizard and configuring, we have to connect the Dynamixel and properly power it up. You can have a look at the pin details at this link: `http://emanual.robotis.com/docs/en/dxl/ax/ax-12a/#connector-information`.

Unlike other RC servos, AX-12 is an intelligent actuator, having a microcontroller that can monitor every parameter of a servo and customize all of them. It has a geared drive, and the output of the servo is connected to a servo horn. We can connect any link to this servo horn. There are two connection ports behind each servo, and each port has pins such as VCC, GND, and data. The ports of the Dynamixel are daisy-chained, so we can connect one servo to another.

The main hardware component interfacing Dynamixel with the PC is called a USB-to-Dynamixel adapter. This is a USB-to-serial adapter that can convert USB into RS232, RS 484, and TTL. In AX-12 motors, data communication is done using TTL. There are three pins in each port. The data pin is used to send to and receive from AX-12, and power pins are used to power the servo. The input voltage range of the AX-12A Dynamixel is from 9V to 12V. The second port in each Dynamixel can be used for daisy-chaining. We can connect up to 254 servos using such chaining.

To work with Dynamixel, we should know some more things. Let's have a look at some of the important specifications of the AX-12A servo. The specifications are taken from the servo manual:

- Weight : 54.6g (AX-12A)
- Dimension : 32mm * 50mm * 40mm
- Resolution : 0.29°
- Gear Reduction Ratio : 254 : 1
- Stall Torque : 1.5N.m (at 12.0V, 1.5A)
- No load speed : 59rpm (at 12V)
- Running Degree : 0° ~ 300°, Endless Turn
- Running Temperature : -5°C ~ +70°C
- Voltage : 9 ~ 12V (Recommended Voltage 11.1V)
- Command Signal : Digital Packet
- Protocol Type : Half duplex Asynchronous Serial Communication (8bit,1stop,No Parity)
- Link (Physical) : TTL Level Multi Drop (daisy chain type Connector)
- ID : 254 ID (0~253)
- Communication Speed : 7343bps ~ 1 Mbps
- Feedback : Position, Temperature, Load, Input Voltage, etc.
- Material : Engineering Plastic

AX-12A specifications

The Dynamixel servo can communicate with the PC at a maximum speed of 1 Mbps. It can also provide feedback about various parameters, such as its position, temperature, and current load. Unlike RC servos, this can rotate up to 300 degrees, and communication is mainly done using digital packets.

You can have a look at how to power and connect the Dynamixel to a PC using the following two links: `http://emanual.robotis.com/docs/en/parts/interface/usb2dynamixel/` and `http://emanual.robotis.com/docs/en/parts/interface/u2d2/`.

Setting up the USB-to-Dynamixel driver on the PC

We have already discussed that the USB-to-Dynamixel adapter is a USB-to-serial converter with an FTDI chip on it. We have to install a proper FTDI driver on the PC to detect the device. The driver is required for Windows but not for Linux, because FTDI drivers are already present in the Linux kernel. If you install the RoboPlus software, the driver may already be installed along with it. If it is not, you can manually install it from the RoboPlus installation folder.

Plug the USB-to-Dynamixel into the Windows PC, and check **Device Manager** (right-click on **My Computer** and go to **Properties | Device Manager**). If the device is properly detected, you'll see something like this:

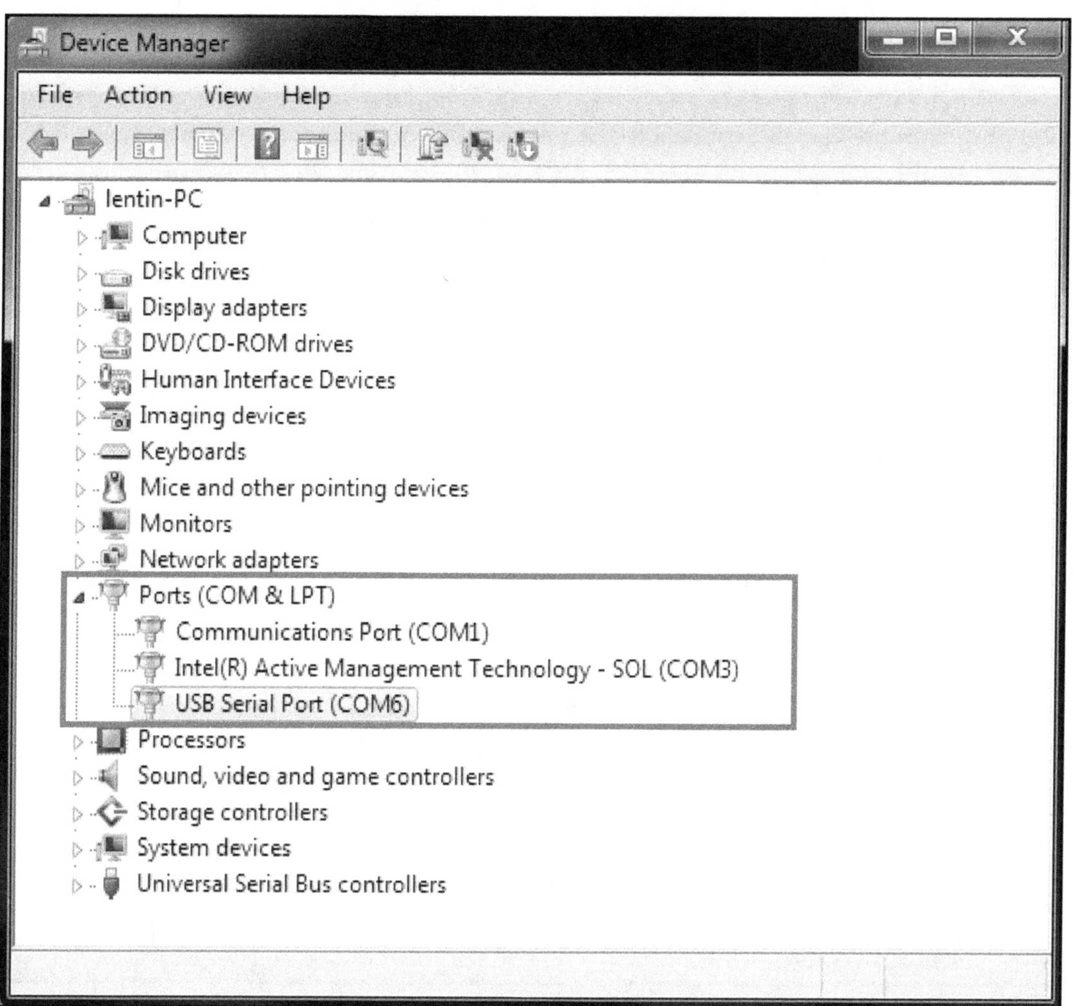

The COM port of the USB-to-Dynamixel

If you are getting a COM port for the USB-to-Dynamixel, you can start the Dynamixel manager from RoboPlus. You can connect to the serial port number from the list and click on the **Search** button to scan for Dynamixel, as shown in the next screenshot. Select the COM port from the list, and connect to the port marked **1**. After connecting to the COM port, set the default baud rate to 1 Mbps, and click on the **Start searching** button:

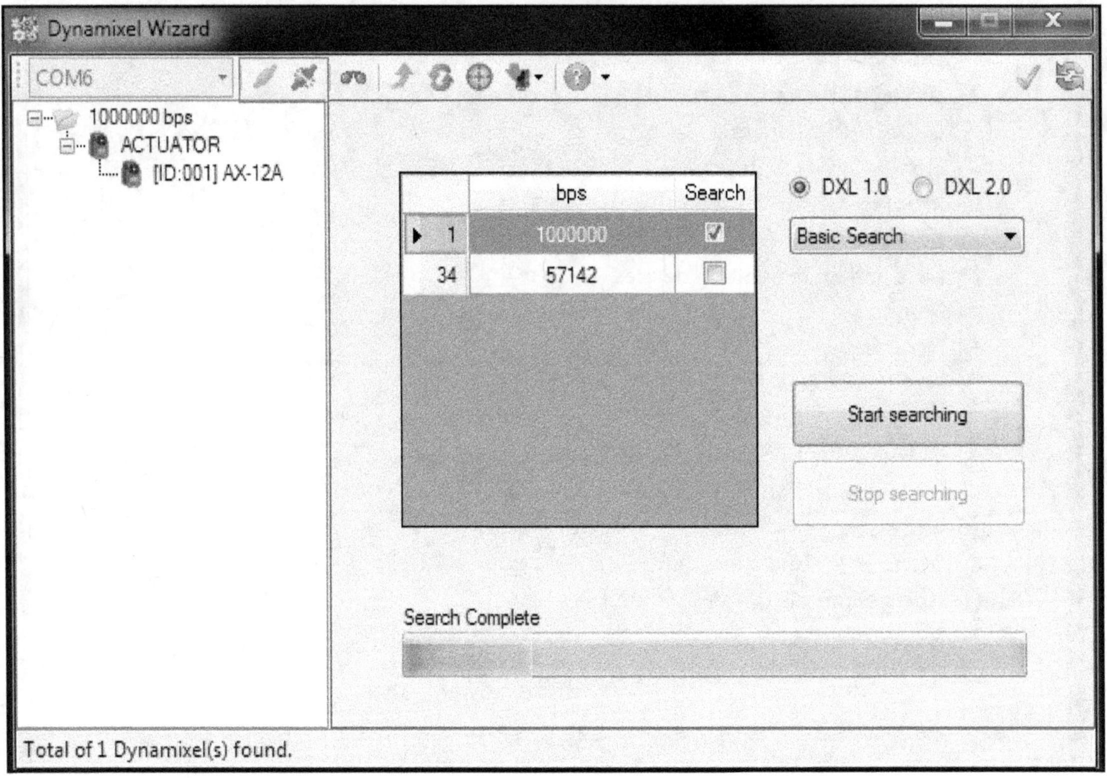

The COM port of the USB-to-Dynamixel

If you are getting a list of servos in the left-hand side panel, it means that your PC has detected a Dynamixel servo. If the servo is not being detected, you can perform the following steps to debug:

1. Make sure that the supply and connections are correct using a multimeter. Make sure that the servo LED on the back is blinking when the power is on; if it does not come on, it can indicate a problem with the servo or power supply.
2. Upgrade the firmware of the servo using the Dynamixel manager. The wizard is shown in the next set of screenshots. While using the wizard, you may need to power off the supply and turn it back on to detect the servo.

3. After detecting the servo, you have to select the servo model and install the new firmware. This may help you to detect the servo in the Dynamixel manager if the existing servo firmware is outdated:

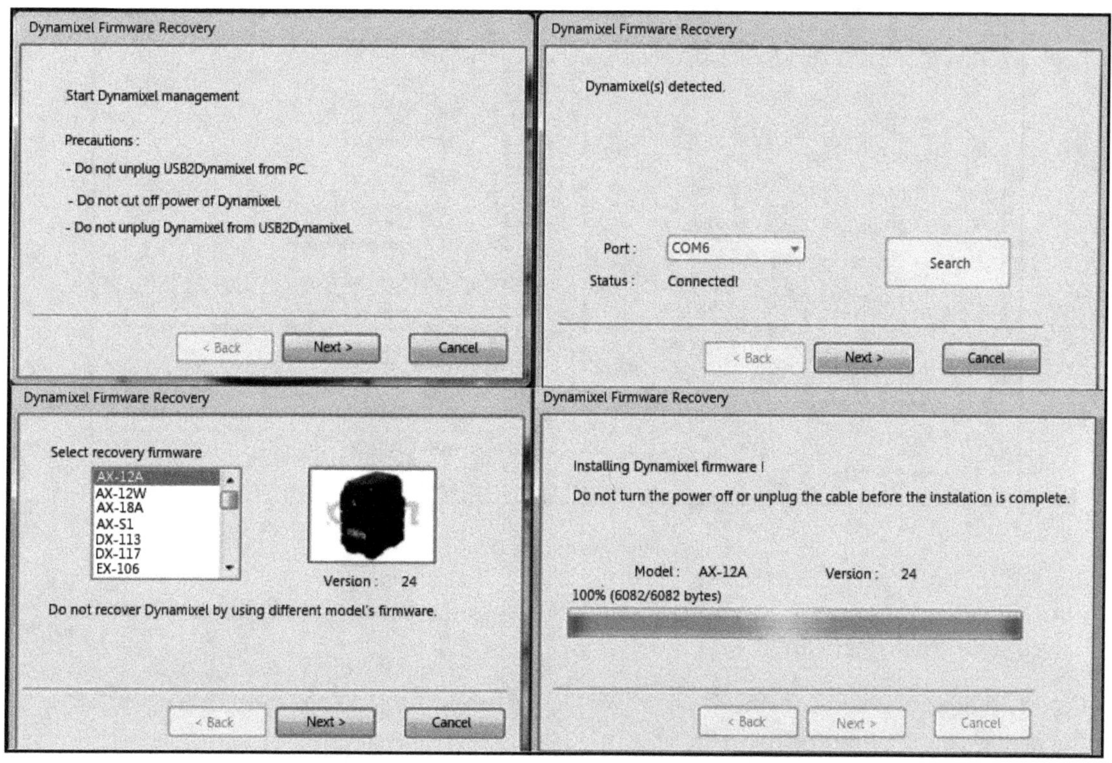

The Dynamixel recovery wizard

If the servos are listed in the Dynamixel Manager, click on one, and you will see its complete configuration. We have to modify some values inside the configuration for our current face-tracker project. Here are the parameters:

- **ID**: `1`
- **Baud Rate**: `1`
- **Moving Speed**: `100`
- **Goal Position**: `512`

The modified servo settings are shown in the following screenshot:

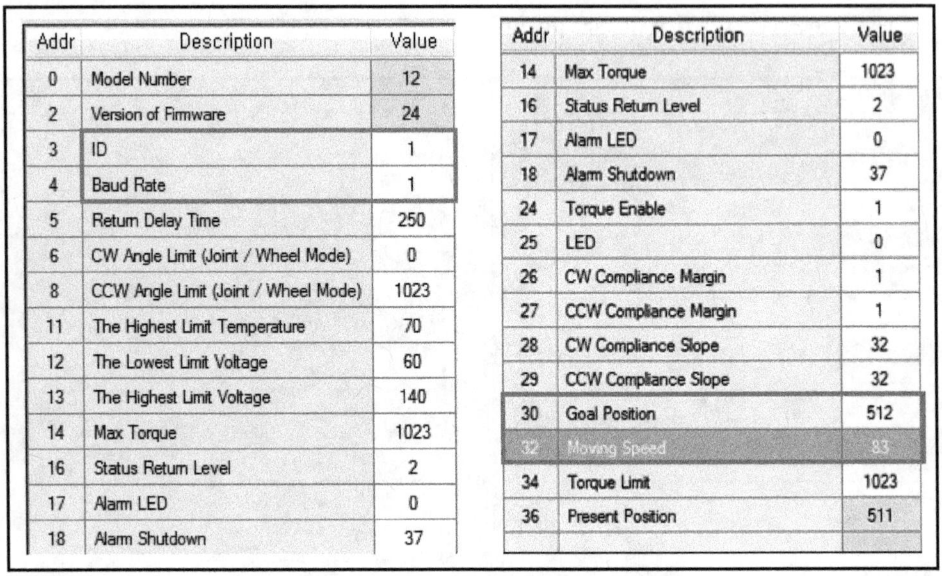

Addr	Description	Value
0	Model Number	12
2	Version of Firmware	24
3	ID	1
4	Baud Rate	1
5	Return Delay Time	250
6	CW Angle Limit (Joint / Wheel Mode)	0
8	CCW Angle Limit (Joint / Wheel Mode)	1023
11	The Highest Limit Temperature	70
12	The Lowest Limit Voltage	60
13	The Highest Limit Voltage	140
14	Max Torque	1023
16	Status Return Level	2
17	Alarm LED	0
18	Alarm Shutdown	37

Addr	Description	Value
14	Max Torque	1023
16	Status Return Level	2
17	Alarm LED	0
18	Alarm Shutdown	37
24	Torque Enable	1
25	LED	0
26	CW Compliance Margin	1
27	CCW Compliance Margin	1
28	CW Compliance Slope	32
29	CCW Compliance Slope	32
30	Goal Position	512
32	Moving Speed	83
34	Torque Limit	1023
36	Present Position	511

Modified Dynamixel firmware settings

After adjusting these settings, you can check whether the servo is working well or not by changing its **Goal Position**.

Nice! You are done configuring Dynamixel—congratulations! What's next? We want to interface Dynamixel with ROS.

Interfacing Dynamixel with ROS

If you successfully configured the Dynamixel servo, then it will be very easy to interface Dynamixel with ROS running on Ubuntu. As we've already discussed, there is no need for an FTDI driver in Ubuntu because it's already built into the kernel. The only thing we have to do is install the ROS Dynamixel driver packages.

The ROS Dynamixel packages are available at the following link: `http://wiki.ros.org/dynamixel_motor`.

You can install the Dynamixel ROS packages using the commands we'll look at now.

Installing the ROS dynamixel_motor packages

The ROS `dynamixel_motor` package stack is a dependency for the face tracker project, so we can install it to the `ros_project_dependencies_ws` ROS workspace by following these steps:

1. Open a Terminal and switch to the `src` folder of the workspace:

```
$ cd ~/ros_project_dependencies_ws/src
```

2. Clone the latest Dynamixel driver packages from GitHub:

```
$ git clone https://github.com/arebgun/dynamixel_motor
```

3. Use `catkin_make` to build the entire packages of the Dynamixel driver. If you can build the workspace without any errors, you are done with meeting the dependencies of this project.

Congratulations! You are done with the installation of the Dynamixel driver packages in ROS. We have now met all of the dependencies required for the face tracker project.

So, let's start working on face-tracking project packages.

Creating face tracker ROS packages

Let's start creating a new workspace for keeping the entire ROS project files for this book. You can name the workspace `chapter_12_ws` and follow these steps:

1. Download or clone the source code of this book from GitHub using the following link:

```
$ git clone
https://github.com/PacktPublishing/ROS-Robotics-Projects-SecondEdit
ion.git
```

2. Now, you can copy two packages, named `face_tracker_pkg` and `face_tracker_control`, from the `chapter_12_ws/` folder into the `src` folder of the `chapter_12_ws` workspace you created.
3. Compile the package using `catkin_make` to build the two project packages.

You have now set up the face tracker packages on your system.

What if you want to create your own package for tracking? Follow these steps:

1. First, delete the current packages that you copied to the `src` folder.

Note that you should be in the `src` folder of `chapter_12_ws` while creating the new packages, and there should not be any existing packages from this book's GitHub code.

2. Switch to the `src` folder:

```
$ cd ~/chapter_12_ws/src
```

3. The next command will create the `face_tracker_pkg` ROS package with the main dependencies, such as `cv_bridge`, `image_transport`, `sensor_msgs`, `message_generation`, and `message_runtime`. We are including these packages because these packages are required for the proper working of the face tracker package. The face tracker package contains ROS nodes for detecting faces and determining the centroid of the face:

```
$ catkin_create_pkg face_tracker_pkg roscpp rospy cv_bridge
image_transport sensor_msgs std_msgs message_runtime
message_generation
```

4. Next, we need to create the `face_tracker_control` ROS package. The important dependency of this package is `dynamixel_controllers`. This package is used to subscribe to the centroid from the face tracker node and control the Dynamixel in a way that the face centroid will always be in the center portion of the image:

```
$ catkin_create_pkg face_tracker_pkg roscpp rospy std_msgs
dynamixel_controllers message_generation
```

You have now created the ROS packages on your own.

What's next? Before starting to code, you will have to understand some concepts of OpenCV and its interface with ROS. Also, you need to know how to publish ROS image messages. So, let's master these concepts in the next section.

The interface between ROS and OpenCV

Open Source Computer Vision (**OpenCV**) is a library that has APIs to perform computer vision applications. The project was started Intel in Russia, and later on, it was supported by Willow Garage and Itseez. In 2016, Itseez was acquired by Intel.

For more information, refer to the following:

- **The OpenCV website**: `http://opencv.org/`
- **Willow Garage**: `http://www.willowgarage.com/`

OpenCV is a cross-platform library that supports most operating systems. Now, it also has an open source BSD license, so we can use it for research and commercial applications. The OpenCV version interfaced with ROS Melodic is 3.2. The 3.x versions of OpenCV have a few changes to the APIs from the 2.x versions.

The OpenCV library is integrated into ROS through a package called `vision_opencv`. This package was already installed when we installed `ros-melodic-desktop-full` in `Chapter 1`, *Getting Started with ROS*.

The `vision_opencv` metapackage has two packages:

- `cv_bridge`: This package is responsible for converting the OpenCV image data type (`cv::Mat`) into ROS image messages (`sensor_msgs/Image.msg`).
- `image_geometry`: This package helps us to interpret images geometrically. This node will aid in processing such as camera calibration and image rectification.

Out of these two packages, we are mainly dealing with `cv_bridge`. Using `cv_bridge`, the face tracker node can convert ROS `Image` messages from `usb_cam` into the OpenCV equivalent, `cv::Mat`. After converting into `cv::Mat`, we can use OpenCV APIs to process the camera image.

Here is a block diagram that shows the role of `cv_bridge` in this project:

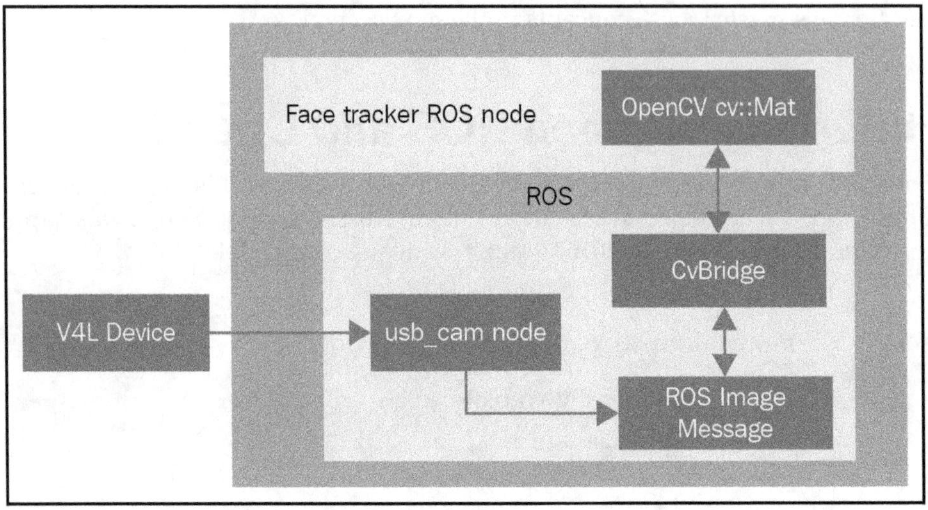

The role of cv_bridge

Here, `cv_bridge` is working between the `usb_cam` node and face-tracking node. We'll learn more about the face-tracking node in the next section. Before that, it would be good if you get an idea of its workings.

Another package we are using to transport ROS `Image` messages between two ROS nodes is `image_transport` (`http://wiki.ros.org/image_transport`). This package is always used to subscribe to and publish image data in ROS. The package can help us to transport images in low bandwidth by applying compression techniques. This package is also installed along with the full ROS desktop installation.

That's all about OpenCV and the ROS interface. In the next section, we are going to work with the first package of this project: `face_tracker_pkg`.

Working with the face-tracking ROS package

We have already created or copied the `face_tracker_pkg` package to the workspace and have discussed some of its important dependencies. Now, we are going to discuss what this package does exactly!

This package consists of an ROS node called `face_tracker_node`, which can track faces using OpenCV APIs and publish the centroid of the face to a topic. Here is a block diagram of the workings of `face_tracker_node`:

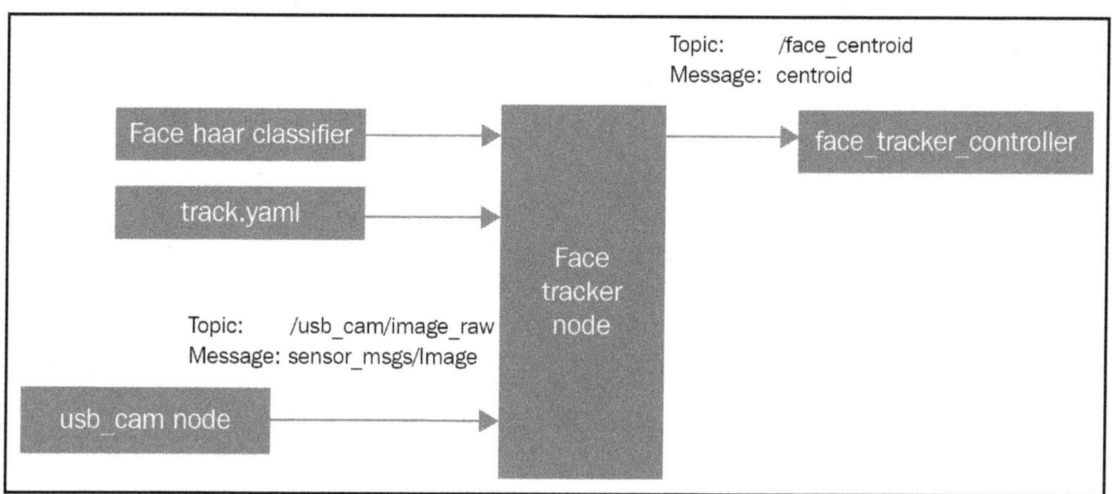

Block diagram of face_tracker_node

Let's discuss the things connected to `face_tracker_node`. One of the sections that may be unfamiliar to you is the face Haar classifier:

- **The face haar classifier**: The Haar feature-based cascade classifier is a machine learning approach for detecting objects. This method was proposed by Paul Viola and Michael Jones in their paper, *Rapid object detection using a boosted cascade of simple features*, in 2001. In this method, a cascade file is trained using a positive and negative sample image, and after training, that file is used for object detection:
    - In our case, we are using a trained Haar classifier file along with OpenCV source code. You will get these Haar classifier files from the OpenCV `data` folder (`https://github.com/opencv/opencv/tree/master/data`). You can replace the desired Haar file according to your application. Here, we are using the face classifier. The classifier will be an XML file that has tags containing features of a face. Once the features inside the XML match, we can retrieve the **Region Of Interest** (**ROI**) of the face from the image using the OpenCV APIs. You can check the Haar classifier of this project from `face_tracker_pkg/data/face.xml`.

- `track.yaml`: This is an ROS parameter file having parameters such as the Haar file path, input image topic, output image topic, and flags to enable and disable face tracking. We are using ROS configuration files because we can change the node parameters without modifying the face tracker source code. You can get this file from `face_tracker_pkg/config/track.xml`.
- `usb_cam` node: The `usb_cam` package has a node publishing the image stream from the camera to ROS `Image` messages. The `usb_cam` node publishes camera images to the `/usb_cam/raw_image` topic, and this topic is subscribed to by the face tracker node for face detection. We can change the input topic in the `track.yaml` file if we require.
- `face_tracker_control`: This is the second package we are going to discuss. The `face_tracker_pkg` package can detect faces and find the centroid of the face in the image. The centroid message contains two values, *X* and *Y*. We are using a custom message definition to send the centroid values. These centroid values are subscribed by the controller node and move the Dynamixel to track the face. The Dynamixel is controlled by this node.

Here is the file structure of `face_tracker_pkg`:

```
.
├── CMakeLists.txt
├── config
│   └── track.yaml
├── data
│   └── face.xml
├── include
│   └── face_tracker_pkg
├── launch
│   ├── start_dynamixel_tracking.launch
│   ├── start_tracking.launch
│   └── start_usb_cam.launch
├── msg
│   └── centroid.msg
├── package.xml
└── src
    └── face_tracker_node.cpp

7 directories, 9 files
```

The file structure of face_tracker_pkg

Let's see how the face-tracking code works. You can open the CPP file at `face_tracker_pkg/src/face_tracker_node.cpp`. This code performs face detection and sends the centroid value to a topic.

We'll look at and understand some code snippets.

Understanding the face tracker code

Let's start with the header file. The ROS header files we are using in the code are here. We have to include ros/ros.h in every ROS C++ node; otherwise, the source code will not compile. The remaining three headers are image-transport headers, which have functions to publish and subscribe to image messages at a low bandwidth. The cv_bridge header has functions to convert between OpenCV ROS data types. The image_encoding.h header has the image-encoding format used during ROS-OpenCV conversions:

```
#include <ros/ros.h>
#include <image_transport/image_transport.h>
#include <cv_bridge/cv_bridge.h>
#include <sensor_msgs/image_encodings.h>
```

The next set of headers is for OpenCV. The imgproc header consists of image-processing functions, highgui has GUI-related functions, and objdetect.hpp has APIs for object detection, such as the Haar classifier:

```
#include <opencv2/imgproc/imgproc.hpp>
#include <opencv2/highgui/highgui.hpp>
#include "opencv2/objdetect.hpp"
```

The last header file is for accessing a custom message called centroid. The centroid message definition has two fields, int32 x and int32 y. These can hold the centroid of the file. You can check this message definition from the face_tracker_pkg/msg/centroid.msg folder:

```
#include <face_tracker_pkg/centroid.h>
```

The following lines of code give a name to the raw image window and face-detection window:

```
static const std::string OPENCV_WINDOW = "raw_image_window";
static const std::string OPENCV_WINDOW_1 = "face_detector";
```

The following lines of code create a C++ class for our face detector. The code snippet creates handles of NodeHandle, which is a mandatory handle for an ROS node; image_transport, which helps to send ROS Image messages across the ROS computing graph; and a publisher for the face centroid, which can publish the centroid values using the centroid.msg file defined by us. The remaining definitions are for handling parameter values from the parameter file, track.yaml:

```
class Face_Detector
  {
    ros::NodeHandle nh_;
```

```
    image_transport::ImageTransport it_;
    image_transport::Subscriber image_sub_;
    image_transport::Publisher image_pub_;
    ros::Publisher face_centroid_pub;
    face_tracker_pkg::centroid face_centroid;
    string input_image_topic, output_image_topic, haar_file_face;
    int face_tracking, display_original_image,  display_tracking_image,
  center_offset, screenmaxx;
```

The following is the code for retrieving ROS parameters inside the `track.yaml` file. The advantage of using ROS parameters is that we can avoid hardcoding these values inside the program and modify the values without recompiling the code:

```
    try{
    nh_.getParam("image_input_topic", input_image_topic);
    nh_.getParam("face_detected_image_topic", output_image_topic);
    nh_.getParam("haar_file_face", haar_file_face);
    nh_.getParam("face_tracking", face_tracking);
    nh_.getParam("display_original_image", display_original_image);
    nh_.getParam("display_tracking_image", display_tracking_image);
    nh_.getParam("center_offset", center_offset);
    nh_.getParam("screenmaxx", screenmaxx);

    ROS_INFO("Successfully Loaded tracking parameters");
    }
```

The following code creates a subscriber for the input image topic and publisher for the face-detected image. Whenever an image arrives on the input image topic, it will call a function called `imageCb`. The names of the topics are retrieved from ROS parameters. We create another publisher for publishing the centroid value, which is the last line of the code snippet:

```
image_sub_ = it_.subscribe(input_image_topic, 1,
&Face_Detector::imageCb, this);
image_pub_ = it_.advertise(output_image_topic, 1);

face_centroid_pub = nh_.advertise<face_tracker_pkg::centroid>
("/face_centroid",10);
```

The next bit of code is the definition of `imageCb`, which is a callback for `input_image_topic`. What it basically does is convert the `sensor_msgs/Image` data into the `cv::Mat` OpenCV data type. The `cv_bridge::CvImagePtr cv_ptr` buffer is allocated for storing the OpenCV image after performing the ROS-OpenCV conversion using the `cv_bridge::toCvCopy` function:

```
    void imageCb(const sensor_msgs::ImageConstPtr& msg)
    {

      cv_bridge::CvImagePtr cv_ptr;
      namespace enc = sensor_msgs::image_encodings;

      try
      {
        cv_ptr = cv_bridge::toCvCopy(msg,
  sensor_msgs::image_encodings::BGR8);
      }
```

We have already discussed the Haar classifier; here is the code to load the Haar classifier file:

```
      string cascadeName = haar_file_face;
      CascadeClassifier cascade;
  if( !cascade.load( cascadeName ) )
      {
        cerr << "ERROR: Could not load classifier cascade" << endl;
      }
```

We are now moving to the core part of the program, which is the detection of the face performed on the converted OpenCV image data type from the ROS `Image` message. The following is the function call of `detectAndDraw()`, which performs the face detection, and in the last line, you can see the output image topic being published. Using `cv_ptr->image`, we can retrieve the `cv::Mat` data type, and in the next line, `cv_ptr->toImageMsg()` can convert this into an ROS `Image` message. The arguments of the `detectAndDraw()` function are the OpenCV `image` and `cascade` variables:

```
detectAndDraw( cv_ptr->image, cascade );
image_pub_.publish(cv_ptr->toImageMsg());
```

Let's understand the `detectAndDraw()` function, which is adapted from the OpenCV sample code for face detection: the function arguments are the input image and cascade object. The next bit of code will convert the image into grayscale first and equalize the histogram using OpenCV APIs. This is a kind of preprocessing before detecting the face from the image. The `cascade.detectMultiScale()` function is used for this purpose (`http://docs.opencv.org/2.4/modules/objdetect/doc/cascade_classification.html`):

```
Mat gray, smallImg;
cvtColor( img, gray, COLOR_BGR2GRAY );
double fx = 1 / scale ;
resize( gray, smallImg, Size(), fx, fx, INTER_LINEAR );
equalizeHist( smallImg, smallImg );
t = (double)cvGetTickCount();
cascade.detectMultiScale( smallImg, faces,
      1.1, 15, 0
      |CASCADE_SCALE_IMAGE,
      Size(30, 30) );
```

The following loop will iterate on each face that is detected using the `detectMultiScale()` function. For each face, it finds the centroid and publishes to the `/face_centroid` topic:

```
for ( size_t i = 0; i < faces.size(); i++ )
{
    Rect r = faces[i];
    Mat smallImgROI;
    vector<Rect> nestedObjects;
    Point center;
    Scalar color = colors[i%8];
    int radius;

    double aspect_ratio = (double)r.width/r.height;
    if( 0.75 < aspect_ratio && aspect_ratio < 1.3 )
    {
        center.x = cvRound((r.x + r.width*0.5)*scale);
        center.y = cvRound((r.y + r.height*0.5)*scale);
        radius = cvRound((r.width + r.height)*0.25*scale);
        circle( img, center, radius, color, 3, 8, 0 );

    face_centroid.x = center.x;
    face_centroid.y = center.y;

        //Publishing centroid of detected face
          face_centroid_pub.publish(face_centroid);

    }
```

To make the output image window more interactive, there are text and lines to alert about the user's face on the left or right or at the center. This last section of code is mainly for that purpose. It uses OpenCV APIs to do this job. Here is the code to display text such as **Left**, **Right**, and **Center** on the screen:

```
        putText(img, "Left", cvPoint(50,240),
FONT_HERSHEY_SIMPLEX, 1,
    cvScalar(255,0,0), 2, CV_AA);
        putText(img, "Center", cvPoint(280,240),
FONT_HERSHEY_SIMPLEX,
    1, cvScalar(0,0,255), 2, CV_AA);
        putText(img, "Right", cvPoint(480,240),
FONT_HERSHEY_SIMPLEX,
    1, cvScalar(255,0,0), 2, CV_AA);
```

Excellent! We're done with the tracker code; let's see how to build it and make it executable.

Understanding CMakeLists.txt

The default `CMakeLists.txt` file made during the creation of the package has to be edited to compile the previous source code. Here is the `CMakeLists.txt` file used to build the `face_tracker_node.cpp` class.

The first two lines state the minimum version of `cmake` required to build this package, and the next line is the package name:

```
cmake_minimum_required(VERSION 2.8.3)
project(face_tracker_pkg)
```

The following line searches for the dependent packages of `face_tracker_pkg` and raises an error if they are not found:

```
find_package(catkin REQUIRED COMPONENTS
  cv_bridge
  image_transport
  roscpp
  rospy
  sensor_msgs
  std_msgs
  message_generation

)
```

This line of code contains the system-level dependencies for building the package:

```
find_package(Boost REQUIRED COMPONENTS system)
```

As we've already seen, we are using a custom message definition called `centroid.msg`, which contains two fields, `int32 x` and `int32 y`. To build and generate C++ equivalent headers, we should use the following lines:

```
add_message_files(
   FILES
   centroid.msg
 )

## Generate added messages and services with any dependencies
listed here
 generate_messages(
   DEPENDENCIES
   std_msgs
 )
```

The `catkin_package()` function is a catkin-provided CMake macro that is required to generate `pkg-config` and CMake files:

```
catkin_package(
  CATKIN_DEPENDS roscpp rospy std_msgs message_runtime
)
include_directories(
  ${catkin_INCLUDE_DIRS}
)
```

Here, we are creating the executable called `face_tracker_node` and linking it to catkin and OpenCV libraries:

```
add_executable(face_tracker_node src/face_tracker_node.cpp)
target_link_libraries(face_tracker_node
   ${catkin_LIBRARIES}
   ${OpenCV_LIBRARIES}
 )
```

Let's now look at the `track.yaml` file.

The track.yaml file

As we discussed, the `track.yaml` file contains ROS parameters, which are required by `face_tracker_node`. Here are the contents of `track.yaml`:

```
image_input_topic: "/usb_cam/image_raw"
face_detected_image_topic: "/face_detector/raw_image"
haar_file_face:
"/home/robot/chapter_12_ws/src/face_tracker_pkg/data/face.xml"
face_tracking: 1
display_original_image: 1
display_tracking_image: 1
```

You can change all of the parameters according to your needs. In particular, you may need to change `haar_file_face`, which is the path of the haar face file (this will be your package path). If we set `face_tracking:1`, it will enable face tracking, otherwise, it will not. Also, if you want to display the original face-tracking image, you can set the flag here.

Launch files

Launch files in ROS can do multiple tasks in a single file. Launch files have an extension of `.launch`. The following code shows the definition of `start_usb_cam.launch`, which starts the `usb_cam` node for publishing the camera image as an ROS topic:

```
<launch>
  <node name="usb_cam" pkg="usb_cam" type="usb_cam_node"
output="screen" >
    <param name="video_device" value="/dev/video0" />
    <param name="image_width" value="640" />
    <param name="image_height" value="480" />
    <param name="pixel_format" value="yuyv" />
    <param name="camera_frame_id" value="usb_cam" />
    <param name="auto_focus" value="false" />
    <param name="io_method" value="mmap"/>
  </node>
</launch>
```

Within the `<node>...</node>` tags, there are camera parameters that can be changed by the user. For example, if you have multiple cameras, you can change the `video_device` value from `/dev/video0` to `/dev/video1` to get the second camera's frames.

The next important launch file is `start_tracking.launch`, which will launch the face-tracker node. Here is the definition of this launch file:

```
<launch>
<!-- Launching USB CAM launch files and Dynamixel controllers -->
  <include file="$(find
face_tracker_pkg)/launch/start_usb_cam.launch"/>

<!-- Starting face tracker node -->
   <rosparam file="$(find face_tracker_pkg)/config/track.yaml"
command="load"/>

   <node name="face_tracker" pkg="face_tracker_pkg"
type="face_tracker_node" output="screen" />
</launch>
```

It will first start the `start_usb_cam.launch` file to get ROS image topics, then load `track.yaml` to get the necessary ROS parameters, and then load `face_tracker_node` to start tracking.

The final launch file is `start_dynamixel_tracking.launch`; this is the launch file we have to execute for tracking and Dynamixel control. We will discuss this launch file at the end of this chapter, after discussing the `face_tracker_control` package. Let's now learn how to run the face tracker node.

Running the face tracker node

Let's launch the `start_tracking.launch` file from `face_tracker_pkg` using the following command:

```
$ roslaunch face_tracker_pkg start_tracking.launch
```

Note that you should connect your webcam to your PC.

If everything works fine, you will get output like the following; the first one is the original image, and the second one is the face-detected image:

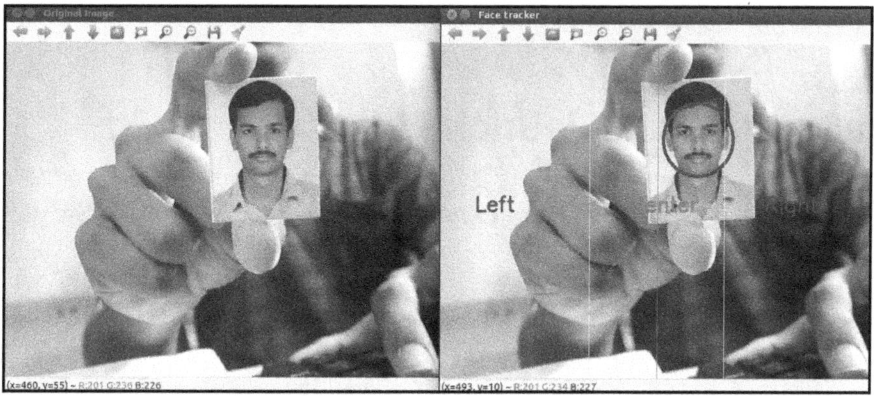

Face-detected image

We have not enabled Dynamixel now; this node will just find the face and publish the centroid values to a topic called `/face_centroid`.

So, the first part of the project is done—what's next? It's the control part, right? Yes, so next, we are going to discuss the second package, `face_tracker_control`.

The face_tracker_control package

The `face_tracker_control` package is the control package used to track a face using the AX-12A Dynamixel servo.

Given here is the file structure of the `face_tracker_control` package:

```
.
├── CMakeLists.txt
├── config
│   ├── pan.yaml
│   └── servo_param.yaml
├── include
│   └── face_tracker_control
├── launch
│   ├── start_dynamixel.launch
│   └── start_pan_controller.launch
├── msg
│   └── centroid.msg
├── package.xml
└── src
    └── face_tracker_controller.cpp

6 directories, 8 files
```

File organization in the face_tracker_control package

We'll look at the use of each of these files first.

The start_dynamixel launch file

The `start_dynamixel` launch file starts Dynamixel control manager, which can establish a connection to a USB-to-Dynamixel adapter and Dynamixel servos. Here is the definition of this launch file:

```
<!-- This will open USB To Dynamixel controller and search for
servos -->
<launch>
    <node name="dynamixel_manager" pkg="dynamixel_controllers"
    type="controller_manager.py" required="true"
  output="screen">
        <rosparam>
            namespace: dxl_manager
            serial_ports:
                pan_port:
                    port_name: "/dev/ttyUSB0"
                    baud_rate: 1000000
                    min_motor_id: 1
                    max_motor_id: 25
                    update_rate: 20
        </rosparam>
    </node>
<!-- This will launch the Dynamixel pan controller -->
  <include file="$(find
face_tracker_control)/launch/start_pan_controller.launch"/>
</launch>
```

We have to mention `port_name` (you can get the port number from the kernel logs using the `dmesg` command). The `baud_rate` parameter we configured was 1 Mbps, and the motor ID was `1`. The `controller_manager.py` file will scan from servo ID `1` to `25` and report any servos being detected.

After detecting the servo, it will start the `start_pan_controller.launch` file, which will attach an ROS joint position controller for each servo.

The pan controller launch file

As we can see from the previous subsection, the pan controller launch file is the trigger for attaching the ROS controller to the detected servos. Here is the definition for the `start_pan_controller.launch` file, which starts the pan joint controller:

```
<launch>
    <!-- Start tilt joint controller -->
    <rosparam file="$(find face_tracker_control)/config/pan.yaml"
 command="load"/>
    <rosparam file="$(find
face_tracker_control)/config/servo_param.yaml" command="load"/>

    <node name="tilt_controller_spawner"
pkg="dynamixel_controllers" type="controller_spawner.py"
          args="--manager=dxl_manager
                --port pan_port
                pan_controller"
          output="screen"/>
</launch>
```

The `controller_spawner.py` node can spawn a controller for each detected servo. The parameters of the controllers and servos are included in `pan.yaml` and `servo_param.yaml`.

The pan controller configuration file

The pan controller configuration file contains the configuration of the controller that the controller spawner node is going to create. Here is the `pan.yaml` file definition for our controller:

```
pan_controller:
    controller:
        package: dynamixel_controllers
        module: joint_position_controller
        type: JointPositionController
    joint_name: pan_joint
    joint_speed: 1.17
    motor:
        id: 1
        init: 512
        min: 316
        max: 708
```

In this configuration file, we have to mention the servo details, such as ID, initial position, minimum and maximum servo limits, servo moving speed, and joint name. The name of the controller is `pan_controller`, and it's a joint position controller. We are writing one controller configuration for ID `1` because we are only using one servo.

The servo parameters configuration file

The `servo_param.yaml` file contains the configuration of `pan_controller`, such as the limits of the controller and the step distance of each movement; also, it has screen parameters such as the maximum resolution of the camera image and the offset from the center of the image. The offset is used to define an area around the actual center of the image:

```
servomaxx: 0.5    #max degree servo horizontal (x) can turn
servomin: -0.5    # Min degree servo horizontal (x) can turn
screenmaxx: 640    #max screen horizontal (x)resolution
center_offset: 50 #offset pixels from actual center to right and
left
step_distancex: 0.01 #x servo rotation steps
```

Let's now look at the face tracker controller node.

The face tracker controller node

As we've already seen, the face tracker controller node is responsible for controlling the Dynamixel servo according to the face centroid position. Let's dissect the code of this node, which is placed at `face_tracker_control/src/face_tracker_controller.cpp`.

The main ROS headers included in this code are as follows:

```
#include "ros/ros.h"
#include "std_msgs/Float64.h"
#include <iostream>
```

Here, the `Float64` header is used to hold the position value message to the controller.

The following variables hold the parameter values from `servo_param.yaml`:

```
int servomaxx, servomin,screenmaxx, center_offset, center_left,
center_right;
float servo_step_distancex, current_pos_x;
```

The following message headers of `std_msgs::Float64` are for holding the initial and current positions of the controller, respectively. The controller only accepts this message type:

```
std_msgs::Float64 initial_pose;
std_msgs::Float64 current_pose;
```

This is the publisher handler for publishing the position commands to the controller:

```
ros::Publisher dynamixel_control;
```

Switching to the `main()` function of the code, you can see the following lines of code. The first line is the subscriber of `/face_centroid`, which has the centroid value, and when a value comes to the topic, it will call the `face_callback()` function:

```
ros::Subscriber number_subscriber =
node_obj.subscribe("/face_centroid",10,face_callback);
```

The following line will initialize the publisher handle in which the values are going to be published through the `/pan_controller/command` topic:

```
dynamixel_control = node_obj.advertise<std_msgs::Float64>
("/pan_controller/command",10);
```

The following code creates new limits around the actual center of the image. This will help to get an approximated center point of the image:

```
center_left = (screenmaxx / 2) - center_offset;
center_right = (screenmaxx / 2) + center_offset;
```

Here is the callback function executed while receiving the centroid value coming through the `/face_centroid` topic. This callback also has the logic for moving the Dynamixel for each centroid value.

In the first section, the `x` value in the centroid is checking against `center_left`, and if it is on the left, it just increments the servo controller position. It will publish the current value only if the current position is within the limit. If it is in the limit, then it will publish the current position to the controller. The logic is the same for the right side: if the face is on the right side of the image, it will decrement the controller position.

When the camera reaches the center of the image, it will pause there and do nothing, and that is what we want. This loop is repeated, and we will get continuous tracking:

```
void track_face(int x,int y)
{
    if (x < (center_left)){
```

```
        current_pos_x += servo_step_distancex;
        current_pose.data = current_pos_x;
      if (current_pos_x < servomaxx and current_pos_x > servomin ){
         dynamixel_control.publish(current_pose);
      }
      }
  else if(x > center_right){
  current_pos_x -= servo_step_distancex;
  current_pose.data = current_pos_x;
    if (current_pos_x < servomaxx and current_pos_x > servomin ){
          dynamixel_control.publish(current_pose);
  }
      }

     else if(x > center_left and x < center_right){
  ;
  }
  }
```

We will now create the `CMakeLists.txt` file.

Creating CMakeLists.txt

Like the first tracker package, there is no special difference in the control package; the difference is in the dependencies. Here, the main dependency is `dynamixel_controllers`. We are not using OpenCV in this package, so there's no need to include it. The necessary changes are highlighted as follows:

```
...
    project(face_tracker_control)
    find_package(catkin REQUIRED COMPONENTS
      dynamixel_controllers
      roscpp
      rospy
      std_msgs
      message_generation
    )

...

    catkin_package(
      CATKIN_DEPENDS dynamixel_controllers roscpp rospy std_msgs
    )

...
```

We can now test the face tracker control package.

Testing the face tracker control package

We have seen most of the files and their functionalities. So, let's test this package first. By following these steps, we can ensure that it is detecting the Dynamixel servo and creating the proper topic:

1. Before running the launch file, we will have to change the permission of the USB device, or it will throw an exception. The following command can be used to get permissions on the serial device:

   ```
   $ sudo chmod 777 /dev/ttyUSB0
   ```

   Note that you must replace `ttyUSB0` with your device name; you can retrieve it by looking at kernel logs. The `dmesg` command can help you to find it.

2. Start the `start_dynamixel.launch` file using the following command:

   ```
   $ roslaunch face_tracker_control start_dynamixel.launch
   ```

```
 * /servomin: -0.5
 * /step_distancex: 0.01

NODES
  /
    dynamixel_manager (dynamixel_controllers/controller_manager.py)
    tilt_controller_spawner (dynamixel_controllers/controller_spawner.py)

auto-starting new master
process[master]: started with pid [6997]
ROS_MASTER_URI=http://localhost:11311

setting /run_id to 6b4d648e-62c8-11e6-ac5f-00177c2e2869
process[rosout-1]: started with pid [7010]
started core service [/rosout]
process[dynamixel_manager-2]: started with pid [7027]
process[tilt_controller_spawner-3]: started with pid [7028]
[INFO] [WallTime: 1471252362.231754] pan_port controller_spawner: waiting for controller_manager dxl_manager to startup in global namespace...
[INFO] [WallTime: 1471252362.661902] pan_port: Pinging motor IDs 1 through 25...
[INFO] [WallTime: 1471252364.696276] pan_port: Found 1 motors - 1 AX-12 [1], initialization complete.
[INFO] [WallTime: 1471252364.951534] pan_port controller_spawner: All services are up, spawning controllers...

[INFO] [WallTime: 1471252364.979589] Controller pan_controller successfully started.
[tilt_controller_spawner-3] process has finished cleanly
log file: /home/robot/.ros/log/6b4d648e-62c8-11e6-ac5f-00177c2e2869/tilt_controller_spawner-3*.log
```

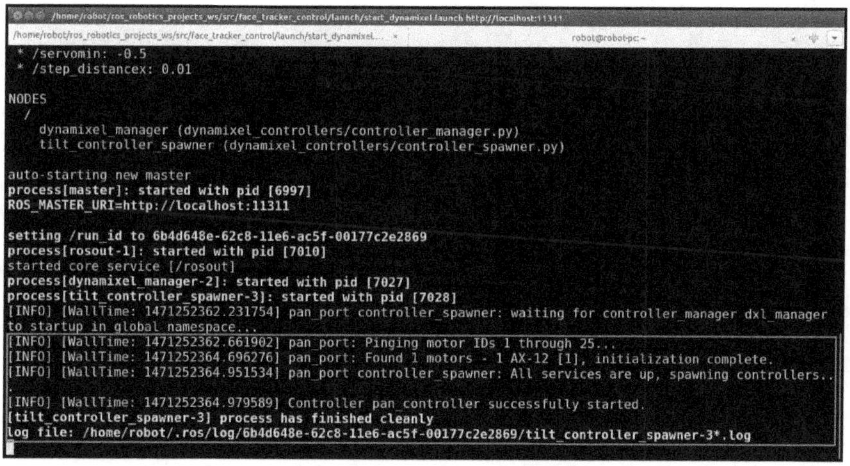

Finding Dynamixel servos and creating controllers

If everything is successful, you will get a message as shown in the preceding screenshot.

If any errors occur during the launch, check the servo connection, power, and device permissions.

The following topics are generated when we run this launch file:

```
robot@robot-pc: ~
/home/robot/ros_robotics_projects_ws/src/face_tracker_control/launch/start_dynamixel....
robot@robot-pc:~$ rostopic list
/diagnostics
/motor_states/pan_port
/pan_controller/command
/pan_controller/state
/rosout
/rosout_agg
robot@robot-pc:~$
```

Face tracker control topics

Now, we need to bring the nodes together.

Bringing all of the nodes together

Next, we'll look at the final launch file, which we skipped while covering the `face_tracker_pkg` package, and that is `start_dynamixel_tracking.launch`. This launch file starts both face detection and tracking using Dynamixel motors:

```
<launch>
<!-- Launching USB CAM launch files and Dynamixel controllers -->
  <include file="$(find
face_tracker_pkg)/launch/start_tracking.launch"/><include
file="$(find
face_tracker_control)/launch/start_dynamixel.launch"/>
<!-- Starting face tracker node -->

<node name="face_controller" pkg="face_tracker_control"
type="face_tracker_controller" output="screen" />

</launch>
```

Let's look at setting up the hardware in the next section.

Fixing the bracket and setting up the circuit

Before doing the final run of the project, we have to do something on the hardware side. We have to fix the bracket to the servo horn and fix the camera to the bracket. The bracket should be connected in such a way that it is always perpendicular to the center of the servo. The camera is mounted on the bracket, and it should be pointed toward the center position.

The following shows the setup I did for this project. I simply used tape to fix the camera to the bracket. You can use any additional material to fix the camera, but it should always be aligned to the center first:

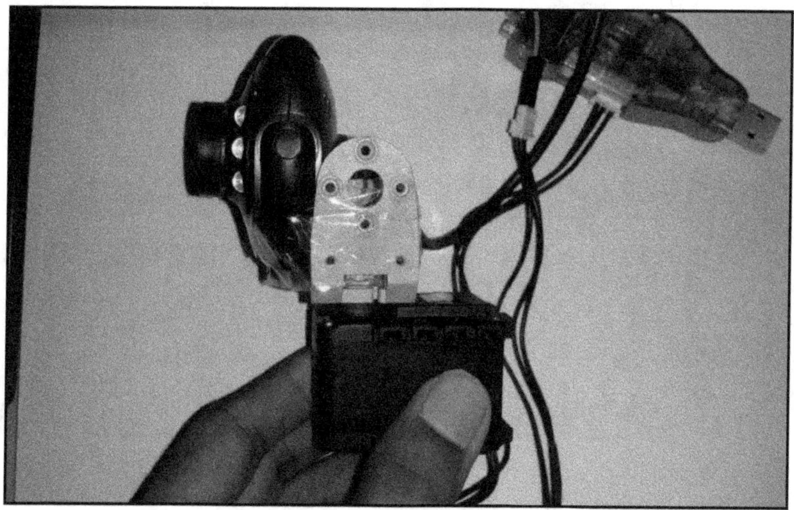

Fixing the camera and bracket to the AX-12A

If you are done with this, then you are ready to go for the final run of this project.

The final run

Here, I hope that you have followed all of the instructions properly; here is the command to launch all of the nodes for this project and start tracking using Dynamixel:

```
$ roslaunch face_tracker_pkg start_dynamixel_tracking.launch
```

You will get the following windows, and it would be good if you could use a photo to test the tracking because you will get continuous tracking of the face:

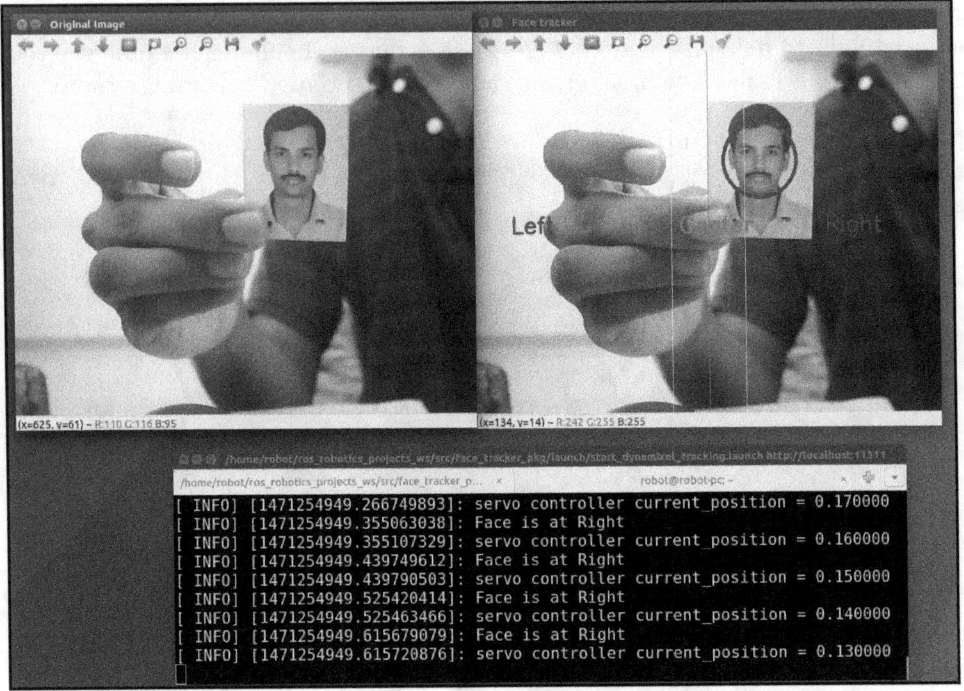

Final face tracking

In the preceding screenshot, you can see the Terminal message that says the image is on the right and the controller is reducing the position value to achieve the center position.

Summary

This chapter was about building a face tracker using a webcam and Dynamixel motor. The software we used was ROS and OpenCV. Initially, we saw how to configure the webcam and Dynamixel motor, and after configuration, we built two packages for tracking. One package was for face detection, and the second package was a controller that can send a position command to Dynamixel to track a face. We then discussed the use of all of the files inside the packages and did a final run to demonstrate the complete workings of the system.

Other Books You May Enjoy

If you enjoyed this book, you may be interested in these other books by Packt:

Mastering ROS for Robotics Programming
Lentin Joseph

ISBN: 978-1-78355-179-8

- Create a robot model of a Seven-DOF robotic arm and a differential wheeled mobile robot
- Work with motion planning of a Seven-DOF arm using MoveIt!
- Implement autonomous navigation in differential drive robots using SLAM and AMCL packages in ROS
- Dig deep into the ROS Pluginlib, ROS nodelets, and Gazebo plugins
- Interface I/O boards such as Arduino, Robot sensors, and High end actuators with ROS
- Simulation and motion planning of ABB and Universal arm using ROS Industrial
- Explore the ROS framework using its latest version

ROS Robotics By Example - Second Edition
Carol Fairchild, Dr. Thomas L. Harman

ISBN: 978-1-78847-959-2

- Control a robot without requiring a PhD in robotics
- Simulate and control a robot arm
- Control a flying robot
- Send your robot on an independent mission
- Learning how to control your own robots with external devices
- Program applications running on your robot
- Extend ROS itself
- Extend ROS with the MATLAB Robotics System Toolbox

Leave a review - let other readers know what you think

Please share your thoughts on this book with others by leaving a review on the site that you bought it from. If you purchased the book from Amazon, please leave us an honest review on this book's Amazon page. This is vital so that other potential readers can see and use your unbiased opinion to make purchasing decisions, we can understand what our customers think about our products, and our authors can see your feedback on the title that they have worked with Packt to create. It will only take a few minutes of your time, but is valuable to other potential customers, our authors, and Packt. Thank you!

Index

D

E

F

G

H

I

J

K

L

M

N

O

P

Q

R

S

T

U

V

W

X

Y

Z

www.ingramcontent.com/pod-product-compliance
Lightning Source LLC
LaVergne TN
LVHW081510050326
832903LV00025B/1432